Plant Biochemical Regulators, edited by Harold W. Gausman
Maximizing Crop Yields, N. K. Fageria
Transgenic Plants: Fundamentals and Applications, edited by Andrew Hiatt
Soil Microbial Ecology: Applications in Agricultural and Environmental Management, edited by F. Blaine Metting, Jr.
Principles of Soil Chemistry: Second Edition, Kim H. Tan
Water Flow in Soils, edited by Tsuyoshi Miyazaki
Handbook of Plant and Crop Stress, edited by Mohammad Pessarakli
Genetic Improvement of Field Crops, edited by Gustavo A. Slafer
Agricultural Field Experiments: Design and Analysis, Roger G. Petersen
Environmental Soil Science, Kim H. Tan
Mechanisms of Plant Growth and Improved Productivity: Modern Approaches, edited by Amarjit S. Basra
Selenium in the Environment, edited by W. T. Frankenberger, Jr., and Sally Benson
Plant–Environment Interactions, edited by Robert E. Wilkinson
Handbook of Plant and Crop Physiology, edited by Mohammad Pessarakli
Handbook of Phytoalexin Metabolism and Action, edited by M. Daniel and R. P. Purkayastha
Soil–Water Interactions: Mechanisms and Applications, Second Edition, Revised and Expanded, Shingo Iwata, Toshio Tabuchi, and Benno P. Warkentin
Stored-Grain Ecosystems, edited by Digvir S. Jayas, Noel D. G. White, and William E. Muir
Agrochemicals from Natural Products, edited by C. R. A. Godfrey
Seed Development and Germination, edited by Jaime Kigel and Gad Galili
Nitrogen Fertilization in the Environment, edited by Peter Edward Bacon
Phytohormones in Soils: Microbial Production and Function, W. T. Frankenberger, Jr., and Muhammad Arshad
Handbook of Weed Management Systems, edited by Albert E. Smith
Soil Sampling, Preparation, and Analysis, Kim H. Tan
Soil Erosion, Conservation, and Rehabilitation, edited by Menachem Agassi
Plant Roots: The Hidden Half, Second Edition, Revised and Expanded, edited by Yoav Waisel, Amram Eshel, and Uzi Kafkafi
Photoassimilate Distribution in Plants and Crops: Source–Sink Relationships, edited by Eli Zamski and Arthur A. Schaffer
Mass Spectrometry of Soils, edited by Thomas W. Boutton and Shinichi Yamasaki
Handbook of Photosynthesis, edited by Mohammad Pessarakli
Chemical and Isotopic Groundwater Hydrology: The Applied Approach, Second Edition, Revised and Expanded, Emanuel Mazor
Fauna in Soil Ecosystems: Recycling Processes, Nutrient Fluxes, and Agricultural Production, edited by Gero Benckiser

Additional Volumes in Preparation

Soil and Plant Analysis in Sustainable Agriculture and Environment, edited by Teresa Hood and J. Benton Jones, Jr.

Agricultural Biotechnology, edited by Arie Altman

CHEMICAL AND ISOTOPIC GROUNDWATER HYDROLOGY
The Applied Approach
Second Edition, Revised and Expanded

Emanuel Mazor
Weizmann Institute of Science
Rehovot, Israel

Marcel Dekker, Inc.　　　New York•Basel•Hong Kong

First edition © 1991 by Halsted Press, a Division of John Wiley & Sons, Inc., New York

ISBN: 0-8247-9803-1

The publisher offers discounts on this book when ordered in bulk quantities. For more information, write to Special Sales/Professional Marketing at the address below.

This book is printed on acid-free paper.

Copyright © 1997 by Marcel Dekker, Inc. All Rights Reserved.

Neither this book nor any part may be reproduced or transmitted in any form or by any means, electronic or mechanical, including photocopying, microfilming, and recording, or by any information storage and retrieval system, without permission in writing from the publisher.

Marcel Dekker, Inc.
270 Madison Avenue, New York, New York 10016

Current printing (last digit):
10 9 8 7 6 5 4 3 2 1

PRINTED IN THE UNITED STATES OF AMERICA

PREFACE

Groundwater is a vital resource in steadily increasing demand by man, but man threatens its quality and mishandles the available quantity. For optimal management the groundwater resource has to be well understood, in terms of both its general behavior and the specific systems exploited. Chemical and isotopic hydrology are tailored to these challenges, and the hydrochemist has a key role as a consultant to groundwater planners, developers, and managers in the frame of the public establishment, research institutes, and the private sector.

This book deals with data processing and phenomenological modeling, based on maximum field observations and field measurements. Phenomenological conceptual models have no dimensional limitations and can accommodate any number of parametric data and dynamic variations. They are useful for:

Understanding the natural regimes of groundwater systems in specific case study areas.

Understanding the consequences of anthropogenic intervention.

Reaching conclusions and recommendations related to the management of the water resource.

Establishing boundary conditions that lead to conceptual models and can help to formulate mathematical models with a reduced number of assumptions and unknowns.

Field hydrochemists are explorers who plan hydrochemical studies, conduct field measurements and observations, collect samples for laboratory analyses, process maximum available data, formulate phenomenological and mathematical models, and communicate the results to the scientific community and to decision makers. This book addresses these topics, emphasizes the processing of data, discusses a large number of published case studies, and preaches for independent interpretations. Because most field hydrologists bring their samples to specialized laboratories for analysis, the laboratory work is not dealt with here.

Physical, chemical, isotopic, and dating methods are presented via the basic observations, stressing the applied aspects.

This book is based partly on a previous book titled *Applied Chemical and Isotopic Groundwater Hydrochemistry* (Mazor, 1991, Open University Press and Halsted Press) and may be regarded as a new revised, updated, and extended edition. Five new chapters have been added dealing with groundwater dating by ^{36}Cl and radiogenic ^4He, paleohydrology, sustainable development of groundwater, and groundwater aspects of urban and statewide planning. Exercises have been added, along with discussed answers.

Emanuel Mazor

CONTENTS

PREFACE iii

1. INTRODUCTION—CHEMICAL AND ISOTOPIC ASPECTS OF HYDROLOGY 1

    1.1 The Concealed Nature of Groundwater 1
    1.2 Groundwater Composition 1
    1.3 Information Encoded into Water During the Hydrological Cycle 3
    1.4 Questions Asked 5
    1.5 Phenomenological Conceptual Models and Mathematical Models 7
    1.6 Economic Considerations of Field Hydrochemistry 8
    1.7 Environmental Issues 10
    1.8 Summary Exercises 10

2. BASIC HYDROLOGICAL CONCEPTS 12

    2.1 The Aerated Intake Zone 12
    2.2 The Saturated Zone and the Water Table 14
    2.3 Aquifers and Aquicludes 15
    2.4 Recharge 17
    2.5 Discharge 18
    2.6 Evapotranspiration 18

2.7	Permeability, Impermeability, and the Selection of Representative Values	19
2.8	Through-Flow Systems: Unconfined and Confined	22
2.9	Karstic Systems: Paths of Preferred Flow	25
2.10	Flow Velocity and Groundwater Age	26
2.11	Flow of Water: Basic Observations	28
2.12	Two Basic Experiments of Water Flow	30
2.13	A Model of L-Shape Through-Flow Paths and Zones of Stagnation	31
2.14	Deep Basin Compartments of Pressurized Water, Petroleum, and Geothermal Fluids	39
2.15	The Traditional Model of U-Shape Flow Paths	40
2.16	Discussion of the U-Shape Flow Paths Model	42
2.17	Summary of Modes of Groundwater Flow	45
2.18	Summary Exercises	45

3. GEOLOGICAL DATA 47

3.1	Lithology and Its Bearing on Water Composition	47
3.2	Properties of Geological Materials and Their Bearing on Recharge and Groundwater Storage	49
3.3	Layering and Its Control on Groundwater Flow	51
3.4	Folded Structures and Their Bearing on Flow Direction, Confinement, and Entrapment	52
3.5	Faults Controlling Groundwater Flow	54
3.6	Intrusive Bodies Influencing Groundwater Flow	55
3.7	Hydrochemical and Isotopic Checks of Geological and Hydrological Models	56
3.8	Summary Exercises	61

4. PHYSICAL PARAMETERS 62

4.1	Water Table Measurements	62
4.2	Interpretation of Water Table Data	63
4.3	Gradient and Flow Direction	71
4.4	The Need for Complementing Data to Check Deduced Gradients and Flow Directions	71
4.5	Velocities and Pumping Tests	73
4.6	Chemical and Physical Measurements During Pumping Tests	73
4.7	Temperature Measurements	75
4.8	Tracing Groundwater by Temperature—A Few Case Studies: The Mohawk River	78

4.9	Cold and Hot Groundwater Systems	83
4.10	Discharge Measurements and Their Interpretative Value	84
4.11	Summary Exercises	89

5. ELEMENTS, ISOTOPES, IONS, UNITS, AND ERRORS — 90

5.1	Elements	90
5.2	Isotopes	90
5.3	Atomic Weight	92
5.4	Ions and Valences	92
5.5	Ionic Compounds	94
5.6	Concentration Units	94
5.7	Reproducibility, Accuracy, Resolution, and Limit of Detection	96
5.8	Errors and Significant Figures	97
5.9	Checking the Laboratory	98
5.10	Evaluation of Data Quality by Data Processing Techniques	99
5.11	Putting Life into a Dry Table	100
5.12	Evaluation of Calculated Reproducibilities and Reaction Errors	103
5.13	Summary Exercises	104

6. CHEMICAL PARAMETERS: DATA PROCESSING — 106

6.1	Data Tables	106
6.2	Fingerprint Diagrams	108
6.3	Composition Diagrams	113
6.4	Major Patterns Seen in Composition Diagrams	113
6.5	Establishing Hydraulic Interconnections	116
6.6	Mixing Patterns	121
6.7	End Member Properties and Mixing Percentages	122
6.8	Water-Rock Interactions and the Types of Rocks Passed	126
6.9	Water Composition: Modes of Description	128
6.10	Compositional Time Series	129
6.11	Some Case Studies	131
6.12	Saturation and Undersaturation with Regard to Halite and Gypsum: The Hydrochemical Implications	136
6.13	Sea-Derived (Airborne) Ions: A Case Study Demonstrating the Phenomenon	136
6.14	Sources of Chlorine in Groundwater	138
6.15	Evapotranspiration Index: Calculation Based on Chlorine as a Hydrochemical Marker	139

6.16	CO$_2$-Induced Water-Rock Interactions Revealed by Chlorine Serving as a Hydrochemical Marker	140
6.17	Exchange of Sodium Versus Calcium and Magnesium Revealed by Chlorine Serving as a Hydrochemical Marker	141
6.18	Summary Exercises	142

7. PLANNING HYDROCHEMICAL STUDIES — 145

7.1	Representative Samples	145
7.2	Data Collection During Drilling	147
7.3	Depth Profiles	149
7.4	Data Collection During Pumping Tests	149
7.5	Importance of Historical Data	150
7.6	Repeated Observations or Time-Data Series	151
7.7	Search for Meaningful Parameters	152
7.8	Sampling for Contour Maps	153
7.9	Sampling Along Transects	153
7.10	Reconnaissance Studies	154
7.11	Detailed Studies	154
7.12	Pollution-Related Studies	155
7.13	Summary: A Planning List	155
7.14	Summary Exercises	156

8. CHEMICAL PARAMETERS: FIELD WORK — 158

8.1	Field Measurements	158
8.2	Smell and Taste	159
8.3	Temperature	160
8.4	Electrical Conductance	160
8.5	pH	161
8.6	Dissolved Oxygen	162
8.7	Alkalinity	163
8.8	Sampling for Dissolved Ion Analyses	164
8.9	Sampling for Isotopic Measurements	164
8.10	Preservation of Samples	165
8.11	Efflorescences	166
8.12	Equipment List for Field Work	166

9. STABLE HYDROGEN AND OXYGEN ISOTOPES — 168

9.1	Isotopic Composition of Water Molecules	168
9.2	Units of Isotopic Composition of Water	169

9.3	Isotopic Fractionation During Evaporation and Some Hydrological Applications	169
9.4	The Meteoric Isotope Line	172
9.5	Temperature Effect	176
9.6	Amount Effect	177
9.7	Continental Effect	179
9.8	Altitude Effect	180
9.9	Tracing Groundwater with Deuterium and ^{18}O: Local Studies	183
9.10	Tracing Groundwater with Deuterium and ^{18}O: A Regional Study	190
9.11	The Need for Multisampling	190
9.12	Sample Collection for Stable Hydrogen and Oxygen Isotopes and Contact with Relevant Laboratories	194
9.13	Summary Exercises	194

10. TRITIUM 196

10.1	The Radioactive Heavy Hydrogen Isotope	196
10.2	Natural Tritium Production	197
10.3	Man-Made Tritium Inputs	198
10.4	Tritium as a Short-Term Age Indicator	201
10.5	Tritium as a Tracer of Recharge and Piston Flow: Observations in Wells	203
10.6	The Special Role of Tritium in Tracing Intermixing of Old and Recent Waters and Calculation of the Extrapolated Properties of the End Members	209
10.7	Tritium, Dissolved Ions, and Stable Isotopes as Tracers for Rapid Discharge Along Fractures: The Mont Blanc Tunnel Case Study	210
10.8	The Tritium-^3He Groundwater Dating Method	214
10.9	Sample Collection for Tritium Measurements	215
10.10	Summary Exercises	215

11. RADIOCARBON AND ^{13}C 217

11.1	The Isotopes of Carbon	217
11.2	Natural ^{14}C Production	217
11.3	Man-Made ^{14}C Dilution and Addition	218
11.4	^{14}C in Groundwater: An Introduction to Groundwater Dating	219
11.5	Lowering of ^{14}C Content by Reactions with Rocks	222

11.6	^{13}C Abundances and Their Relevance to ^{14}C Dating of Groundwater	224
11.7	Application of $\delta^{13}C$ to Correct Observed ^{14}C Values for Changes Caused by Interactions with Carbonate Rocks	225
11.8	Direction of Down-Gradient Flow and Groundwater Age Studied by ^{14}C: Case Studies	228
11.9	Flow Discontinuities Between Adjacent Phreatic and Confined Aquifers Indicated by ^{14}C and Other Parameters	236
11.10	Mixing of Groundwaters Revealed by Joint Interpretation of Tritium and ^{14}C Data	248
11.11	Piston Flow Versus Karstic Flow Revealed by ^{14}C Data	250
11.12	Sample Collection for ^{14}C and ^{13}C Analyses and Contact with Relevant Laboratories	253
11.13	Summary Exercises	253

12. CHLORINE-36   255

12.1	The Radioactive Isotope of Chlorine and Its Production	255
12.2	A Potential Tool for Groundwater Dating	255
12.3	Processes Controlling the ^{36}Cl Concentration in Groundwater	256
12.4	Groundwater Dating by ^{36}Cl	259
12.5	A Number of Case Studies	261
12.6	Summary	268
12.7	Summary Exercises	269

13. NOBLE GASES   270

13.1	Rare, Inert, or Noble?	270
13.2	Atmospheric Noble Gases and Their Dissolution in Water	271
13.3	Groundwater as a Closed System for Atmospheric Noble Gases	272
13.4	Studies on Atmospheric Noble Gas Retention in Groundwater Systems: Cold Groundwater	274
13.5	Further Checks on Atmospheric Noble Gas Retention: Warm Groundwater	280
13.6	Identification of the Nature of the Unsaturated Recharge Zone: A Porous Medium or a Karstic System	282
13.7	Depth of Circulation and Location of the Recharge Zone	286
13.8	Paleotemperatures	287
13.9	Sample Collection for Noble Gas Measurements	291
13.10	Summary Exercises	292

14. HELIUM-4 AS A TOOL FOR GROUNDWATER DATING — 293

- 14.1 Sources of Radiogenic ^{4}He Dissolved in Groundwater — 293
- 14.2 Parameters Determining Helium Concentration in Groundwater — 294
- 14.3 Case Studies and Observed Trends in the Distribution of Helium Concentrations in Groundwater Systems — 295
- 14.4 Mantle Helium — 298
- 14.5 The Helium-Based Groundwater Age Equation — 298
- 14.6 Comparison of Helium-Based Water Ages and Hydraulically Calculated Ages — 299
- 14.7 Does Helium Diffuse Through the Upper Crust? — 300
- 14.8 Helium Concentration Increases with Depth: Groundwater as a Sink of Helium — 300
- 14.9 Effect of Mixing of Groundwaters of Different Ages on the Observed Helium Concentration — 302
- 14.10 Helium Losses or Gains Due to Phase Separation Underground or During Ascent to the Sampling Point: Modes of Identification and Correction — 304
- 14.11 Internal Checks for Helium Ages — 304
- 14.12 The Issue of Very Old Groundwaters — 307
- 14.13 Conclusions — 308
- 14.14 Summary Exercises — 309

15. INTERRELATIONS BETWEEN GROUNDWATER DATING, PALEOCLIMATE, AND PALEOHYDROLOGY — 310

- 15.1 Scope of the Problem — 310
- 15.2 The Need to Reconstruct Paleo-Input Values — 311
- 15.3 Stagnant Systems as Fossil Through-Flow Systems — 312
- 15.4 Screening of Data for Samples of "Last Minute" Mixing of Groundwaters of Different Ages — 312
- 15.5 Paleohydrological Changes — 313
- 15.6 Paleoclimate Changes — 317
- 15.7 Selecting Sites for Noble Gas Based Paleotemperature Studies — 318
- 15.8 Stages of Data Processing — 318
- 15.9 Summary Exercises — 319

16. DETECTING POLLUTION SOURCES — 320

- 16.1 Scope of the Problem — 320
- 16.2 Detection and Monitoring of Pollutants: Some Basic Rules — 321

	16.3	Groundwater Pollution Case Studies	323
	16.4	Summary of Case Studies	355
	16.5	Summary Exercises	355

17. SUSTAINABLE DEVELOPMENT OF GROUNDWATER — 356

    17.1 The Concept — 356
    17.2 Man and Groundwater: A Historical Perspective — 356
    17.3 The Concern over Groundwater Quality: A Lever for the Demand of Clean Nationwide Management — 357
    17.4 The Value of the Environment: An Essential Parameter Is Economic Calculations — 358
    17.5 Recharge and Its Controls — 358
    17.6 Overpumping Policy: The Through-Flow Systems — 360
    17.7 Optimization of Pumping Rates in the Through-Flow System — 361
    17.8 Optimal Utilization of the Groundwater Resource — 362

18. GROUNDWATER ASPECTS OF URBAN AND STATEWIDE DEVELOPMENT PLANNING AND MANAGEMENT — 366

    18.1 Surface Coverage by Roads and Buildings — 366
    18.2 Street Runoff Mitigation — 367
    18.3 Raising of the Water Table Due to Sewage Infiltration — 367
    18.4 Location of Urban Landfills, Sewage Treatment Plants, and Industrial Zones — 368
    18.5 Irrigation Optimization — 369
    18.6 Agricultural and Urban Use of Treated Effluents — 369
    18.7 Groundwater Vulnerability Maps: A Questionable Approach — 370
    18.8 Monitoring Networks: Basic Principles — 371
    18.9 Active Data Banks — 372

19. HYDROCHEMIST'S REPORTS — 374

    19.1 Why Reports? — 374
    19.2 Types of Reports — 374
    19.3 Internal Structure of Reports — 375

20. ANSWERS — 383

REFERENCES — 399

INDEX — 411

1
INTRODUCTION—CHEMICAL AND ISOTOPIC ASPECTS OF HYDROLOGY

1.1 The Concealed Nature of Groundwater

Groundwater is a concealed fluid that can be observed only at intermittent points at springs and in boreholes. Aquifers and flow paths are traditionally deduced by information retrieved from drilling operations, for example, rock cuttings, and occasionally cores, but these are poor small-scale representatives of the relatively large systems involved. This feature is familiar to geologists who try to predict the lithological and stratigraphic column to be passed by a new borehole—the outcome reveals time and again a higher spatial complexity and variability than predicted by interpolation of data available from existing boreholes. Prediction of hydraulic barriers, as well as prediction of preferred flow paths, based on knowledge of the regional geology, are shorthanded as well. Water properties have been measured traditionally, but these were mostly limited to the depth of the water table, or the hydraulic head, temperature, and salinity. Modern chemical and isotope hydrology encompasses many more parameters and deals with the integrative interpretation of all available data. The concealed nature of groundwater becomes gradually exposed.

1.2 Groundwater Composition

Water molecules are built of two elements: hydrogen (H) and oxygen (O). The general formula is H_2O. Different varieties of atoms with different masses, called isotopes, are presented in water, including light hydrogen, ^1H (most common); heavy hydrogen, called deuterium, ^2H or D (rare); light oxygen, ^{16}O

(common); ^{17}O (very rare); and heavy oxygen, ^{18}O (rare). Thus, several types of water molecules occur, the most important being $H_2^{16}O$—light, most common water molecule; and $HD^{16}O$ and $H_2^{18}O$—heavy, rare water molecules.

The major terrestrial water reservoirs, the oceans, are well mixed and of rather uniform composition. Upon evaporation, an isotopic separation occurs because the light molecules are more readily evaporated. Thus, water in clouds is isotopically light compared to ocean water. Upon condensation, heavier water molecules condense more readily, causing a "reversed" fractionation. The degree of isotopic fractionation depends on the ambient temperature and other factors which are discussed in Chapter 9.

Water contains dissolved salts, dissociated into cations (positively charged ions) and anions (negatively charged ions). The most common dissolved cations are sodium (Na^+), calcium (Ca^{2+}), magnesium (Mg^{2+}), and potassium (K^+). Most common anions are chloride (Cl^-), bicarbonate (HCO_3^-), and sulfate (SO_4^{2-}), discussed in section 5.5. The composition of groundwater, that is, the concentration of the different ions, varies over a wide range of values.

Water also contains dissolved gases. A major source is air, contributing nitrogen (N_2), oxygen (O_2), and noble gases (Chapter 13)—helium (He), neon (Ne), argon (Ar), krypton (Kr), and xenon (Xe). Other gases, of biogenic origin, are added to the water in the ground, for example, carbon dioxide (CO_2), methane (CH_4), and hydrogen sulfide (H_2S). Biogenic processes may consume the dissolved oxygen.

Cosmic rays interact with the upper atmosphere and produce a variety of radioactive isotopes, three of which are of special hydrological interest: (1) ^3H, better known as tritium (T), the heaviest hydrogen isotope, (2) carbon-14 (^{14}C), the heaviest isotope of carbon, and (3) chlorine-36 (^{36}Cl), a rare isotope of chlorine. The rate of radioactive decay is expressed in terms of half-life, that is, the time required for atoms in a given reservoir to decay to half their initial number. The three radioactive isotopes—tritium, ^{14}C, and ^{36}Cl—are incorporated into groundwater. They decay in the saturated zone (section 2.2), providing three semiquantitative dating tools—for periods of a few decades, to 25,000 years, and 10^5 to 10^6 years, respectively (sections 10.4, 11.4, and 12.2), The three isotopes also have been produced by nuclear bomb tests, complicating the picture but providing additional information.

Rocks contain uranium and thorium in small concentrations, and their radioactive decay results in the production of radiogenic helium-4 (^4He). The helium reaches the groundwater and is dissolved and stored. With time, radiogenic helium is accumulated in groundwater, providing an independent semiquantitative dating method (section 13.9) useful for ages in the range of 10^4 to 10^8 years.

1.3 Information Encoded into Water During the Hydrological Cycle

The hydrological cycle is well known in its general outline: ocean water is evaporated, forming clouds that are blown into the continent, where they gradually condense and rain out. On hitting the ground, part of the rain is turned into runoff, flowing back into the ocean, part is returned to the atmosphere by evapotranspiration, and the rest infiltrates and joins groundwater reservoirs, ultimately returning to the ocean as well. Hydrochemical studies reveal an ever-growing amount of information that is encoded into water during this cycle. It is up to the hydrochemist to decipher this information and to translate it into terms usable by water management personal (Fig. 1.1).

Rain-producing air masses are formed over different regions of the oceans, where different ambient temperatures prevail, resulting in different degrees of isotopic fractionation during evaporation and cloud formation. This explains observed variations in the isotopic composition of rains reaching a region from different directions or at different seasons. In terms of this chapter, we can say that water is encoded with isotopic information even in the first stages of the water cycle—cloud formation.

Wind carries water droplets of seawater which dry and form salt grains. These windborne salts, or sea spray, reach the clouds and are carried along with them. Sea spray also reaches the contingents as aerosols that settle on surfaces and are washed into the ground by rain.

Rainwater dissolves atmospheric N_2 and O_2 and the rare gases, as well as CO_2. In addition, radioactive tritium, ^{14}C, and ^{36}Cl are incorporated into the hydrological cycle and reach groundwater.

Man-made pollutants reach rainwater through the air. The best known is acid rain, which contains sulfur compounds. A large variety of other pollutants of urban life and industry are lifted into the atmosphere and washed down by rain. As clouds move inland, or rise up mountains, their water condenses and gradually rains out. Isotopic fractionation increases the deuterium and ^{18}O content of the condensed rain, depleting the concentration of these isotopes in the vapor remaining in the cloud. As the cloud moves on into the continent or up a mountain, the rain produced becomes progressively depleted in the heavy isotopes. Surveys have been developed into a source of information on distance of recharge from the coast and recharge altitude, parameters that help identify recharge areas (sections 9.7 and 9.8).

Upon contact with the soil, rainwater dissolves accumulated sea spray and dust, as well as fertilizers and pesticides. Fluid wastes are locally added, providing further tagging of water.

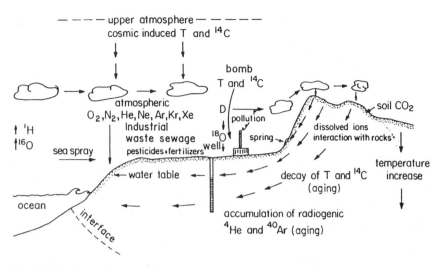

Fig. 1.1 Information coded into water along its surface and underground path: right at the beginning, water in clouds is tagged by an enrichment of light hydrogen (^{1}H) and oxygen (^{16}O) isotopes, separated during evaporation. Cloud water equilibrates with the atmosphere, dissolving radioactive tritium (T) and radiocarbon (^{14}C) produced naturally and introduced by nuclear bomb tests. Also, atmospheric gases are dissolved, the most important being oxygen (O_2), nitrogen (N_2), and the noble gases (He, Ne, Ar, Kr, and Xe). Gas dissolution is dependent on temperature and altitude (pressure). Sea spray reaches the clouds, along with urban and industrial pollutants. As clouds produce rain, heavy hydrogen (D) and heavy oxygen (^{18}O) isotopes are preferentially enriched, leaving isotopically depleted water in the cloud. Thus, inland and in mountainous regions rains of different isotopic compositions are formed. Upon hitting the ground, rainwater dissolves accumulated sea spray aerosols, dust particles, pesticides, and fertilizers. These are carried to rivers or introduced into the ground with infiltrating water. Sewage and industrial wastes are introduced into the ground as well, mixing with groundwater. Underground, water is disconnected from new supplies of tritium and ^{14}C, and the original amounts decay with time. On the other hand, radiogenic helium (^{4}He) and argon (^{40}Ar), produced in rocks, enter the water and accumulate with time. As water enters the soil zone it becomes enriched with soil CO_2 produced by biogenic activity, turning the water into an acid that interacts with rocks and introducing dissolved ions into the water. The deeper water circulates the warmer it gets due to the existing geothermal gradient.

Infiltrating water, passing the soil zone, becomes loaded with CO_2 formed in biogenic processes. The water turns into a weak acid that dissolves soil components and rocks. The nature of these dissolution processes varies with soil and rock types, climate, and drainage conditions. Dissolution ceases when ionic saturation is reached, but exchange reactions may continue (section 6.8).

Introduction

Daily temperature fluctuations are averaged out in most climates at a depth of 40 cm below the surface, and seasonal temperature variations are averaged out at a depth of 10–15 m. At this depth the prevailing temperature is commonly close to the average annual value. Temperature of shallow groundwater equaling that of the rainy season indicates rapid recharge through conduits. In contrast, if the temperature of shallow groundwater equals the average annual temperature, this indicates retardation in the soil and aerated zone until temperature equilibration has been reached. Further along the flow path groundwater gets heated by the geothermal heat gradient, providing information on the depth of circulation or underground storage (sections 4.7–4.9).

While water is stored underground, the radioactive isotopes of tritium, ^{14}C, and ^{36}C decay and the concentrations left indicate the age of the water. At the same time, radiogenic ^{4}He accumulates, providing another age indicator. A summary of the information encoded into water during its cycle is given in Fig. 1.1.

1.4 Questions Asked

Many stages of the water cycle are described by specific information implanted into surface water and groundwater. Yet, field hydrochemists have limited access to the water, being able to measure and sample it only at single points—wells and springs. Their task is to reconstruct the complete water history. A list of pertinent topics is given below.

1.4.1 Water Quality

Because groundwaters differ from one place to another in the concentration of their dissolved salts and gases, it is obvious that they also differ in quality. Water of high quality (low salt content, good taste) should be saved for drinking, irrigation, and a number of specific industries. Poorer water (more salts, poor taste) may be used, in order of quality, for domestic purposes (other than drinking and cooking), certain types of agricultural applications, stock raising, and industrial consumption. The shorter water resources are, the more important become the management manipulations that make sure each drop of water is used in the best way.

1.4.2 Water Quantity

As with any other commodity, for water the target is "the more the better." But, having constructed a well, will we always pump it as much as possible? Not at all. Commonly, a steady supply is required. Therefore, water will be pumped at a reasonable rate so the water table will not drop below the pump inlet and water will flow from the host rocks into the well at a steady rate.

Young water is steadily replenished, and with the right pumping rate may be developed into a steady water supply installation. In contrast, pumping old water resembles mining—the amounts are limited. The limitation may be noticed soon or, as in the Great Artesian Basin of Australia, the shortage may be felt only after substantial abstraction, but in such cases the shortage causes severe problems due to local overdevelopment. Knowledge of the age of water is thus essential for groundwater management. The topic of water dating is discussed in Chapters 10, 11, 12, and 13.

1.4.3 Hydraulic Interconnections

Does pumping in one well lower the output of an adjacent well? How can we predict? How can we test? Everyone knows that water flows down-gradient, but to what distances is this rule effective? And is the way always clear underground for water to flow in all down-gradient directions? The answers to these questions vary from case to case and therefore a direct tracing investigation is always desired but rarely feasible. Compositional, age, and temperature similarities may confirm hydraulic interconnections between adjacent springs and wells, whereas significant differences may indicate flow discontinuities (sections 2.8, 4.4, and 6.5).

1.4.4 Mixing of Different Groundwater Types

Investigations reveal an increasing number of case studies in which different water types (e.g., of different composition, temperature, and/or age) intermix underground. The hydrochemist has to explore the situation in each study area. We are used to regarding groundwater as being of a single (homogeneous) type, which it often is. In such cases a quality analysis represents the whole water body, and similarly, all other parameters measured reflect the properties of the entire system. But when fresh recent groundwater is mixed with varying quantities of saline old groundwater, the hydrochemist's life becomes tough and interesting. The first task is to find out the properties of the end members of the intermixing waters and their origins (sections 6.6 and 6.7). Then it is possible to recommend proper exploitation policies, balancing the need for the better water type with the demand for large water quantities.

1.4.5 Depth of Circulation

Understanding a groundwater system includes knowledge of the depth of circulation. Temperature can supply this information: the temperature of rocks is observed to increase with depth and the same holds true for water kept in rock systems—the deeper water is stored, the warmer is it when it issues from a spring or well (section 4.7).

Introduction

1.4.6 Location of Recharge

Every bottle of water collected at a spring or well contains information on the recharge location. Water molecules are separated during precipitation from clouds—molecules with heavier isotopes being rained out more efficiently than molecules with lighter isotopes. As a result, clouds that are blown to high mountains precipitate "light rain." Study of the isotopic composition of groundwaters provides clues to the recharge altitude (section 9.8) and, hence, provides constraints on the possible location of recharge.

1.4.7 Sources of Pollution

The major topics the hydrochemist has to deal with regarding pollution are

Early detection of the arrival of pollutants into groundwater in order to provide warning in due time, allowing for protective action.
Identification of pollutant input sources.

This brings us back to the problems of groundwater intermixing and hydraulic interconnections. It also brings to mind the urgent need for basic data to be collected on nondisturbed systems so that charges caused by contamination can be detected at an early stage.

1.5 Phenomenological Conceptual Models and Mathematical Models

The ultimate goal of a hydrochemical study is the construction of a conceptual model describing the water system in three spatial dimensions with its evolution in time. To achieve this goal a large number of observations and measurements are required to find out the site-specific setups. Thus, at the basic stage of investigation a phenomenological approach is required (section 16.1). The phenomenological conceptual model is mainly qualitative, recognizing recharge areas, topographic relief, terminal bases of drainage, modes of water up take, lithological compositions, geological structures, the number and nature of aquifers and aquiclude, karstic features and other preferred flow paths, storage capacities, flow directions and approximate flow velocities, degrees of confinement, identification of compartments of complete stagnation (entrapment), depths of groundwater flow and storage, response to rain, flood, and snowmelt events, water ages, and sources of pollution. A phenomenological conceptual model is best summarized in diagrams (section 18.3).

The conceptual model is needed to plan further stages of investigation, to forecast system behavior, and to provide the necessary base for mathematical modeling. Phenomenological models are not only a scientific necessity but are also useful in dealing with clients or financing authorities. A preliminary conceptual model may provide a useful introduction to research proposals, clarifying the suggested working plan and budgetary needs. Finally, the conceptual model is a must for final reports (section 19.3).

A mathematical model is the outcome of a set of mathematical operations applied to sets of data and based on premises. The appeal of mathematical models is their quantitative aspect. However, mathematical models suffer from technical restrictions to accommodate the three spatial dimensions, ranges of measured parameters, changes of the groundwater regime with season, climatic fluctuations, and response to human intervention (section 16.2). Thus, mathematical modeling necessitates simplifications, for example, application of single hydraulic conductivity coefficients selected to represent entire study areas or large sections therein (section 2.7), assumption of lithological homogeneity, application of simplified relief and water table configurations so they may be described by simple mathematical equations, and other boundary conditions. Assumptions are valid as long as they are backed by a phenomenological model that is based on real field data. Otherwise, they are camouflaged unknowns, bringing the mathematical calculations to a stage in which the number of unknowns is larger than the number of relevant equations. Hence, the mathematical model should be based of a phenomenological model.

1.6 Economic Considerations of Field Hydrochemistry

Applied water research is nearly always conducted with a restricted budget. Traditionally the larger sums go to the construction of weirs and other discharge measuring devices in surface water studies and the drilling of new wells in groundwater studies. Little money is allocated to hydrochemical investigations. The ratio of outcome to cost is, however, in many cases in favor of hydrochemical studies.

New drillings are traditionally conducted to check water occurrences, to establish the depth of water tables, and to be equipped for pumping new niches of known aquifers. Inclusion of hydrochemical studies as an integral part of the drilling operation may double or triple the amount and quality of information gained, with only a small increase in costs. A number of examples are given below.

1.6.1 Measurements During Drilling, Providing Insight into the Vertical Distribution of Water Bodies

Blind drilling is common, aiming at a specific water horizon or aquifer. It is the hydrochemist's task to convince the drilling organization that drilling should be stopped at intervals, water removed so that new water enters the well, and the following measurements carried out:

Water level
Temperature
Electrical conductivity (indicating salinity)
Samples collected for chemical and isotopic analyses.

This will provide vital information on the vertical distribution of water horizons and their properties (section 7.2). Admittedly, arresting a drilling operation costs money, but the gained information is indispensable because water layers may otherwise be missed. Savings may be made by conducting hydrochemical measurements and sampling at planned breaks in the drilling operation on weekends and holidays, or during unplanned breaks due to mechanical troubles.

1.6.2 Initial and Current Databases

Whenever a new well is completed, the abstracted aquifer should be studied in detail, including water table and temperature measurements and complete laboratory analysis of dissolved ions and gases, stable isotopes, and age indicators such as tritium, ^{14}C, and ^{36}Cl. Analysis for suspected pollutants, for example, fertilizers, pesticides, pollutants from local industries, and domestic sewage, should be carried out as well. This wealth of data is needed to provide answers to the questions raised in section 1.4.

Pumping changes the underground pressure regime, and fluids begin to move in new directions. Thus, information that was not collected at the very beginning of a well operation is lost and may not be obtained later. The hydrologic evolution of flow patterns induced by human activity in exploited water systems can be established by frequent comparisons to the initial situation (section 7.5).

As can be sensed, hydrochemical investigations produce an enormous number of observations and measured results. This is the strength of this discipline, but the flood of data has to be well registered in active and user-friendly data banks. Much thinking and care should be devoted to the administration of data banks to ensure they do not turn into data sinks (sections 18.8 and 18.9).

1.6.3 Periodically Repeated Hydrochemical Measurements (Time Series)

There are questions that cannot be answered by the mere placement of new wells. Information is needed on the:

Number of water types that are intermixed underground.
Seasonal variations in water quality and quantity.
Influence of pumping operations in adjacent wells.

This information may be obtained by conducting periodically repeated measurements in selected wells and springs (sections 6.10 and 7.6). The costs of repeated measurements are small in comparison to drilling, and they provide information not otherwise available, significantly increasing the understanding of the water system in terms needed for its management.

The economical value of hydrochemical studies also has wider aspects, as they may reduce the number of new drillings needed or improve their location. The price of hydrochemical measurements is small compared to drilling expenses, but they should be minimized by careful planning and proper selection of the parameters measured at each stage of the study, and proper selection of water sampling frequencies. These topics are discussed in Chapter 7.

1.7 Environmental Issues

Intervention of man in groundwater regimes is intensive in densely populated areas, and quality deterioration is manifold, examples being:

Encroachment of saline groundwater as a result of local lowering of water tables in unconfined systems, and reduction of water pressures in confined and stagnant systems (section 2.17 and 14.6).
Encroachment of seawater in coastal wells as a result of overpumping and an extreme lowering of the water table (section 17.6).
Reduction of local recharge due to urbanization-related sealing of the ground and formation of street runoff (section 18.2).
Pollution as a result of irrigation, industry, sewage, fuel spillage, etc.

The common thread in these water deterioration processes is their being cases of intermixing of natural waters of different qualities, or mixing of uncontaminated water with contaminated effluents. hence, the need (1) to be familiar with the methodology of mixing identification and reconstruction of the properties of the involved end members (section 6.7), and (2) establishment of initial databases (sections 18.8 and 18.9).

1.8 Summary Exercises

Exercise 1.1: "All the rivers flow into the sea . . ." stated Ecclesiastes. This remarkable observation was made more than 2000 years ago. What is remarkable about it? Which basic concepts of hydrology is it describing?

Exercise 1.2: "... and the sea does not fill up" continues Ecclesiastes. The ancient observer raises a fundamental research problem—he draws attention to what seems to be a violation of a universal law, which he must have known. What law is it?

Exercise 1.3: The above citations make it clear that the concept of the water cycle was not known to the ancient philosophers. Which segments of the water cycle were concealed from the ancient scientists?

Exercise 1.4: The introductory chapter is devoted to the encoding of information into water along the hydrological cycle. At which point is all this information erased?

Exercise 1.5: What kinds of information are inscribed into water underground?

2
BASIC HYDROLOGICAL CONCEPTS

2.1 The Aerated Intake Zone

A typical well passes through two zones: an upper zone, usually of soil-covered rock, that contains air and water in pores and fissures, and a lower zone, often of rock, that contains only water in pores and fissures. This separation into two parts is observed everywhere, and the two zones have been named the aerated (air-containing) zone and the *saturated* (water saturated) *zone*. The top of the saturated zone (where water is hit in a well) is called the water table (Fig. 2.1). The water table is described either in terms of its depth from the surface or as the altitude above sea level (section 2.2).

The aerated zone is also called the vadose zone (from Latin vadosus, shallow). Its thickness varies from zero (swamps) to several hundred meters (in regions of elevated and rugged topography and in arid climates), but it is commonly 5–25 m. In most cases the aerated zone has two parts: soil at the top and a rocky section beneath.

The soil, formed by weathering and biogenic processes, varies in thickness from zero (bare rock surface) to a few meters, reaching maximum thickness in alluvial valleys. The soil zone is characterized by the plants that grow in it, their roots penetrating 10–40 cm, but certain plants, mainly trees, have roots that grow as deep as 20–30 m. Roots may penetrate the lower rocky part of the aerated zone, occasionally reaching the water table. Plants respire and liberate CO_2 into the soil pores. Bacteria decompose plant remains in the soil, adding to the CO_2 of the soil air. This biogenic CO_2 dissolves in infiltrating water, producing carbonic acid, that slowly but continually dissolves minerals

Basic Hydrological Concepts

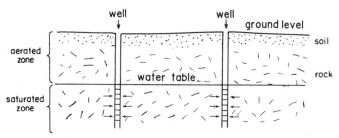

Fig. 2.1 A schematic section through typical wells: at the upper part the soil and rocks contain air and water in pores and fissures, forming the aerated zone. Below occur rocks with only water in their pores and fissures, forming the saturated zone. The top of the saturated zone is the water table, recognizable in wells as the depth at which water is encountered.

and rock components. This process creates soil and provides recharging water with dissolved salts (section 6.8).

Intake of water in the aerated zone is either by infiltration into the soil cover or, on bare rock surfaces, by infiltration into intergranular pores (as in sandstone), fissures and joints (as in igneous rocks or quartzite), or dissolution conduits and cavities (limestone, dolomite, gypsum, rock salt). Only pores and fissures that are interconnected, or communicate, are effective to infiltration.

Water infiltration in a sandy soil may be described as water movement in a homogeneous granular medium (or as in a sponge). Water descends in this setting in downward moving fronts, or in a piston-like flow (Fig. 2.2). In such a mode of flow a layering exists, the deeper water layers being older than the shallower ones. Measurements of tritium and other contaminants reveal that such piston flows proceed at velocities of a few centimeters to a few meters a year. This velocity depends on the porosity of the soil, the degree of pore

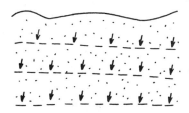

Fig. 2.2 Piston flow of water infiltrating into a homogeneous granular soil (sponge-type flow). A layered structure is formed, lower water layers being older than shallow ones. Penetration velocities are observed to range from a few centimeters to a few meters per year.

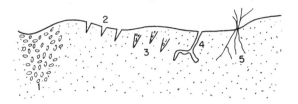

Fig. 2.3 Enhanced recharge through parts of high conductance in soil, alluvium, and nonconsolidated rocks: (1) gravel in buried streambeds; (2) open desiccation cracks; (3) old desiccation cracks filled with coarse material; (4) animal holes; (5) plant roots.

interconnection, the amount and mode of distribution of local precipitation, and drainage conditions. In contrast to the sponge-type flow, recharge into the ground is often rapid, through conduits and zones of preferred transmissivity. Examples of such conduits are open desiccation cracks, old desiccation cracks that have been filled up with coarse material, living and decayed roots, animal holes or bioturbations, and gravel-packed channels of buried streambeds (Fig. 2.3). Karstic terrains (section 2.9) provide extreme examples of recharge via conduits. In most cases water descends through the aerated zone in a combined mode, that is, through intergranular pores and through conduits that cross the granular medium (Fig. 2.4).

2.2 The Saturated Zone and the Water Table

Infiltrating water settles by gravity, reaching a depth at which all voids in the ground are filled with water. This portion of the ground is called the saturated zone. The water in this zone forms a continuous medium interwoven with the solid matter system.

The water in the saturated zone settles with a smooth horizontal upper face, called the water face, or water table. By analogy, water poured into a vessel settles with a smooth horizontal upper face—a miniature water table. The depth

Fig. 2.4 Infiltration of recharge water in a combined mode: part of the water moves slowly through the granular (mainly soil) portion of the aerated zone and part moves fast through open conduits in zones of high transmission.

Basic Hydrological Concepts

of the water table, or the water level, is expressed either in relation to the land surface or as the altitude above sea level. The depth of the water table varies from zero, in marshy areas, to several hundred meters, in dry mountainous terrains. A depth of a few tens of meters is common.

Measurements of the water table are of prime interest for management purposes and for the establishment of flow directions. Study of water table fluctuations is most valuable: rapid response to rain, flood events, or snowmelt indicates fast recharge via conduits. Delayed response indicates slow infiltration through a porous medium. Changes in the water table, caused by pumping in adjacent wells, establishes the existence of hydraulic interconnections between the respective wells (sections 4.1 and 4.2).

The slope of the water table is defined by three major factors: the base of the drainage system (sea, lake, river), the topography of the land surface, and the occurrence of impermeable rock beds.

The saturated zone usually contains strata of varying permeabilities, forming aquifers and aquicludes, described in the following sections.

2.3 Aquifers and Aquicludes

2.3.1 *Aquifers*

Rock beds at the saturated zone that host flowing groundwater are called aquifers (from Latin: *aqua*, water; *ferre*, to bear or carry). Aquifer rocks contain the water in voids—pores and fissures. The size and number of voids and the degree of interconnection between those pores and fissures define the qualities of the aquifers. The same properties define infiltration efficiency and capacity of intake of recharge water. These properties are discussed below for different rocks.

Conglomerates. These rock deposits are made up of rock pebbles cemented to different degrees. If poorly cemented, conglomerates are extremely efficient water intake systems; they form aquifers with large storage volumes and water flows relatively fast through them (provided gradients are steep enough and drainage conditions are good). Often conglomerates are cemented by calcite, iron oxides, or silica. Cementation reduces the water-carrying capacity of an aquifer, and in extreme cases conglomerates are so tightly cemented that they act as aquicludes.

Sandstone. Sandstone is in most cases composed of quartz grains with pores constituting 10–50% of the rock volume. The pores are interconnected, providing high infiltration efficiencies and providing sandstone aquifers with high storage capacities and rapid water through-flow. Cementation reduces the pore space, but cemented rocks tend to develop mechanically formed fissures that let water through.

Limestone and dolomite. These carbonate rocks are often well crystallized and can be poor in interconnected pores. However, limestone and dolomite tend to fracture under tectonic stress, and through such fractures groundwater can move. Water can have two completely different effects on the fractures: either the water fills up the fissures by precipitation of carbonates or, under high flow conditions, the water dissolves the rock and causes further opening. Dissolution of limestone and dolomite may thus create surface and subsurface conduits that enhance intake of recharge water and form high-conducting aquifers with large water storage capacities. Karstic systems are the result of extreme dissolution activity in hard carbonate rocks (section 2.9).

Igneous rocks. Granite and other intrusive igneous rocks have intrinsically well-crystallized structures with no empty pores. However, being rigid they tend to fracture and may have a limited degree of infiltration capacity and aquifer conductivity. Weathering in humid climates may result in extensive formation of soil, but little dissolution opening of fractures. Thus, granitic rocks may form very poor to medium quality aquifers.

Basalt and other extrusive lava rocks are similar to granitic rocks in having no pores and tending to fracture. In addition, lava rocks often occur in bedded structures, alternating with paleosoils and other types of conducting materials, for example, rough lava flow surfaces. Thus, basaltic terrains may have medium lateral flow conductivities. Weathering of basaltic rocks and paleosoils produces clay minerals (section 6.8) that tend to clog fractures. Tuff (a rock formed by fragmental volcanic ejecta) is porous when formed, but is readily weathered into clay-rich impervious soils.

Metamorphic rocks. Quartzite, schist, gneiss, and other metamorphic rocks and metasediments vary in their hydrological properties, but in most cases are fractured and transmit water. In rainy zones weathering may produce clays that occasionally clog the fractures.

The rocks discussed in this section have intrinsically medium to good infiltration efficiency, aquifer conductance, and storage capacities. These features may be developed to different degrees as a result of secondary processes such as tectonic fracturing and chemical dissolution opening on the one hand, and clogging by sedimentation and weathering products on the other hand. Thus, close study of rock types and of accompanying features are essential parts of hydrological investigations.

2.3.2 Aquicludes

Rock strata that prevent passage of groundwater are called aquicludes (from Latin: *aqua*, water; *claudere*, to close). Aquicludes are important components of groundwater systems because they seal the aquifers and prevent water from infiltrating to great depths. Aquicludes are essential to the formation of springs

Basic Hydrological Concepts

and shallow accessible aquifers. Aquiclude rocks have low water conductivity caused by a lack of interconnected voids or conduits. A number of common aquiclude-forming rocks are discussed below.

Clay. Clay minerals have several characteristics that make them good aquicludes:

When consolidated, they have effectively no interconnected pores.
They swell, closing desiccation fractures or tectonic fractures.
They are plastic—another property that makes clays effective sealing agents.

The effectiveness of clay aquicludes depends on the type of clay; montmorillonite being perhaps the best. The thickness of the clay aquiclude is of prime importance—a clay bed 1 m thick may be a significant barrier to underground water movement, but water may slowly leak through. The thicker the aquiclude, the higher is its sealing efficiency. Clay aquicludes with a thickness of several hundred meters are known.

Shales. Shales are derivatives of clays, formed in slow diagenetic processes. Shales have little or negligible swelling capacity and have medium plasticity. They form effective aquicludes at thicknesses of at least a few meters to tens of meters.

Igneous rocks. Large igneous bodies, such as stocks or thick sills, often act as aquicludes because they lack interconnected fractures or dissolution conduits. Dikes and sills may act as aquicludes if they are either very fresh and not fractured or are weathered into clay-rich rocks.

2.4 Recharge

The term recharge relates to the water added to the groundwater system, that is, to the saturated zone. Thus, *recharge* is the balance between the amount of water that infiltrates into the ground and evapotranspiration losses.

The location of recharge areas is of prime importance in modern hydrochemistry, as such areas have to be protected in order to preserve groundwater quality. Urbanization, agriculture, and almost any other kind of human activity, may spoil the quality of recharge water. Limitations on the use of fertilizers and pesticides in recharge areas are often enforced, as well as limitations on mining and other earth-moving activities. Since these limitations have financial impacts, an increasing number of cases are being taken to court and hydrochemists may be asked to deliver expert opinions involving the identification of areas of active recharge.

The following are a few features characterizing recharge regions:

Depth of water table is at least a few meters below the surface.
Water table contours show a local high.

Significant seasonal water table variations are noticed.
Local groundwater temperatures are equal to, or several degrees colder, than the average ambient annual temperature.
Effective water ages are very recent (a few months to a few years).
Salt content is low in most cases in nonpolluted areas (less than about 800 mg/l total dissolved ions).
Surface is covered by sand, permeable soil, or outcrops of permeable rocks.

2.5 Discharge

The term discharge relates to the emergence of groundwater at the surface as springs, water feeding swamps and lakes, and water pumped from wells. Discharge is the output of groundwater. The rate of discharge is measured in units of volume per time, for example, cubic meters per hour (m^3/h) (section 4.10). The bulk of the groundwater is drained into terminal bases of drainage, which in general are the seas. Most of this recharge is concealed, but can be detected indirectly, for example, by the presence of fresh water in shallow wells at the seashore or in offshore boreholes.

Discharge areas away from the terminal base of drainage warrant identification in relation to drainage operations, water exploration, and water quality preservation.

Several features characterize discharge areas:

Water table levels mark, in most cases, a high in the water table contour map.
Water temperatures are a few degrees above the average annual ambient temperature, reflecting warming by the thermal gradient while circulating at depth (section 4.8).
Waters are either fresh or saline (if they have passed through salinizing rocks) (section 3.1).

2.6 Evapotranspiration

Only part of the water infiltrating into the ground keeps moving downward into the saturated groundwater zone. An important part of infiltrated water is transferred back into the atmosphere. Two major mechanisms are involved: evaporation and transpiration, together called evapotranspiration.

Evaporation is a physical process, caused by heat energy input, providing water molecules with kinetic energy that transfers them from an interpore liquid phase into a vapor phase. Evaporation depends on local temperature, humidity, wind, and other atmospheric parameters, as well as soil properties. A maximum value is provided by the extent of evaporation from an open water body; this varies from place to place in the range of 1–2 m/year. Taken at face value, this would imply that no effective recharge occurs in areas that have an

Basic Hydrological Concepts

average annual precipitation rate lower than the potential evaporation rate measured in an open water body. This is not the case, however, as effective recharge is known in regions that have as little as 70 mm average annual rainfall. This indicates that water infiltrated into the ground is partially protected from evaporation. In fact, evaporation occurs mainly from the uppermost soil surface and, when this is dry, further evaporation depends on capillary ascent of water, which is a slow and only partially efficient process. It seems that water that has infiltrated to a depth of 3 m is "safe" from capillary ascent and evaporation. From this discussion it becomes clear that the relative contribution of precipitation to recharge is proportional to the amount of rain falling in each rain event. Intensive rain events push water deep into the ground, contributing to recharge, whereas sporadic rains only wet the soil and the water is then lost by subsequent evaporation.

The mode of rainfall and degree of evaporation influence groundwater composition. High degrees of evaporation result in enrichment of the heavier stable isotopes of hydrogen and oxygen (sections 9.3 and 9.6), and in higher concentrations of dissolved salts.

Transpiration is the process by which plants lose water, mainly from the surfaces of leaves. Plants act as pumps, their roots extracting water from the soil and the leaves transpiring it into the atmosphere. Thus, the depth of the transpiration effect on soil moisture is defined by the depth of a plant's root system. The latter is most intensive to 40 cm and drops to almost nil at depths greater than 2 m. Roots of certain plants, mainly trees, are occasionally observed to penetrate to 20–30 m.

2.7 Permeability, Impermeability, and the Selection of Representative Values

2.7.1 Permeability Definition

A large sector of hydrology deals with through-flow systems, that is, groundwater systems that are recharged. A term is needed that expresses the ability of different rock types to let water infiltrate and flow through. This term is the permeability coefficient, k.

Henry Darcy, a French engineer, conducted experiments in 1856 in which he measured the flow velocity of water through a sand-filled tube. He found that the flow velocity (V) is directly proportional to the difference in hydraulic head (Δh), and is inversely proportional to the flow distance (Δl):

$$V = k\,(\Delta h/\Delta l)$$

where $\Delta h/\Delta l$ is the hydraulic gradient. The coefficient of proportionality, k, depends of the rock properties and is case specific. Hence, k is called the coefficient of permeability.

A convenient way to define the coefficient of permeability, in order to compare its value for different types of rocks, is to determine its empirical value for hydraulic gradients of 45°, that is, for $\Delta h/\Delta l = 1$. In such cases $k = V$.

The permeability coefficient (k) has the units of velocity, that is, distance/time. It is determined either in laboratory experiments or derived from pumping tests. Both methods are semiquantitative, but are still highly informative, as the values observed for common rocks span more than seven orders of magnitude. A variety of units are in use—m/day being a common one. The following are a few of the average permeability or hydraulic conductivity values floating around in the literature, expressed in m/day:

shale, 10^{-7}; clay, 10^{-6}; sandstone, 10^{-2}; limestone, 1; sand, 10; and gravel, 10^2.

2.7.2 Impermeability—Discussion of the Concept

Metal, glass, or ceramic vessels are generally agreed to be impermeable. Yet, when it comes to rocks many hydrologists believe that there exist no impermeable natural materials, and all rocks are permeable, although to different degrees. This approach may be examined in light of the following analysis. The accepted permeability values, listed above, assign seven orders of magnitude difference between the permeability of clay or shale and that of sandstone or fractured limestone. Thus, the resistance to flow exerted by 1 m of clay or shale is equivalent to the resistance exerted by 10,000 km of sandstone or fractured limestone (check the calculation!). Or, in a different example: if water flows a certain distance through limestone or sandstone in 1 year, it will take the water 10 millions years (!) to flow an equivalent distance through clay or shale. These examples sure grant clay and shale the title of impermeability.

The question of whether certain rocks are impermeable or possess a very low permeability is not just a matter of semantics, it is of principle importance to the understanding of hydrological systems, as discussed in sections 2.7 and 2.10–2.13. A permeability of 10^{-7} m/day cannot be directly measured. It is assigned to clay by interpolation of indirect measurements, and this is done with prejudice—believing a priori that clay is permeable. In this regard one may follow the physicists—they do apply absolute terms, such as opaque, rigid, or impermeable. Why should not hydrologists do the same in the appropriate cases? Finally, there are examples of impermeable rock materials:

Fluid inclusions are common in mineral crystals that are 10^7–10^8 years old. The millimeter thick walls of these crystals demonstrate the existence of impermeable rock materials.
Oil and gas deposits are trapped for long periods, stored in permeable rocks that are engulfed by impermeable rocks (no one will claim that these are through-flow systems with active oil or gas recharge).

Basic Hydrological Concepts

Along the same lines, groundwater is expected to exist in traps as well, engulfed by impermeable rocks. Related arguments and ample observations are discussed in sections 2.12 and 2.14.

2.7.3 Selection of Representative Permeability Coefficients

Permeability is observed to vary enormously in what looks like homogeneous rock bodies, and it varies much more between the different rock types that constitute a studied system. Disregarding this large range of different prevailing permeabilities, mathematical models accommodate only single permeability values for entire case studies or for large sections of studied systems. Thus, representative values are selected in the frame of a whole list of model simplifications. Such simplifications can lead to misinterpretations and therefore should be avoided. If, however, it is decided to select a representative permeability coefficient for a model, the selection should at least be done correctly.

The procedure adopted to select the representative permeability coefficient is rarely explained in published case studies. It seems that a common approach is to apply a weighted representative value. Let us examine this procedure in light of an example of a system that is composed of 90% fractured limestone (a coefficient of 1 m/day according to the accepted tables) and 10% of shale (10^{-7} m/day). The average weighted k will be

$$k = 0.9 \times 1 + 0.1 \times 10^{-7} = 0.9 \text{ m/day}$$

The calculated value of 0.9 m/day is practically equal to the value of the high-conducting limestone and does not express at all the presence of the flow-resisting shale. Such a selected value is totally wrong, and will lead to the calculation of an erroneously high flow velocity, a too short travel time, and a meaningless young water age (section 2.10).

In order to properly represent the flow-resistance of the shale, a harmonic mean seems much more plausible, or in our example,

$$k \text{ representative} = (90 + 10) : (90/10 + 10/10^{-7}) = 1.0 \times 10^{-4}) \text{ m/day}$$

This representative k reflects the ability of even thin beds of clay and shale to stop through-flow.

The discussed example provides a glimpse into the pitfalls that await the hydrologist in the selection of a representative permeability value—the different modes of calculation can vary over almost five orders of magnitude! It seems that a large list of modeled case studies should be revised in light of the necessity to properly select the applied k values. In most cases the application of the harmonic mean calculation provides significantly lower k values, resulting in significantly higher water ages (section 2.10), which is in good agreement

2.7.4 Porosity

Water is stored in rocks mainly in pores. The *effective porosity* of a rock is the volume percentage of the rock that may contain water in pores. The values (Table 2.1) are high for nonconsolidated granular rocks—soils, clays, silts, and gravels. The porosity is low for crystallized rocks such as limestone, dolomite, and most igneous rocks. Movement of water through rocks is through interconnected pores, fissures, and conduits. *Permeability* is a measure of the ease of flow through rocks (hydraulic conductivity is another term). Permeability relates to the pores and voids that are interconnected; it differs from porosity, which is the total pore volume. Clay has a high content of water stored in the molecular lattice, but these water sites have practically no interconnections. Thus, in spite of its high capacity for containing water, clay has a low permeability, making it a most efficient aquiclude. In contrast, sandstone (nonconsolidated) has a high permeability, providing excellent aquifer properties.

2.8 Through-Flow Systems: Unconfined and Confined

2.8.1 Phreatic (Unconfined, Free Surface) Aquifers

Phreatic aquifers have free communication with the aerated zone. The synonym *free surface aquifer* relates to the free communication between the aquifer and the vadose zone. An example is shown in Fig. 2.5. The term phreatic originates from the Greek word for a well.

Phreatic aquifers are the most exploited type of aquifer and most of the hydrochemist's work is performed on them. Phreatic aquifers are the collectors of infiltrating recharge water, and this process is well reflected in the chemical, physical, and isotopic parameters of the aquifer's water: water table fluctuations indicate the nature of recharge intake (sections 4.1 and 4.2), water table gradients indicate flow directions (section 4.3), water temperature indicates the nature of infiltration and depth of circulation (section 4.7), and tritium and radiocarbon concentrations reflect water ages and modes and velocity of flow (sections 10.7 and 11.8).

Table 2.1 Rock porosities, volume percent

Soils	50–60%	Fine sand	30–35%
Clay	40–55%	Pebbles	30–40%
Silt	40–50%	Sandstone	10–20%
Coarse sand	35–40%	Limestone	1–20%

Basic Hydrological Concepts

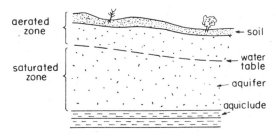

Fig. 2.5 Basic components of a phreatic groundwater system: intake outcrops, an aerated zone, the water table, the saturated zone that constitutes a water-bearing aquifer, and impermeable rock beds of the aquiclude that seal the aquifer at its base.

2.8.2 Confined Aquifers and Artesian Flow

Confined aquifers are water-bearing strata that are sealed at the top and the bottom by aquiclude rocks of low permeability (Fig. 2.6). Confined aquifers are commonly formed in folded terrains (section 3.4) and have a phreatic section, where the aquifer rock beds are exposed to recharge infiltration, and a confined section, where the aquifer rock beds are isolated from the landscape surface by an aquiclude (Fig. 2.6).

The water in the saturated zone of the phreatic section of a confined system exerts a hydrostatic pressure that causes water to ascend in wells. In fact, a confined aquifer can often be identified by the observation that water ascends

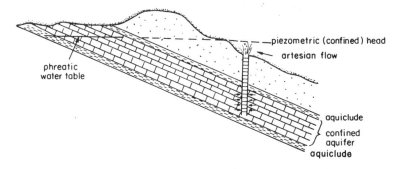

Fig. 2.6 Components of a confined aquifer with through-flow: tilted, or folded, water-bearing rock strata, sealed at the top and the base by aquicludes. Each active confined system also has a phreatic section at outcrops of the aquifer rocks. The level of the water table in the phreatic section defines the piezometric head in the confined section. Water ascends in boreholes drilled into confined aquifers. Water reaches the surface in artesian flow in boreholes that are drilled at altitudes lower than the piezometric head.

in a borehole to a level higher than the level at which the water was first struck. In extreme cases the water ascends to the surface, constituting an artesian well. This phenomenon of water ascending in a well and flowing by itself was first described in 1750 in the area of Artois, a province in northern France. The term artesian is applied to self-flowing wells and to aquifers supporting such wells. The term semiartesian describes wells in which water ascends above the depth at which the water was struck but does not reach the surface.

The level water reaches in an artesian well reflects its pressure, called the *piezometric*, or *confined, water head* (Fig 2.6). In boreholes drilled at altitudes that are lower than the piezometric head, water will reach the surface in a jet (or wellhead pressure) with a pressure that is proportional to the difference between the altitude of the wellhead and the piezometric head. The piezometric head is slightly lower than the water level in the relevant phreatic section of the system, due to the flow resistance of the aquifer. Confined aquifers often underlay as phreatic aquifer, a shown in Fig. 2.7. The nature of such groundwater systems may be revealed by data measured in boreholes and wells: the water levels in wells 1 and 2 of Fig. 2.7 did not rise after the water was encountered, and both wells reached a phreatic aquifer. Well 3 is artesian, and the drillers' account should include the depth in which the water was struck and the depth and nature of the aquiclude. The hydraulic interconnection between well 1 and well 3 may be established by:

Checking their water heads.
The resemblance of the relevant aquifer rocks.
Agreement of the measured dips at the outcrop and the depth at which the water was encountered at well 3.

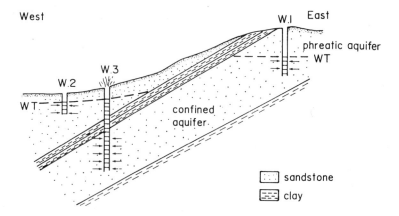

Fig. 2.7 A confined aquifer underlying a phreatic aquifer. The nature of such a system may be established by parameters measured at boreholes and wells (see text).

Basic Hydrological Concepts

Flow of groundwater in confined aquifers is determined by the water head gradient and by the degree to which the system is drained. In extreme cases water may be trapped in confined aquifers. The topic of water movement in confined aquifers is best studied by means of chemical and isotopic parameters (section 6.5).

2.8.3 Stagnant Aquifers (Groundwater Traps)

The discussion so far has dealt with through-flow systems that have active recharge and discharge. However, there are stagnant aquifers in which static groundwater is stored. This topic is discussed in detail in section 2.12.

2.9 Karstic Systems: Paths of Preferred Flow

Hard calcareous rocks, that is, limestone and dolomite, develop specific features noticeable on the land surface: dissolution fissures, cavities, caves, and sinkholes in which runoff water disappears. On the slopes of such terrains occur springs with high discharges that often undergo a marked seasonal cycle. In such terrains intake of recharge water and underground flow are in open conduits (sections 2.1 and 2.2). A classical area of such conduit-controlled water systems has been described in the Karst Mountains of Yugoslavia, and from there originated the term *karstic*, pertaining to groundwater flow in dissolution conduits. Some basic features of karstic systems are depicted in Fig. 2.8. Sinkholes are inlets to conduits in which runoff water can move rapidly into deeper parts of the system. Karstic springs occur at the base of local drainage basins, commonly major river beds or the seashore. Abandoned spring outlets remain "hanging" above new drainage bases, forming caves that can sometimes be rather deep. Dissolution conduits are formed by the chemical interaction of CO_2 charged water with calcareous rocks (section 6.8).

Karstic systems are often formed in generations. For example, during the Pleistocene, when the oceans had a lower water level, low-level karstic systems were formed in coastal areas. With the rise of sea level, higher karstic systems evolved.

The karstic nature of recharge intake zones is often recognizable from specific geomorphic features, but this is not always the case., Karstic systems are occasionally concealed, and karstic outlets may occur tens of kilometers from the intake areas. Furthermore, paleokarstic conduits may be in operation in regions that at present have a semiarid climate. Thus, indicators for karstic flow are needed. In small-scale studies, direct tracing of karstic underground interconnections has been successfully carried out with dyes, spores and radioactive tracers. The limitations are:

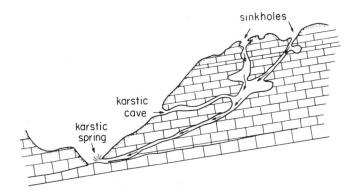

Fig. 2.8 Components of a karstic system: sinkholes and fissures acting as inlets for rapid intake of runoff water; dissolution channels; water discharging in springs at local bases of drainage, for example, river beds; caves exposed at former dissolution channels formed at former (higher) drainage bases.

The dependence on sinkholes as tracer inlets.
The logistics of tracing experiments that limits them to distances from hundreds of meters to at most a few kilometers.

Hydrological and hydrochemical indicators for karstic flow include:

Rapid flow, revealed by very young groundwater ages (months to a few years).
Water temperatures that maintain the intake temperatures (e.g., snowmelt temperatures).
Significant seasonal discharge fluctuations.
Light stable isotope components, reflecting composition of high-altitude precipitation and indicating lack of evaporation.
Chemical composition reflecting contact with limestone and/or dolomite (section 6.8).

2.10 Flow Velocity and Groundwater Age

The term *flow velocity* sounds simple: the distance the water moves in a unit of time. In this sense it is also handled by Darcy's experiment:

$V = k\,(\Delta h/\Delta l)$

In the experiment the water flows in a tube in which the water enters in one end, flows along the tube, and leave at the other end. The impermeable walls of the tube turn it into a well-defined *one-dimensional* system, in which the point of recharge, the flow path, and the point of discharge are rigidly defined.

Basic Hydrological Concepts

The term *groundwater age* sounds simple as well: the time that has passed since a water parcel was recharged into the saturated water system. Or, in terms of travel time, the time that has passed since the water entered the ground until it reaches the point at which it is observed, for example, at a spring or a well. In a simplistic way one could compute the age of water (t) in Darcy's experiment by dividing the length of the flow path (Δl) by the measured flow velocity (V):

$$t = \Delta l / V$$

Again, we have to remember that the above equation holds true for the one-dimensional setup of flow through a vertical tube.

Most hydrology textbooks apply a tilted-tube version of the Darcy experiment in order to relate it to aquifers. The analogy is possibly relevant only to tilted confined through-flow systems of the kind drawn in Fig. 2.7. However, such setups are scarce, since tilted rock systems rarely have adequate recharge intake areas, they do not serve as conduits, as they are often blocked by tectonically squeezed plastic rocks and as a result of compaction, and their lower end is often buried beneath the base of drainage (Fig. 2.10). Thus, many confined systems host stagnant (trapped) groundwater (section 2.11), for which the Darcian equations are not geared (Mazor and Native, 1992, 1994; Mazor, 1995; Mazor et al., 1995 and Fridman et al., 1995).

The bulk of groundwater through-flow takes place in unconfined systems. Precipitation falls over the entire surface of such systems and recharge is fed into them at every point. As a result, there is a constant increase in the water flux along the flow path: the base flow of the system accumulates water along its course, and as a result the flow velocity must increase as well. Hence, no singular flow velocity exists in an unconfined system.

Similarly, water has no singular age in unconfined systems. Water recharged at the upper reaches of a system has a long flow path, and would attain a relatively high age along the flow path, but it is constantly mixed with more re-

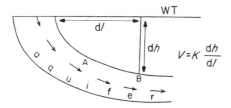

Fig. 2.9 Darcy's law applied to a groundwater system: the velocity (V) at which water flows in the aquifer rocks is proportional to the coefficient of permeability (k) and the gradient ($\Delta h/\Delta l$). This is a greatly simplified approach, and a number of assumptions have to be checked with hydrochemical methods (see text).

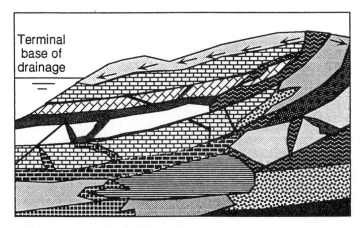

Fig. 2.10 Rock beds in a subsidence basin. The part above the terminal base of drainage, for example, the sea, functions as a through-flow system (arrows). The deeper rock beds are fossil through-flow systems that host stagnant groundwater as they are (1) covered by impermeable rocks, (2) bisected by plastic impermeable rocks that have been squeezed into stretch joints in the competent rock beds and in between bedding plane thrusts, and (3) placed in a zone of zero hydraulic potential.

cent water that is recharged along the downflow course. Thus, unconfined systems are characterized by constant mixing of waters of different ages. Hence, there is no room for hydraulic (Darcian) calculation, as it is built only for singular water age values.

This brief discussion of the flow regime of an unconfined system emphasizes its three-dimensional nature and the large variety of water fluxes, velocities, and ages that prevail in each case study. The unconfined groundwater regime differs fundamentally from the one-dimensional tube of Darcy's experiment, with its particular properties.

A completely different approach to the question of groundwater age is taken in the isotopic groundwater dating methods, using tritium, ^{14}C, ^{36}Cl, and ^{4}H (sections 10.4, 11.4, 12.2, and 14.5). The short-range isotopic dating methods are informative for through-flow systems, and the long-range isotopic dating methods are especially useful for stagnant groundwaters.

2.11 Flow of Water: Basic Observations

The direction of flow of water is controlled by two major factors:

Earth's gravitational field.
The degrees of hydraulic freedom.

Basic Hydrological Concepts

Four basic modes of water flow are observable in daily life (Fig. 2.11):

Free fall, as in water that is released from a tilted bottle or as in rain. All directions are hydraulically free (three-dimensional degrees of hydraulic freedom) and the water is free to flow vertically under the force of gravity (Fig. 2.11a).

Lateral overflow, as in water reaching a filled tub. The water that fills the tub blocks the downflow direction of the added water (two-dimensional degrees of hydraulic freedom) and the latter overflows, that is, it flows laterally toward the bases of drainage (Fig. 2.11b). The observation that the water flows laterally indicates its surface (water table) is slightly tilted by the critical angle of flow. The latter is defined by the water viscosity. The water present in the deeper part of the tub does not flow—it is stagnant, bounded in a "dead volume," situated under zero hydraulic potential difference. Small eddies create a transitional mixing zone (Fig. 2.11b).

Forced flow, as in water flowing in a tube or a pipe. The impermeable walls of the pipe restrict the flow to the course of the pipe (one-dimensional degree of hydraulic freedom), with only one inflow point and one outflow point (Fig. 2.11c).

Stagnation (zero flow), as in water in a closed bottle. The impermeable walls of the bottle completely block the mobility of the water (zero degrees of hydraulic freedom) and it stays at rest (Fig. 2.11d).

The flow of groundwater is discussed in these terms in the following section.

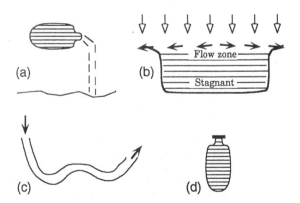

Fig. 2.11 Four basic modes of flow of water: (a) free fall, three-dimensional degrees of hydraulic freedom; (b) lateral overflow, two-dimensional degrees of hydraulic freedom; (c) forced flow (one dimension of hydraulic freedom); and (d) stagnation (zero degrees of hydraulic freedom).

2.12 Two Basic Experiments of Water Flow

Let us expose an empty aquarium to rain. The vessel will fill up and water will overflow laterally toward the "bases of drainage." The observation that the water is flowing indicates that the water surface, or water table, is slightly inclined (Fig. 2.12) by the critical angle of flow. This angle is determined by properties of the water, mainly viscosity, which in turn is determined by salinity and temperature. If the vessel is exposed long enough to rain, a steady state is reached at which new rain flows out laterally, whereas water at the bottom of the vessel remains at rest, forming a zone of stagnation ("dead volume"). Small eddies cause some mixing of water from the flowing zone with water of the upper part of the stagnation zone, forming a transition or mixing zone.

As a second experiment an aquarium is filled with sand and then exposed to rain (Fig. 2.13). Water infiltrates through the sand, forming a saturated zone and a water table that rises until lateral overflow toward the edges of the vessel occurs. This critical angle of flow is slightly steeper in this experiment, as the water has to overcome the resistance to flow caused by the sand. At a steady state the rain infiltrates vertically down through the heaped sand relief, performing free flow until the water table is reached, at which time the infiltrating water switches to lateral flow. The presence of the sand impedes the formation of eddies, and the mixing zone is significantly thinner as compared to the previ-

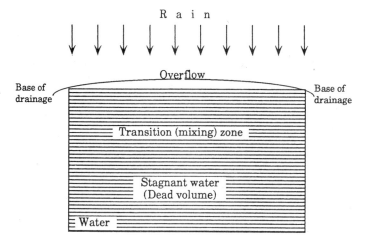

Fig. 2.12 A basic experiment in water flow: an aquarium exposed to rain. The vessel fills up and the additional rainwater overflows. At a steady state, all the arriving rainwater flows laterally toward the bases of drainage, and the water at the bottom of the vessel is stagnant (dead volume).

Basic Hydrological Concepts

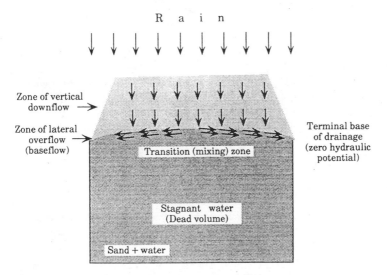

Fig. 2.13 A second basic experiment in water flow: an aquarium filled and heaped with sand, exposed to rain. At steady state the rain infiltrates along vertical downflow paths (free flow) until a zone of lateral flow (overflow) is reached. The water at the deeper part is stagnant. Eddies create a thin transition or mixing zone.

ous experiment, leaving a larger portion of the water beneath the base of drainage in a stagnant state.

The phenomenon of lateral overflow is of basic importance in groundwater systems: in steady-state conditions, typical for unconfined groundwater systems, the rocks beneath the level of the terminal base of drainage are filled with water that is under no potential energy gradient, blocking the downward flow direction for newly arriving water. Thus, the hydraulic degree of freedom is only two-dimensional at the level of the base of drainage, causing lateral flow.

2.13 A Model of L-Shape Through-Flow Paths and Zones of Stagnation

Groundwater regimes are best understood by discussing entire groundwater drainage basins, that is, from the principal water divide to the terminal base of drainage, which in most cases is the sea (Fig. 2.14).

2.13.1 Unconfined Permeable Systems

At the first stage of the discussion let us assume that all the rocks in the system are highly permeable to a great depth beneath the level of the terminal base

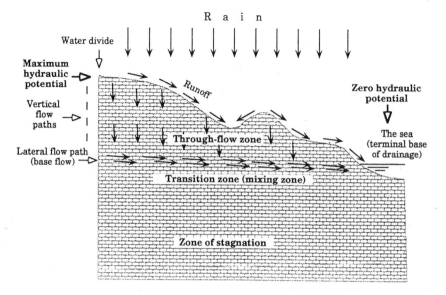

Fig. 2.14 An entire groundwater system, from the water divide to the terminal base of drainage, built of permeable rocks. The following patterns of water motion are recognizable: (1) a through-flow zone with vertical flow paths that join a lateral flow path toward the terminal base of drainage; (2) a transition (mixing) zone; and (3) a zone of stagnation occurring beneath the level of the terminal base of drainage (zero hydraulic potential).

of drainage—the entire system is unconfined. The groundwater regime may be discussed by its following components:

Runoff. Precipitation falls over the surface of the system, partitioning between infiltrating water and runoff. The runoff gets focused into the valleys and riverbeds, turning the latter into areas of increased infiltration rates and thus increased recharge rates (Fig. 2.14).

Infiltration and evapotranspiration. Infiltrating water is partially returned to the atmosphere by evaporation of water that is retarded on the surface and in the soil zone. Another part of the infiltrating water is returned to the atmosphere by plants by way of transpiration. Both effects, together called evapotranspiration, are effective to a depth of 1–3 m. In extra heavy rain or snowmelt events infiltrating water moves beyond this depth and is then on its way into the saturated zone. The portion of water that enters the saturated zone is the recharge.

Evapotranspiration returns distilled water into the atmosphere. The salts, dissolved in the original precipitation water, remain in the soil zone, and at the end of dry seasons some of these salts accumulate on the surface. The next rain event redissolves these salts, and eventually they are brought into the saturated zone. Thus, recharge water is richer in atmospherically derived salts than the original precipitation. The factor by which the chlorine concentration in the recharge water is greater than its concentration in the respective precipitation serves as a semiquantitative tool to estimate the amount of water transferred to the atmosphere by evapotranspiration (section 6.15).

Recharge. The portion of infiltrating water that reaches the saturated zone is the recharge. The following observations indicate occurrence of recharge: (1) a rise of the water table following rainy seasons, (2) measurable concentrations of tritium and other contaminants, indicating recharge during the last decades, (3) the continuous supply of water in shallow wells that sustain abstraction over hundreds and thousands of years.

Zone of through-flow. The term *through-flow* pertains to the section of the water cycle at which water is recharged, then flows through voids in the rocks, and eventually is discharged at a terminal base of drainage. This definition leaves out other modes of groundwater motion, which are not connected to ongoing recharge, for example, diffusion and migration due to compaction. Through-flow occurs in the zone of hydraulic potential differences, that is, between the topographic relief and the terminal base of drainage at which the hydraulic potential is zero (Fig. 2.14).

The term *terminal base of drainage* is applied to emphasize that the level of final drainage is meant, not an intermediate base of drainage such as a lake, a valley, or a river. The terminal base of drainage is in general the sea.

Two major flow path directions prevail in the through-flow zone: vertical downflow and lateral overflow, *or* base flow (Figs. 2.13 and 2.14). The lateral flow path is defined by the location and altitude of the terminal base of drainage, and the slope of the lateral flow path is determined by the critical angle of flow. Recharged groundwater flows vertically down until it reaches the lateral flow zone. The situation is simple in coastal plains—the recharge water flows vertically down (infiltrates) through the aerated zone and the zone of lateral flow coincides with the water table. In mountainous terrains the zone of vertical downflow is to greater depths.

The vertical downflow paths and the lateral flow zone take the shape of the letter L, hence the name "zone of L-shape through-flow paths," which differs from the U-shape flow paths model discussed in section 2.15.

Zone of groundwater stagnation. At depths below sea level all the rock systems of the continents are filled with water to their full capacity—they are saturated. Being below sea level, the water stored in these rocks is under no hydraulic potential difference, and therefore this water does not flow—it is static, or stagnant (Fig. 2.14). This situation is similar to that of water stored in a tub, it cannot flow out, and hence, is stagnant. Additional water reaching the tub overflows. The same is observed in the sand-filled aquarium experiment (Fig. 2.13): after steady state is reached, all the new rainwater infiltrates down to the level of the rim and flows out, whereas the deeper water remains static.

A closer look at the zone of lateral base flow (overflow). Overflowing water flows laterally by the critical angle. This angle is determined by the water viscosity, which in turn is dependent on the temperature and concentration of dissolved ions. Groundwater flows laterally toward the terminal base of drainage at a critical angle that is determined by the hydraulic conductivity, or permeability, of the rocks (k) the water viscosity, which depends on the temperature (T), and the concentration of dissolved ions (i):

Critical angle of flow = water table gradient = $f\{k, T, i\}$

This topic warrants more experimental and theoretical work.

Nonpumped wells in coastal plains reveal water tables that are at sea level at the seashore, while the water table is observed to be higher inland, portraying a water table surface that is inclined seaward by a gradient of 1–3 m/km. These water table gradients seem to reflect critical angles of flow that vary because of variations in the rock conductivity (k). In coastal plains the observed water table reflects the regional base flow and the gradient is the critical angle of flow.

The zone of lateral base flow accumulates recharge water introduced over the surface. The flux, or amount of water flowing at each section, increases from the principle water divide in the downflow direction and reaches a maximum value at the drainage front. Thus, the zone of lateral base flow has a certain thickness that is expected to grow toward the base of drainage.

Through-flow–stagnation transition zone (mixing zone). Eddies cause some mixing of water from the through-flow zone with water in the underlying stagnation zone, resulting in the formation of a mixing or transition zone. Chemical, isotopic, and dating parameters are useful in determining the location and depth of the transition zone by tracing mixtures reflected by these parameters, based on the compositional differences of the water from the two zones, and conditioned on access to adequate wells at various depths. A thickness of up to 50 m is assumed for a typical transition zone.

A closer look at the zone of vertical downflow paths. Local recharge flows vertically down until the zone of lateral flow is reached. In coastal plains the vertical flow zone is fairly thin, on the order of a few meters to a few tens of meters, and it clearly coincides with the aerated zone, and the water table signifies the zone of lateral base flow. In mountainous regions the vertical flow paths are longer, and usually the higher the topographic relief is, the thicker is the zone of vertical downflow. The following observations testify to the existence of vertical downflow paths:

During the construction of the 11-km long Mont Blanc tunnel, water was sampled at nearly 70 points of emergence (Fontes et al., 1979). The results are discussed in detail in section 10.7 and are portrayed in Figs. 10.13 and 10.14. The chemical composition reflected the overlying rocks, and the stable hydrogen and oxygen isotopes reflected the overlying topography (altitude effect), bringing the researchers to the conclusion that recharge water descended through fractures in a vertically downflow direction. The height of the covering mountains is about 2300 m above the tunnel, which in turn, is about 1300 m above sea level. Thus, in this part of the Alps the vertical flow zone has a thickness of at least 2300 m and the zone of lateral flow is below 1300 masl. The Mont Blanc tunnel study revealed additional information: tritium was found in all water samples, in concentrations that indicated rapid downflow—in a matter of months to a few years. In addition, the tritium and chlorine concentrations varied significantly between adjacent points of water collection, indicating that the downflow occurs in fractures and other vertical flow paths that have restricted hydraulic interconnections. Flow of waters with different properties has since been studied in other tunnels, providing evidence for vertical downflow paths in mountainous regions.

Flow of water from fractures is often encountered in mines at different depths. Significant variations in the water properties suggest locations in the downflow zone. In contrast, in mines located in the lateral flow zone or in stagnant systems, uniform water properties are expected.

Mountain slopes and escarpments are often with no springs, although they are recharged by precipitation. This is a clear indication that in such regions the recharge water flows vertically down until it meets a lateral flow zone at greater depth (Fig. 2.15).

Karstic shafts manifest vertical downflow paths.

Hydraulic connectivity in the vertical and lateral flow zones. Hydraulic connectivity is poor in the vertical downflow zone and significantly higher in a lateral base flow zone.

Downflow paths in fractures and in porous material have restricted hydraulic interconnections, as has been well demonstrated in the mentioned study of water at the Mont Blanc tunnel. In mountainous terrains this results in some ambiguity as to the configuration of the water table. Recharge water flows down in a wide range of velocities, varying among fractures and between fracture flow and porous flow. Furthermore, karstic systems (section 2.9) contain in the dry season air pockets that fill up in the high flow season (Herzberg and Mazor, 1979). Disconnected setups of the downflow paths reveal:

Different heights of the water table in neighboring wells.
Different seasonal fluctuations of the water table in adjacent wells.

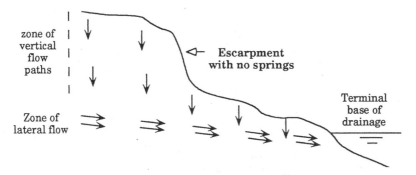

Fig. 2.15 Many escarpments and mountain slopes have no springs, especially in semi-arid regions. This shows that in such regions the vertical downflow paths of the local recharge water are relatively deep, meeting a lateral base flow zone that is situated deeper than the base of the escarpment.

Selectivity in the propagation of contaminants.
Different water properties in nearby wells and springs.

The hydraulic connectivity is much better along the flow paths of the lateral flow zone. This is best demonstrated on coastal plains: groundwater flows from the mountain foothills toward the sea, and along these flow paths local recharge water is added and intermixed.

Evidences for the existence of hydraulically isolated confined groundwater systems. Adjacent wells in deep artesian systems are often observed to differ in (1) their lithological sequences, indicating that abstraction is from different aquifers that are either stacked one on the other or are laterally adjacent; (2) the chemical composition of the water, indicating that efficient hydraulic barriers prevent the different waters from intermixing; (3) the isotopic composition of the water, indicating recharge under different climatic regimes; (4) the concentration of the isotopic age indicators, demonstrating recharge at different times; (5) the frequently observed high isotopic ages (more than 25,000 years, indicated by a lack of measurable ^{14}C, and more than 10^6 years, indicated by a lack of atmospherically produced ^{36}Cl), indicating cutoff from on-going recharge, and (6) the hydraulic pressures, indicating that the various wells tap distinct aquifers. Arguments of this kind have been used to demonstrate the existence of trapped groundwaters in the Great Artesian Basin, Australia (Mazor and Nativ, 1992, 1994; Mazor, in press), the Dead Sea Rift Valley (Mazor et al., 1995), and the Ramon National Geological Park, Israel (Fridman et al., 1995).

Basic Hydrological Concepts

Fossil through-flow systems. Trapped groundwater usually occurs in confined systems that are deeply buried in subsidence basins and rift valleys, as demonstrated by the list of observations discussed in the previous chapter. These stagnant systems often sustain self-flowing (artesian) wells. The water encountered in these wells has isotopic compositions that establish with certainty a meteoric origin. Hence, the trapped water systems are fossil through-flow systems that underwent subsidence, followed by accumulation of sealing rocks and the formation of new through-flow systems that eventually were buried as well. Continental sediments, accumulated in most basins, are characterized by frequent lithological changes, vertically and laterally, including facies changes and interfingering. These environments make room for a multitude of separated aquifers, a tendency that is further accentuated by tectonic deformations. The age of groundwaters trapped in sedimentary basins is expected to be between the age of the hosting rocks and the age of the subsidence stage that caused the hydraulic isolation.

The four basic modes of water flow, discussed in the previous chapter, are well observed in groundwater systems: free flow (three-dimensional degrees of hydraulic freedom) in the vertical downflow zone; overflow (two-dimensional degrees of hydraulic freedom) in the zone of lateral base flow; forced flow (one-dimensional degree of hydraulic freedom) in fractures and karstic channels; and stagnation (zero degrees of hydraulic freedom) in groundwater traps. These basic modes of groundwater flow are the essence of the presented model of L-shape through-flow paths and zones of stagnation.

2.13.2 Effect of Hydraulic Barriers in the Through-Flow and Stagnation Zones

The through-flow zone has so far been described in a simplified mode, assuming all the hosting rocks are homogeneously permeable. Deviations from the simplified L-shape of the flow path is caused by the presence of hydraulic barriers, such as clay and shale, that may in certain places block the downflow and create local perched water systems and springs (Fig. 2.16) or cause steps in the path of the lateral flow zone. But the overall L-shape is generally preserved, as the water of perched systems finds pathways to resume the vertical downflow direction.

2.13.3 Through-Flow and Stagnation Zones: The Global Picture

Figure 2.17 portrays a global picture: through-flow systems prevail above sea level, stagnant trapped systems dominate beneath sea level (along with petro-

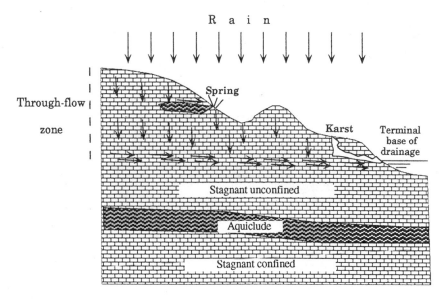

Fig. 2.16 A lense of impermeable rock creating a local perched water table and a spring; an aquiclude sealing off a stagnant zone, turning it into a confined system; and karstic features.

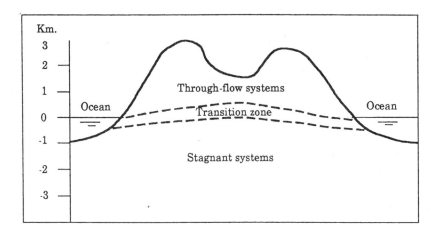

Fig. 2.17 A global picture. At depths above sea level through-flow systems prevail, near sea level exists a transition zone with both through-flow and stagnant systems, and beneath sea level prevail stagnant groundwater systems.

Basic Hydrological Concepts

leum traps), and a transition zone exists in between. The volume of water stored in stagnant systems seems to match the volume stored in through-flow systems.

2.14 Deep Basin Compartments of Pressurized Water, Petroleum, and Geothermal Fluids

A large number of articles deal with different aspects of "abnormal pressures" observed in fluid-containing rock compartments (Martinsen, 1994). Recently, there appeared a comprehensive collection of 30 articles dealing with basin compartments and seals (edited by Ortoleva, 1994). The topic is presented by Ortoleva and Wescott (1994):

> In the early 1970s, Powley and Bradely recognized that basins worldwide exhibited an unexpected degree of hydrologic segregation. They observed regions of a sedimentary basin that were isolated from their surroundings by a relatively thin envelope of low-permeability rock and had an interior of sufficiently high permeability to maintain a consistent internal hydrostatic fluid pressure gradient. They named these regions fluid (or pressure) compartments. Pressure compartments have several remarkable features. First, their internal fluid pressures may greatly exceed or be significantly less than any regional topographically controlled hydrologic head or drain. Thus, their internal pressure cannot be explained by connection to another similarly pressured source or sink region. Next, their bounding rock often does not follow a single stratigraphic unit but transverses several units. This contradicts conventional thinking, wherein low permeability beds such as shale, salt or bedded anhydride serve to separate geopressured regions above and below but there are no recognized lateral boundaries. Furthermore, this surrounding shell was such an efficient barrier to fluid flow that it could retain very large potentiometric pressure heads over geologic time—i.e., the hundreds-million-year time frame. They thereby termed the surrounding rock a "seal"—i.e., a relatively thin region of rock at the top, bottom, and all sides that can retain pressure gradients on time scales very much longer than that required for dissipating nonhydrostatic gradients within the interior. The fact that these compartments often contained large reserves of oil and gas made them very interesting.

The size of identified pressurized compartments is reported to vary from small-scale structures, 1–10 km across, to mega-scale compartment complexes of hundreds of kilometers across; they have been reported from depths of 2000–5000 m; and potentiometric heads were mostly between the local (theoretical) hydrostatic head and the local lithostatic head. Multiple compartments are distributed vertically and laterally. The pressurized compartments are regarded to be closed chemical systems as no through-flow occurs in them.

Examples of large-scale hydraulically isolated compartments were reported (in Ortoleva, 1994) from Mississippi salt basin, Oklahoma, Kansas, Texas, Louisiana, Colorado, Nebraska, New Mexico, California, Nova Scotia, Alberta Basin, Caspian Sea, Black Sea area, Yugoslavia, Guatemala, Australia, Indonesia, and the Persian Gulf. We would like to add the Great Artesian Basin, Australia, as an example of a megacompartment that is divided into many small-scale subcompartments. The Great Artesian Basin is divided into a shallow unconfined system, underlain by tertiary rocks that host medium-quality groundwater, beneath which is the Cretaceous marine rock sequence that hosts mainly artesian saline groundwater, and beneath that is the Jurassic continental rock sequence that hosts fresh artesian groundwater (the main exploited system). Each of these large-scale systems is divided into numerus small-scale compartments, recognizable by their different properties (Mazor 1993, 1995).

Bradley and Powley (1994) suggest, "[T]he primary cause of either pressure or flow is epirogenic movement, with erosion or deposition, which changes the temperature, stress, and pore pressure in the system, which in turn, affect the diagenesis in the system."

The phenomenon of hydraulically sealed, isolated, and pressurized fluid containing compartments has been observed in structures over 2000 m deep studied in connection with oil and gas prospecting. The concept is, in a way, a newcomer to the oil exploration community, and is regarded as an unexpected observation. This point is reflected, for example, in the terms "abnormal pressure" or "overpressure," applied to indicate pressures greater than the originally expected hydrostatic pressure.

2.15 The Traditional Model of U-Shape Flow Paths

Figure 2.18 depicts a cross section of a model that has been much quoted in the hydrological literature from Hubbert (1940) to most recent textbooks and articles. The modeling is restricted to a valley flank in a small drainage basin. The cross section is through half a valley, from a local divide to a nearby low-order valley. The three marked planes are assumed in the model to be impermeable (no water is assumed to flow across the vertical planes that are marked beneath the local divide and beneath the valley, and an impermeable horizontal rock bed is assumed at some depth). The water table is modeled as a moderated replica of the topography, which in turn is assumed to be expressible by a simple mathematical equation. Figure 2.19 shows a more detailed cross section of the modeled area. The system is assumed to be composed of homogeneous permeable rocks. Equipotential lines, calculated by the model, were computed, and lines of force, or flow lines, were deduced perpendicular to the

Basic Hydrological Concepts

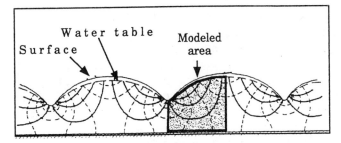

Fig. 2.18 A cross-section of a much-quoted model (following Freeze and Cherry, 1979, who cited Hubbert, 1940). The surface is described as undulating in a mode that can be expressed by a simple mathematical equation, and the water table is assumed to follow topography in a fixed mode. The stippled section describes a water system from a low-order divide to a nearby low-order valley; the thick lines mark there impermeable planes that are an intrinsic part of the U-shape flow paths model, enlarged in Fig. 2.19. The cross section emphasizes topographic undulations, and disregards the location of the terminal base of drainage and the location of the main water divide.

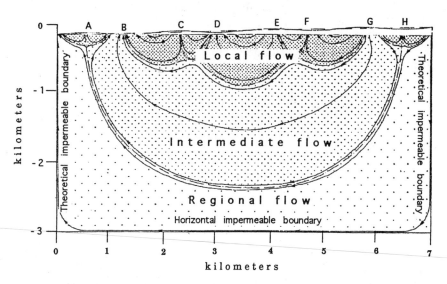

Fig. 2.19 A more detailed cross section of the U-shape model area shown in Fig. 2.18 (condensing several figures of Toth, 1963 and 1995). Groundwater flow paths deduced by the U-shape flow paths model are marked with arrows denoting flow directions. Three flow zones have been concluded: local, intermediate, and regional, with alternating points of discharge (e.g., points A, B, D, F) and points of recharge (points C, E, G, H). The symmetry of the suggested flow lines, centered in the modeled box, reveals that they are a direct outcome of the assumption of the three impermeable flow planes.

former. These model calculations, incorporating Darcy's law, deduced U-shape flow paths that are composed of zones of recharge (points C, E, G, and H in Fig. 2.19) and zones of discharge (points A, B, D, F in Fig. 2.19) closely coinciding with topographic highs and lows, respectively. The resulting flow lines were divided (Toth, 1963) into three flow regions: local, intermediate, and regional. Realistic dimensions for the model were suggested to be a width on the order of 7 km and a depth on the order of 3 km, and in the 1995 paper, Toth suggested the model to represent entire sedimentary basins.

The early U-shape flow paths models assumed homogeneous permeable rocks for the entire system, but this restriction was later removed by the argument that "all rocks are permeable to some degree." Thus, the road was paved to apply the U-shape model to entire basins.

A basic outcome of the U-shape flow paths model and its many derivations was that all groundwater systems were discussed in terms of through-flow, at various flow velocities, and stagnant (static) groundwater was disregarded.

2.16 Discussion of the U-Shape Flow Paths Model

As mentioned, the basic U-shape model, depicted in cross section in Figs. 2.18 and 2.19, assumes three impermeable planes and discusses one-half of a land unit that is hard to identify in nature. Let us examine this approach in the frame of an entire groundwater drainage basin, as shown in Fig. 2.20. The assumption of the three impermeable planes looses ground because the modeled small land cell is part of a larger through-flow system, with recharge inputs from higher altitudes and through-flow toward the terminal base of drainage.

The local, intermediate, and regional flow regions, depicted in Fig. 2.19 as an outcome of the U-shape model, are hydraulically nonlogical. There is no physical process that determines that water recharged at the right end of the modeled half basin (point H in Fig. 2.19) will have the deepest (regional) penetration and discharge at the left corner (point A in Fig. 2.19), that water recharged in the next location (point G in Fig. 2.19) will penetrate to intermediate depths, and in a symmetrical mode discharge at the left of the model (point B in Fig. 2.19), while shallow (local) circulation of water recharge and discharge will occur only at the center of the modeled section (points C–F in Fig. 2.19). The suggested flow paths in Fig. 2.19 are seen to be a freak of the assumed three impermeable planes.

If the principle premise of the model, namely the existence of the three impermeable boundaries, is wrong, then the whole model collapses and the discussion could be terminated as this point. But the notion of deep U-shape flow paths prevails and warrants discussion of the feasibility of its conclusions.

Presentations of the U-shape model in the literature do not specify the location of rain or snowmelt water, and do not discuss the resulting recharge

Basic Hydrological Concepts   43

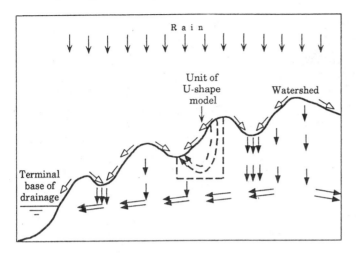

Fig. 2.20 The modeled box of Figs. 2.18 and 2.19, plotted on a regional cross-section from a major watershed to the terminal base of drainage. The assumption of three impermeable planes is groundless, as discussed in the text.

distribution. Commonly, rain falls over the entire area of an unconfined system, as depicted in the cross section shown in Fig. 2.20. Runoff, too, is all over, but with an uneven distribution, as it runs fast from the topographic highs and accumulates and flows relatively slowly at the topographic lows. Thus, recharge takes place all over the surface, but is intensified in the valleys, contrary to the discharge in valleys concluded by the U-shape model (e.g., points B, D, F, G, and H in Fig. 2.19).

Various researchers deduced that the flow regime is most sensitive to the topography: in flat terrains regional flow systems were concluded to prevail, whereas in homocky terrains, numerous shallow flow regions have been concluded to be superimposed on the suggested regional system (Freeze and Witherspoon, 1967, adopted in various following publications). This is another artificial outcome of the U-shape model, as the depth of the flow paths is determined by the level of the terminal base of drainage, and the latter is never discussed in the U-shape model.

A major shortcoming of the U-shape flow paths model is the "forcing" of water down to below the level of the base of drainage, and then "forcing" it up at hypothetical discharge locations (Fig. 2.19). Above the level of the terminal base of drainage, in the through-flow zone, all water particles flow "downhill," under the force of gravity, the through-flow zone occupying the space in which hydraulic potential differences prevail (Figs. 2.14–2.16). Thus, no water particle has to "push away" other particles along its flow path, as they move by themselves, and energy consumption goes to overcome friction alone. In contrast, beneath the level of the terminal base of drainage, no hydraulic potential exists and the water particles do not move by themselves. Hence, to make water flow in a U-path beneath the level of the base of drainage, other water particles have first to be "pumped" or "pushed" up toward a discharge region, and this necessitates a great source of energy, which does not exist. Evoking such a flow means to claim the finding of a perpetual mobile process.

Deep basinwide through-flow models are based on the assumption that water-conducting rock strata, or aquifers, stretch uninterrupted for hundreds of kilometers and serve as through-flow ducts. Through these aquifer ducts water is suggested to flow from recharge areas at marginal outcrops. The aquifer ducts are commonly defined in accordance with regional stratigraphic units (Fig. 2.21). However, stratigraphic units incorporate many alternations of permeable and impermeable rocks, constituting a multitude of potential aquifers (Fig. 2.10). Hence, according to the basinwide through-flow model, a multitude of separate ducts and marginal recharge areas must exist, contrary to field observations. The model of stagnant groundwater traps fits the observations much better (Mazor and Nativ, 1994; Mazor, 1995).

Compaction caused by the weight of overlying rocks is called upon to explain the water pressure in deep-seated artesian aquifers (Mazor, 1995). Compaction counteracts through-flow in deep aquifers, as it causes collapse features, and pressurizes the entrapped water. If through-flow would take place, the open ends of the hypothetical aquifer ducts would serve as pressure vents and no high pressures would be observable. The very fact that the artesian systems are pressurized indicates they are closed systems and not open ones.

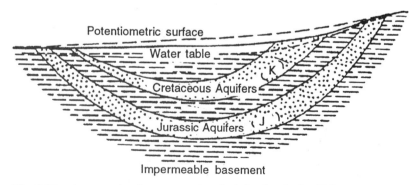

Fig. 2.21 An example of a cross section depicting hundreds of kilometers across subsidence basin, with stratigraphic units assumed to act as through-flow ducts (the Great Artesian Basin, Australia, following Seidel, 1980).

Basic Hydrological Concepts

2.17 Summary of Modes of Groundwater Flow

The discussed phenomenological model of L-shape through-flow paths and zones of stagnation may be summarized in light of the degrees of hydraulic freedom for recharge input and for flow as presented in section 2.11:

Mode of flow	Degrees of hydraulic freedom	Flow and direction
Free flow	Three dimensional	Vertical downflow
Lateral "overflow"	Two dimensional	Lateral base flow
Forced flow	One dimensional	Karstic conduits, joints
Stagnation	Zero dimensional	Groundwater traps

2.18 Summary Exercises

Exercise 2.1: Match the observations to the concluded processes or existing systems:

Observation
1. A well has an upper dry section
2. A well has a lower part filled with water
3. The water table often rises seasonally
4. The water table often falls seasonally
5. The upper soil dries up seasonally
6. Plants depend on rain or irrigation

Process or system
A. Transpiration
B. Recharge
C. Aerated zone
D. Discharge
E. Evaporation
F. Saturated zone

Exercise 2.2: Groundwater is visualized by some people as occurring in underground lakes and rivers. Looking at a well we find in it water—is this an indication that the well taps such a lake?

Exercise 2.3: In a year with twice the average annual rainfall, the resulting recharge is significantly greater than twice the average annual recharge, a phenomenon most evident in arid regions. Why?

Exercise 2.4: In light of the previous question, is the average annual rain a sufficient representative parameter to assess recharge? Is another parameter needed?

Exercise 2.5: In the course of drilling a borehole, water was encountered first at a depth of 20 m, and again at a depth of 54 m. Can you describe the local aquifers in terms of being phreatic or confined? Which additional data would you like to have to further check your conclusions?

Exercise 2.6: The Banias Spring issues at the foot of snow-covered Mt. Hermon. Which measurements are recommended in order to find out whether this is a karstic spring fed mainly by snowmelt?

Exercise 2.7: What is the definition of groundwater age? Where is there a better chance to determine meaningful water ages, in unconfined systems or in groundwater traps?

3
GEOLOGICAL DATA

3.1 Lithology and Its Bearing on Water Composition

The composition of rocks and soils has a direct bearing on the quality of water. Hydrochemically a rough classification into three rock groups seems practical:

Rocks in which fresh groundwater is common, that is, rocks that contribute extremely small amounts of salts to the water.
Carbonate rocks that contribute dissolved matter but maintain good potable quality.
Rocks that enrich the water with significant amounts of dissolved salts, often making them nonpotable.

The lithological parameter is only one of several parameters that control groundwater quality. Other factors include evaporation at the surface prior infiltration, transpiration, wash-down of sea spray, and reducing conditions in the aquifer, connected to H_2S production. Water moves underground and its salt or mineral content is determined by all soil and rock types it passes through. Thus, occasionally, saline water may be encountered in rocks that by themselves do not contribute soluble salts.

3.1.1 Rocks that Preserve High Water Quality

Pure sandstone, rich in quartz grains and with little or no cement, may host very fresh water with little addition to the salts brought in by recharging rain (Table 3.1). Other rocks hosting fresh water are quartzite (sandstone cemented by

Table 3.1 Lithological imprints on groundwater composition

Rock	Groundwater composition
Sandstone	Low salinity (300–500 mg/l); HCO_3^- major anion, Na^+, Ca^{2+}, Mg^{2+} in similar amounts; good taste.
Limestone	Low salinity (500–800 mg/l); HCO_3^- major anion, Ca^{2+} dominant cation; good taste.
Dolomite	Low salinity (500–800 mg/l); HCO_3^- major anion; Mg^{2+} equals Ca^{2+}; good taste.
Granite	Very low salinity (300 mg/l); HCO_3^- major anion, Ca^{2+} and Na^+ major cations; very good taste.
Basalt	Low salinity (400 mg/l); HCO_3^- major anion; Na^+, Ca^{2+}, Mg^{2+} equally important; good taste.
Schist	Low salinity (300 mg/l); HCO_3^- major anion; Ca^{2+} and Na^+ major cations; good taste.
Marl	Medium salinity (1200 mg/l); HCO_3^- and Cl^- major anions, Na^+ and Ca^{2+} major cations; poor taste but potable.
Clay and shale	Often containing rock salt and gypsum. High salinity (900–2000 mg/l); Cl^- dominant anion, followed by SO_4^{2-}; Na^+ major cation; poor taste, occasionally non-potable.
Gypsum	High salinity (2000–4000 mg/l); SO_4^{2-} dominant anion; Ca^{2+} dominant cation, followed by Mg^{2+} or Na^+; bitter, non-potable.

silica), basalt, granite, and other igneous rocks (Table 3.1). The total dissolved salts in the water encountered in these rocks is often in the range of 300–500 mg/l.

3.1.2 Carbonate Rocks

Limestone and dolomite commonly host water with 500–800 mg/l total dissolved salts, mainly $Ca(HCO_3)_2$, in the case of limestone, and $(Ca,Mg)(HCO_3)_2$ in the case of dolomites. The relevant chemical water-rock interactions are discussed in section 6.8. Water encountered in limestone and dolomite is generally tasty and of high quality.

3.1.3 Evaporites: Gypsum and Salt Rock

Gypsum ($CaSO_4 \cdot 2H_2O$) is formed mainly in lagoons (semiclosed shallow water bodies, mainly along oceanic coasts) that are subjected to high degrees of evaporation. Seawater evaporated to one-third of its original volume, precipitates gypsum (or anhydrite—a water-poor variety of gypsum). Groundwater in contact with gypsum rock dissolves it, along with accompanying salts, reaching 1000–2600 mg/l of SO_4, along with Cl, Ca, Mg, and Na, the total dissolved salt content reaching 2000–4000 mg/l (Table 3.1). The taste of such water is bad, mainly bitter. Water with up to 600 mg/l SO_4 can be consumed in an emergency, but higher concentrations will cause thirst rather then quench it. Stock may, however, drink such water.

Rock salt (NaCl) is formed by even higher degrees of evaporation, also in lagoons and closed lakes. Seawater is saturated in NaCl when evaporated to one-tenth of its original volume. Water coming in contact with rock salt dissolves it, becoming rich in NaCl. Water in contact with rock salt is brackish or brine, nonpotable water, with salinities as high (brackish water) or higher (brine) than seawater.

3.1.4 Clay and Shale Hosting Gypsum and Rock Salt

Clay and shale are hydroaluminum silicates that by themselves do not add salts to the water that comes in contact with them. However, clay and shale often contain veins and nodules of gypsum, pyrite, and rock salt. Clay and shale are impermeable and form aquicludes rather than aquifers (sections 2.3 and 2.4), but because of the high solubility of gypsum and especially rock salt, groundwater in contact with clay and shale at the base of aquifers often gets saline and is of poor quality.

3.1.5 Lithological Considerations in Well Location and Design: The Water Quality Aspect

Selection of well sites is done on the basis of several aspects, water quality being a major one. Thus, a thorough knowledge of the lithological section of candidate well sites is of prime importance. Gypsum and rock salt should be avoided in any case. Aquicludes of salinizing rocks, such as clay and shale, have to be avoided, either by terminating the wells or their perforated sections above the clay or shale layer, or by sealing the well at the aquiclude section, avoiding entrance of poor quality water (Fig. 3.1).

3.2 Properties of Geological Materials and Their Bearing on Recharge and Groundwater Storage

Infiltration of recharge water occurs through interconnected pores, open joints and fissures, or combinations of these features (Fig. 3.2). In contrast, rocks rich in pores that are too small or isolated from each other, nonporous rocks, and rocks with no open fissures are inefficient for infiltration and recharge. The same principles hold true for water movement and water storage in aquifers. Rocks with open interconnected pores and open interconnected fissures are best for recharge and storage.

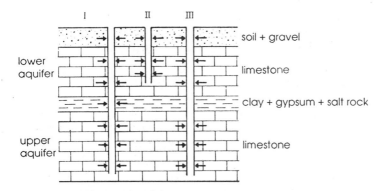

Fig. 3.1 Two aquifers of nonsaliferous rocks, separated by an aquiclude of clay with gypsum and salt rock. (I) fully perforated and producing saline water; (II) stopped at a safe distance above the clay, abstracting good water from the upper aquifer alone; (III) sealed for several meters above and below the clay bed, producing good water from both aquifers.

3.2.1 Materials with Open Interconnected Pores

All granular rocks are rich in interconnected pores. The most common types are sand and sandstone. Their recharge and storage quality increases with the grain size. Cement (carbonate, silica, iron oxide) reduces the pore volume and, in extreme cases, may result in an impermeable sandstone (however, cemented sandstone tends to fracture). Conglomerate (a rock made up of pebbles of rocks) is an extreme case of a granular rock, and if noncemented, has excellent recharge and storage properties.

3.2.2 Materials with Open Interconnected Fissures

All brittle rocks tend to be fractured to different degrees. The joints and fissures may in certain cases be somewhat enlarged by dissolution, but they may

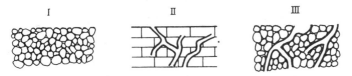

Fig. 3.2 Modes of water infiltration. (I) porous texture with no cement between the grains, high infiltration rate and water storage capacity; (II) nonporous texture with open interconnected fissures, good infiltration and storage capacity; (III) partially porous texture with interconnected fissures, infiltration and water storage are possible in two modes: through noncemented interconnected pores and through fissures.

Geological Data 51

also be clogged with clays produced by weathering. Thus, it is not enough to know the rocks that constitute the terrain in studied areas, you must also know their degree of fracturing and weathering.

An extreme case of open interconnected conduits is manifested by limestone and dolomite in karstic terrains, as described in section 2.9.

3.2.3 Materials with Poorly Connected Pores

Clay and shale have a large number of minute pores, totaling up to 55% of the rock volume. Yet, these pores are poorly interconnected, resulting in very low permeability or impermeability (section 2.7). Clay and shale significantly slow infiltration and serve as aquicludes.

3.2.4 Materials with Isolated Fissures

Metamorphic rocks are rich in open foliation fractures and fissures, but these are isolated or poorly interconnected. Such rocks make poor recharge terrains and poor aquifers. These rocks are occasionally fractured by tectonic processes, improving their infiltration and storage properties.

3.2.5 Studying Rocks for Their Conductivity or Recharge and Storage Capacities

Rock properties should be examined at as many outcrops as possible, looking for friability, cementing, degree and nature of fractures and dissolution conduits, and animal burrowing (Figs. 2.3, 2.4, 2.8, and 3.2). However, one should bear in mind that rocks at exposures are altered by weathering and joint formation due to stress release, and thus may poorly represent the rocks at depth. Similar observations may be conducted on drill cores, but these are expensive and their record is limited in size. Laboratory tests on cores provide semiquantitative data on the nature of rock pores and fissures and conductivities.

Another way to determine the relevant rock properties is to examine the properties of groundwater in springs and wells. Young water ages and seasonal variations of discharge and temperature indicate rapid passage through high conducting rocks, whereas high water ages and constant discharge and temperature indicate retardation and low conductivity, at least along part of the water system.

3.3 Layering and Its Control on Groundwater Flow

Many rock types have a layered structure, individual rock layers varying in thickness from a few centimeters to tens of meters. Layered rocks include marine sediments, most continental sediments, lava flows and volcanic ejecta, and intrusive sills. The hydraulic properties vary from one rock layer to another, often resulting in abrupt changes along the vertical axis. In terms of the permeability coefficient (k) the lateral coefficient (k_x) may significantly differ from the vertical coefficient (K_z). The alternation of aquifers and aquicludes results from the layered structure of different rocks, and the occurrence of springs is often controlled by the layering of rocks. Fissures may be restricted to individual rock layers or cross several rock beds, in which case water flow is improved, mainly in the vertical direction.

3.4 Folded Structures and Their Bearing on Flow Direction, Confinement, and Entrapment

Water flows from high to low points and prefers the path of least resistance. In regions built of bedded rocks, tilting may be a prime factor determining flow direction (besides topographic gradients), as shown in Fig. 3.3. Among folded structures, synclines are important because they often gather water from of neighboring anticlinal structures (Fig. 3.4). Hence, an understanding of the tectonic regime of a well target area is essential. Flow directions deduced from tectonic and topographic arguments should be checked by other methods, as outlined in the following chapters.

Confined aquifers (section 2.8) are rare in tectonically undisturbed regions with horizontal rock beds (Fig. 3.5). Tilting of the aquifer and aquiclude "sandwich" makes room for the formation of confined aquifers. It provides each case with a recharge outcrop section, forming a phreatic aquifer (section 2.8), and a confined section, fed by the former (Fig. 3.6).

Subsidence structures are common. They are formed by the gradual lowering of rock strata, forming basin shapes. Two major processes operate during subsidence: burial of older rock beds, and accumulation of new sediments at the surface forming new rock beds or lenses that eventually get buried as well,

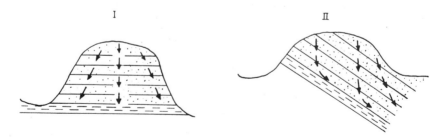

Fig. 3.3 Factors controlling the flow direction of groundwater. (I) porous rock; water direction is determined by topographic gradients alone; (II) a tilted impervious rock bed deviates direction of water flow.

Geological Data

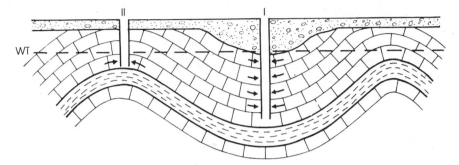

Fig. 3.4 Groundwater accumulating in a syncline. Well I, located at the syncline, has a significantly higher production then well II, located in adjacent anticline. The lowering of the water table, due to pumping, will cut well II dry in a short time. On the other hand, fracturing in the crest of an anticline occasionally makes these locations a better place for water supply.

the water stored in them becoming trapped (Fig. 3.7). Subsidence structures may be narrow, such as rift valleys and intermountain basins (ranges and basins) that are tens of kilometers wide, hundreds of kilometers long, and hundreds to thousands of meters deep, and in other cases subsidence structures may be hundreds to thousands of kilometers across and a few thousand meters deep.

Subsidence structures often host traps of stagnant groundwater bodies. The groundwater entrapment is caused by several processes: (1) burial to beneath sea level, that is, burial into a zone of zero hydraulic potential, and hence a zone of stagnation; (2) coverage by younger sediments that include clay and shale that cut off the groundwater traps from active recharge; (3) frequent changes in lithology (e.g., interfingering, which is common in continental sediments that fill most subsidence structures), hydraulically insulating individual rock bodies, turning them into groundwater traps; and (4) tectonic activity that causes faulting, folding, and thrusting accompanied by squeezing of plastic

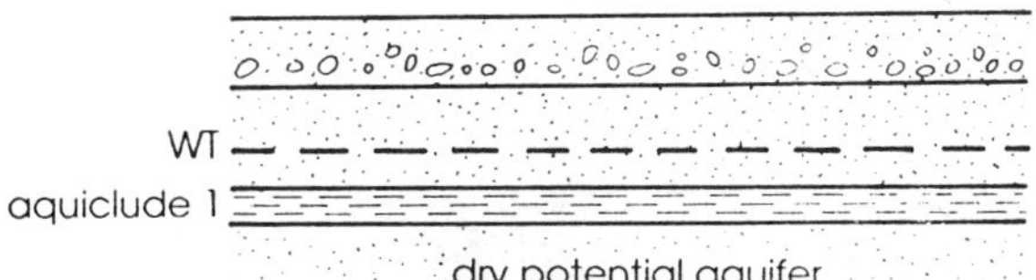

Fig. 3.5 Horizontal rock beds often allow for the formation of only one recharged (phreatic) aquifer, the first aquiclude preventing recharge water from reaching lower potential aquifers.

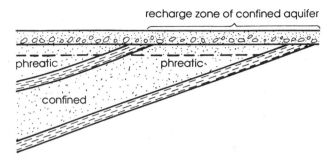

Fig. 3.6 Tectonic tilting of rock strata provides the necessary condition for the formation of a partially confined aquifer. A phreatic section, located at the outcrop of the rock bed, serves as recharge intake area feeding the confined section in cases where the inclined rock bed has a drainage outlet at its lower end enabling through-flow.

impermeable rocks (e.g., clay and shale) into gaps formed in the competent aquifer rocks, causing further hydraulic isolation of water-bearing rock bodies.

3.5 Faults Controlling Groundwater Flow

Faults cause discontinuities in rock sequences and thus control groundwater flow. A fault may set high-conducting aquifer rocks against impervious rocks, resulting in water ascent along the fault zone and formation of springs marking the fault line (Fig. 3.8). Occasionally groundwater continues in its general flow path across a fault, but in different rock beds (Fig. 3.9). Another fault

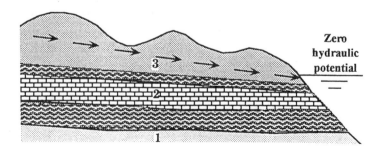

Fig. 3.7 Stagnant confined systems are former (fossil) through-flow systems (1) that got buried beneath the level of the terminal base of drainage, and (2) were covered by new rock systems that eventually got buried. System (3) is presently active as a through-flow system.

Geological Data

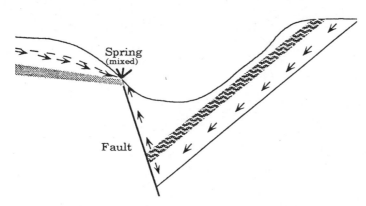

Fig. 3.8 A fault has placed a high conducting aquifer against an igneous rock of low permeability. Water ascends along the fault zone, forming a line of springs. Only part of the faults are open and conduct groundwater flow, others are clogged by compression and/or mineralization.

controlled example is the gathering of groundwater into rock beds of a rift valley (Fig. 3.10).

The above examples look simple, but in reality subterranean structures are often hard to recognize, and physical, chemical, and isotopic tracers are recruited to assist interpretation.

Fault zones may be open and highly conductive, or they may be sealed by mineral precipitation, forming hydraulic barriers.

3.6 Intrusive Bodies Influencing Groundwater Flow

Igneous intrusive bodies occasionally are observed to act as subterranean barriers or dams. Good examples are dikes (Fig. 3.11), sills (Fig. 3.12), and

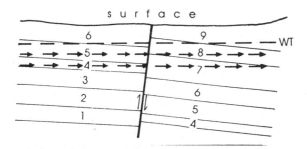

Fig. 3.9 Groundwater changing aquifer beds along the flow path due to strata displacement by a fault.

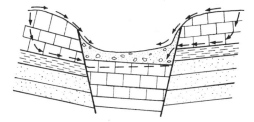

Fig. 3.10 Runoff and groundwater flowing into a rift valley.

stocks. Fresh igneous rocks are nonpermeable, but with time they become fractured and may be somewhat conducting. Clay-rich weathering products may fill such fractures and improve their sealing properties.

In the case of dikes, water gets dammed on the upstream side (Fig. 3.11). When placing a well near a dike it is essential to

Recognize the existence of the dike, although it is often concealed by alluvium. Determine the direction of groundwater flow.

Lack of such information may result in a dry borehole drilled on the downflow side of the dike.

3.7 Hydrochemical and Isotopic Checks of Geological and Hydrological Models

Geological models are an independent entity and an essential part of hydrological models. Geological models deal with rock sequences (candidate aquifers and

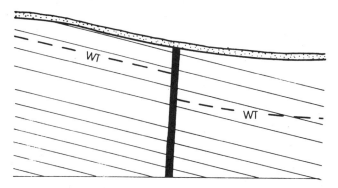

Fig. 3.11 A dike intersecting the flow of groundwater and acting as a subsurface dam. The water table is higher on the upstream side, providing a promising site for drilling a well.

Geological Data

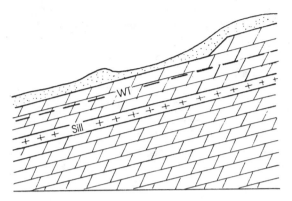

Fig. 3.12 A sill intruded between carbonatic rock beds forming a local aquiclude. Weathering into clay minerals may improve the sealing capacity of a sill.

aquicludes), rock properties (in terms of their potential influence on the quality of hosted groundwater), existence of preferred flow paths (karstic features, open fractures and faults), identification of subsidence structures, and identification of hydraulic barriers (intrusive bodies, and clay or shale squeezed into and around tectonically shaped rock bodies). However, geological data are scarce and of a fragmented nature, seldom leading to a unique model solution. Hence, geological conclusions, or suggestions, should always be checked against hydrochemical and isotopic observations.

Similarly, hydrological models, based on precipitation distribution, topographic relief, location of the terminal base of drainage, hydraulic head data, and geology, are also of a fragmented nature and seldom lead to a unique model solution. The situation may be significantly improved by the application of hydrochemical and isotopic checks, as discussed below.

3.7.1 Concentration of Conservative Dissolved Ions Serving as a Check of Hydraulic Connectivity

Hydraulic connectivity is a basic premise in the construction of groundwater flow paths, calculation of flow velocity, and computation of travel time and water age. Groundwater is suggested to flow along a hydraulic gradient that is deduced from hydraulic head data obtained from measurements in wells and springs (section 4.3). These suggested gradients are valid only if the measured wells are all hydraulically interconnected (section 6.5). In other cases groundwater is assumed to flow in unconfined and confined aquifers, described as continuous and effective ducts, based on geologically deduced configurations. However, the geological data are fragmentary and are often based on mapping or stratigraphic, rather than lithologic, units (section 2.3). Independent methods are needed to check the crucial assumption of hydraulic connectivity in every case study and for every geological and hydrological model.

There are essentially only two sources of chloride in groundwater: sea spray in the form of atmospheric (airborne) salts, and dissolution of halite in the limited number of cases where this mineral is present in the rocks passed by the water. Chloride is a conservative ion; it does not participate in water-rock ion exchange interactions, and once it enters the groundwater there is no process that can remove it (section 6.12). Other conservative ions of interest in the present context are bromide and lithium. Water that comes into contact with halite dissolves it rapidly, and in a matter of days becomes saturated, the water reaching a chloride concentration of more than 180,000 mg/l. Common groundwaters contain chloride in the range of 10–1000 mg/l and are therefore undersaturated with respect to halite, indicating no halite is present in their respective rock systems. Thus, the common source of chloride in groundwater is atmospheric and its concentration is by and large controlled by the amount of water lost by evapotranspiration prior to entry into the saturated zone (section 2.12).

In the saturated zone the chloride concentration stays constant unless mixing of different water types occurs. A chloride concentration that decreases along a suggested groundwater flow direction indicates the suggested flow path model is not valid, as there is no process that can reduce the chloride concentration. In such a case the assumption of hydraulic interconnection between wells that tape the same water body is wrong, and one deals with separate aquifers (Fig. 3.13). An abrupt increase in the observed chloride concentration along a suggested groundwater flow path indicates as well that the assumption of hy-

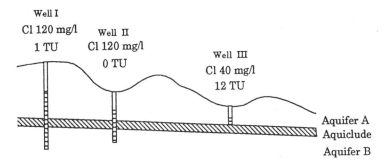

Fig. 3.13 A transect through wells that display a gradually lower water table, yet the chloride concentration is lower in the well with the lower water table. In this case the explanation is that the wells tap different aquifers and are not hydraulically interconnected. Tritium values further clarify the picture: wells I and II tap a confined aquifer, whereas well III taps a phreatic through-flow system.

draulic connectivity is invalidated (Fig. 3.14). Certain springs are fed by a mixture of water from a deep aquifer, ascending along a fault plane, and water from a shallow aquifer (Fig. 3.8). The intermixing waters have different compositions and the mixing ratios vary seasonally. Repeated analysis of a spring that reveals varying composition indicates an outlet that drains two (or more) distinct aquifers.

3.7.2 Concentration of Age Indicators as a Check of Hydraulic Connectivity

Groundwater in unconfined systems receives recharge over the entire surface and, hence, is expected to have a recent age (section 10.4). If a well included in a model of an unconfined system reveals an absence of tritium, the model has to be modified to include a confined aquifer beside the unconfined one (Fig. 3.14). Water in a confined aquifer is expected to have the same age or reveal a certain age increase downflow. However, if the age varies suddenly between adjacent wells, for example, gets younger in the downflow direction, the connectivity assumption is disproved (Fig. 3.13).

3.7.3 Seasonal and Abstraction-Induced Changes in Water Composition as a Check for Interconnection of Two or More Aquifer Systems (Mixing of Groundwaters)

Models based on geological observations have no way to indicate the number of water types involved in a studied system—commonly a single groundwater type is assumed to prevail. Hydraulic models, based on water table and hydrau-

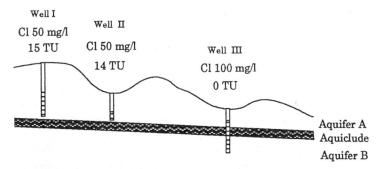

Fig. 3.14 A transect through wells that display a gradually lower water table, yet the chloride concentration is distinctly higher in the well with the lowest water table. This indicates that wells I and II are interconnected, but well III taps a different aquifer. Tritium data provides an additional insight—wells I and II tap a through-flow system, well III taps a confined aquifer.

lic head measurements, are similarly shorthanded and cannot detect intermixing of different groundwater types, a phenomenon that requires hydraulic interconnection between distinct sections of a studied groundwater system. A single analysis of dissolved ions is not helpful either. However, repeated measurements may indicate such mixing cases, as is further discussed in section 6.6. The very fact that the same spring or well reveals different ion concentrations at different times, indicates several (at least two) water types are intermixing in a ratio that varies seasonally, or as a response to abstraction. Thus, the geological and hydrological models have to be modified to allow for an interconnection between distinct sections of the studied system. The topic is further discussed in section 6.6.

3.7.4 Water Composition as a Tool to Identify Types of Rocks Existing in a Studied Groundwater System

The domain of groundwater is concealed, and we have only a limited number of glimpses at the geology, for example, at outcrops, and some information that is retrievable from drillings. As a result, it is hard to portray with certainty the lithology along the flow path. Complementary information is supplied by the chemical composition of groundwater: very fresh water reflects contact with noncemented sandstone, basalt, or granite (section 3.1); water with a significant concentration of calcium and HCO_3 indicates passage through limestone; magnesium along with calcium and HCO_3 indicates passage through dolomite; high SO_4 concentration (greater than 2000 mg/l) indicates contact with gypsum, and high sodium and chlorine concentration (greater than 180,000 mg/l) reflects contact with rock salt (section 3.1).

3.7.5 Composition of Stable Hydrogen and Oxygen Isotopes as a Tool to Distinguish Between On-Going Recharge and Fossil Recharge

The composition of the stable hydrogen and oxygen isotopes of groundwater is controlled by the prevailing climatic and hydrologic conditions (section 1.2), among other factors. Old groundwater that had been recharged under different climatic conditions is often distinguished by an isotopic composition that differs from the presently recharged groundwater (section 9.9). The earth is at present in a relatively warm climatic phase, which means that the paleoclimates were mostly colder, in good correlation with published case studies that in general reveal that old groundwaters have relatively light isotopic compositions.

Geological and hydraulic data leave open the question of whether a given groundwater system is a renewed resource, with ongoing recharge, or a non-renewed resource, representing fossil recharge. In many instances the stable isotopes of the water provide the answer.

3.7.6 Composition of Water in Adjacent Wells as a Tool for Spatial Mapping of Stagnant Aquifers

Geological models are commonly portrayed in stratigraphic units (formations, members, epochs, etc.). These units often have large-scale extensions, leaving the hydrologist with the question of whether a studied setup consists of a single through-flow system or is made of numerous stagnant systems. The problem is especially acute in artesian wells situated in subsidence structures (section 3.4). Adjacent wells with similar water composition are likely to belong to the same stagnant aquifer, whereas wells with significantly different water compositions belong to different systems, stacked one upon the other or laterally adjacent. Thus, ionic and isotopic compositions and concentration of age indicators supply the means to map the spatial distribution and extension of stagnant aquifers, helping to refine the geological and hydrological models.

3.8 Summary Exercises

Exercise 3.1: High-quality drinking water contains dissolved solids (a) below 800 mg/l, (b) above 2000 mg/l, (c) the more the better.

Exercise 3.2: Match the rocks on the left side with the (likely) properties on the right side:

1. Granite   A. Fractured
2. Dolomite   B. Rich in interconnected pores and fissures
3. Basalt   C. Poor in interconnected pores and fissures
4. Limestone   D. Hosting good potable water
5. Clay   E. Salinizing water
6. Chalk   F. Forming good aquifers
7. Marl   G. Forming medium to poor aquifers
8. Sandstone   H. Forming good aquicludes
9. Conglomerate

Example Answer: 1—A, D, and G.

Exercise 3.3: List rock properties that should be noticed and described in the field work, relevant to hydrological and hydrochemical studies.

Exercise 3.4: Study the location of the wells in Figs. 4.13a and 4.13b. How can water properties indicate whether wells I and II are interconnected and that wells III and IV are disconnected?

Exercise 3.5: A water issuing in a spring contains 55 mg Cl/l. What is the likely source of this Cl?

4
PHYSICAL PARAMETERS

4.1 Water Table Measurements

The depth of the water table in a well is measured relative to an agreed upon mark at the top, for example, the edge of the casing. This reference point should be clearly marked to ensure reproducibility of the measurements. The depth of the water table is obtained in meters below the top of the well, or more meaningfully, in meters below the surface, as shown in Fig. 4.1. AC is the depth of the water table below the reference mark at the edge of the well casing (or any other selected mark). This might differ slightly from BC, the depth of the water table below the surface (Fig. 4.1). In the field notebook the depth AC should be used, as it is exact and properly defined. In a final report reference to the surface may be preferred. In any case, one has to specify how the depth is expressed: "below well head" or "below surface."

A variety of simple accurate and professional tools are available to measure water table depths to an accuracy of 0.2 cm. These instruments consist of a wire with a metal tip that is lowered and upon hitting the water an electrical circuit closes, indicating the water table (Fig. 4.2). The wire is marked for depth readings. The described mode of water table measurement can be performed only in an open, unequipped well. A pumped well can be specially equipped for water table measurements, and this is of interest as it reveals the local dynamics of the lowering of the water table due to abstraction (section 4.6).

Regional interpretation of water table data obtained in many wells is done in meters above sea level (masl). Each well head has to be surveyed and its altitude has to be determined. The measured water table depth is subtracted from

Physical Parameters 63

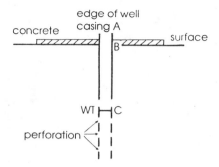

Fig. 4.1 Depth of water table: AC is the depth below the reference mark at the edge of the casing; BC is the depth below the surface.

the well head altitude in order to obtain the water table in absolute altitude units (Fig. 4.3).

4.2 Interpretation of Water Table Data

The depth of the water table is of interest in the management of the well, and helps in identifying the aquifer rocks. The water table altitude is needed for comparison of water tables in neighboring wells to determine the direction of water flow. Data may be processed as water table maps or as transects. These

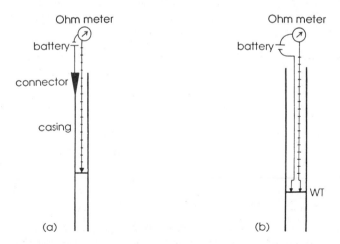

Fig. 4.2 Depth of water table (WT) measurement, based on the closing of an electrical circuit, by the water: (a) one wire, the casing serving as the second wire (provided it is made of metal and unbroken); (b) two wires lowered into the well.

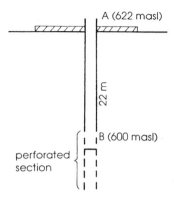

Fig. 4.3 Expressing water table in altitude units (relative to sea level). The measured depth AB is subtracted from the altitude of point A, determined by direct measurement or read from topographic map. In the example, the reference point A has been determined to be 622 masl and the depth of the water table is 22 m. Hence, the water table in this well is 622 - 22 = 600 masl.

modes relate to a singular point in time, that is, the regional water table as measured at a certain date. The next stage of interpretation deals with time variations of the water table, often refereed as water table fluctuations. The processing of water table measurements is herein dealt with in some detail as it is an integral part of the processing of hydrochemical and isotopic data.

4.2.1 Water Table Maps

Water table maps consist of equipotential lines—contours of equal water table altitudes. The contours are drawn to fit measured water table altitudes, as shown in Fig. 4.4. An immediate outcome is the deduction of the dominant direction of groundwater flow (Fig. 4.4b).

A certain degree of ambiguity or subjectivity is inherent in the compilation of water table data. The solutions are not unique. Yet, in most cases, the general picture is clear. The ambiguities are reduced as the number of measured wells increases.

4.2.2 Water Table Transects

Water table transects portray the surface altitudes (topography) and water table altitudes of wells located along a line as shown in Fig. 4.5c. The transect provides a convenient picture of the local topography, depth of groundwater, spacing of existing wells, and inferred direction of groundwater flow.

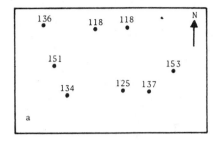 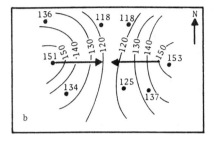

Fig. 4.4 Drawing equipotential lines: (a) a map with well locations and water tables in masl; (b) equipotential lines based on the well data; the arrows show deduced main directions of groundwater flow.

A water table transect may be superimposed on a geological transect, providing insight into the groundwater system, as shown in Fig. 4.6.

4.2.3 Water Table Fluctuations

The water table is dynamic in most systems. It changes in response to rain events, flood events, snowmelt, recharge, and pumping. To decipher the interplay of these ongoing processes, periodic water table measurements are essential. Historical data are in many cases available from local water authorities, which conduct routine water table measurements. Measurement of the depth of the water table at the time of each sample collection is essential in order to couple the chemical and isotopic results with the hydrological data.

Repeated water table measurements in a well may be presented on a hydrograph, as a function of time. In the example given in Fig. 4.7, the low water table may be interpreted as reflecting lack of recharge in the winter, and the rise may reflect snowmelt recharge followed by summer rains. Countless combinations of hydrograph shapes and modes of interpretation are possible. Knowledge of local precipitation and climate is needed for proper interpretation. Rapid response to rain or flood events may indicate conduit-dominated intake (sections 2.1, 2.8, and 3.2), whereas response delayed by weeks or months indicates recharge through homogeneous porous media.

Water table fluctuations are occasionally accompanied by measurable variations in water temperature or composition, providing crucial information on mixing of different water types. Water table measurements are an important tool in tracing recharge. Three cases, reported by Winslow et al. (1965) are discussed in the following sections.

Well no.	1	2	3	4	5	6	7	8	9
surface altitude (masl)	380	374	370	364	359	356	351	344	341
water table (masl)	346	344	341	339	337	334	332	330	330
distance on transect (km)	0	1.2	1.8	3.3	5.1	6.9	7.5	8.9	9.9

(a)

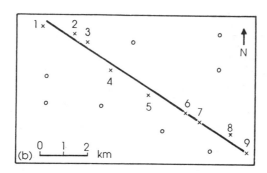

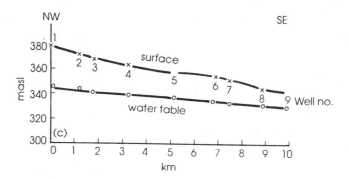

Fig. 4.5 Water table transect: (a) data table; (b) well location; (c) the transect, drawn through the numbered wells (x), having a NW-SE direction (o = other wells, not included in the transect).

4.2.4 Recharge by Different Types of Precipitation

A detailed hydrograph, based on frequent water table measurements at an observation well at the Saratoga National Historic Park, New York, is given in Fig. 4.8, along with the local precipitation (Winslow et al., 1965). The authors offered the following interpretation:

1. Between January and May snowmelt provided some recharge, resulting in a slow rise of the water table.

Physical Parameters

Well no	depth (m)	lithology
1	86	soil (6 m) ; limestone (80 m)
2	90	soil (4 m) ; limestone (86 m)
3	75	soil (5 m) ; limestone (70 m)
4	75	soil (6 m) ; limestone (69 m)
5	46	soil (5 m) ; limestone (41 m)
6	72	soil (10 m) ; limestone (62 m)
7	54	soil (10 m) ; soil + gravel (6 m) ; calcareous sandstone (38 m)
8	58	soil (8 m) ; soil + gravel (14 m) ; calcareous sandstone (36 m)
9	59	soil (5 m) ; soil + gravel (18 m) ; calcareous sandstone (36 m)

(a)

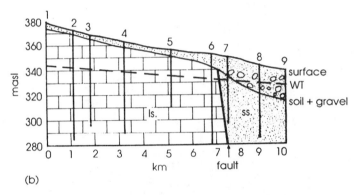

(b)

Fig. 4.6 Geological cross section: (a) data abstracted from geological reports based on drill cuttings; (b) geological section (interpretation), along with the topographic (surface) and water table transects of Fig. 4.5.

2. In April began the rainy season, augmenting the continuing snowmelt, causing a water table rise.
3. In June began the principal growing season, causing increased transpiration losses that lowered the water table.
4. In September 4.7 inches of rain fell during a 4-day hurricane and the water table rose slightly.
5. During the cold month of December only snow fell and the water table continued to decline.

This detailed tracing of recharge history, reflected in the water table fluctuations, is unique in its completeness and reflects direct local recharge and the lack of pumping in the Saratoga Historic Park terrain.

This case study demonstrates how rewarding a simple set of data can be, a topic to be further discussed in section 7.6 in relation to time-data series.

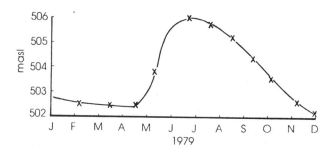

Fig. 4.7 A hydrograph based on repeated measurements of the water table in an observation well. Possible interpretation: winter months reflect restricted recharge, whereas starting in May snowmelt contributions are noticed, followed by summer rains.

4.2.5 Recharge by a River

Recharge of a well by the Mohawk River, New York, has been demonstrated by Winslow et al. (1965) by studying hydrographs of the well and the river (Fig. 4.9):

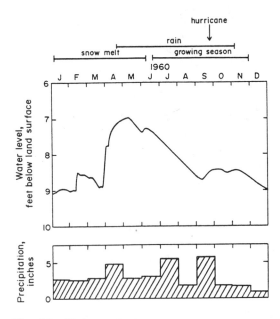

Fig. 4.8 Hydrograph of an observation well at the Saratoga National Historic Park, New York (after Winslow et al., 1965), and local precipitation graph. Interpretation is discussed in the text.

Physical Parameters

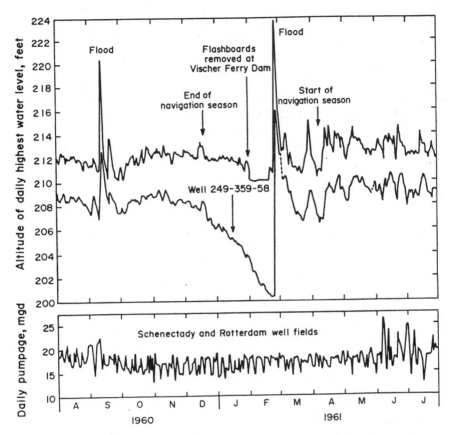

Fig. 4.9 Hydrograph of the Mohawk River and water table variations in an adjacent observation well (following Winslow et al., 1965). The perfect match (text) proves recharge from the river. Pumping from adjacent wells (bottom) was steady and could not cause the observed changes of the water table in the well.

1. The water table of the well followed the river during August.
2. In September a sudden rise in the river was caused by a flood event, followed by an immediate rise of the water table in the well.
3. Lowering the river level by the opening of the dam at the end of the navigation season (March) was followed by a sharp decline in the well.
4. A flood at the end of February caused a sharp rise in the well.
5. Stabilization of the river level at the beginning of the new navigation season was followed by a parallel stabilization in the well.

The close response of the well to fluctuations in the river level demonstrates that the well is dominantly recharged by the Mohawk River. Possible interference by pumping of adjacent wells may be ruled out on the basis of the monotonous abstraction, shown at the bottom of Fig 4.9.

4.2.6 Velocity of a Recharge Pulse

Changes in the water table of the Mohawk River and a number of adjacent observation wells is reported in Fig. 4.10, adapted from Winslow et al. (1965). The wells followed the river, with a time lag of 4–12 hours (insert in Fig. 4.10). Two possible explanations for this time lag may be envisaged: (1) arrival of the hydraulic pulse, or (2) arrival of the recharge front (assuming piston flow, section 2.1). To tell the two apart, the time lag observed for these wells by temperature measurements is helpful, as discussed in section 4.8 (see Fig. 4.21). The temperature time lag of, for example, well 58, has been observed to be about 3 months, whereas the water table time lag was only 12 hours. The latter defines the arrival of the hydraulic pulse, whereas the former defines the travel time of the recharge front. The distances given in the insert in Fig. 4.10, divided by the respective time lags, provided the propagation velocity of the hydraulic pulse, which was 300 ± 50 ft/h, or 100 ± 20 m/h.

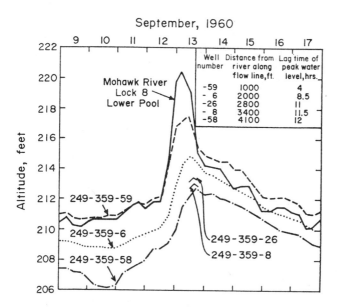

Fig. 4.10 Water table fluctuations measured at wells near the Mohawk River (from Winslow et al., 1965). The wells followed a flood event with a time lag of several hours correlated with the distance from the river (insert table). The distance divided by the time lag provided the propagation velocity of the hydraulic pulse.

Physical Parameters

The case studies discussed above, and depicted in Figs. 4.8–4.10, reveal the importance of repeated measurements, providing evolution with time, and the importance of auxiliary data, such as distribution of local precipitation or discharge in adjacent pumping wells.

4.3 Gradient and Flow Direction

The gradient, or drop in elevation, of the water along its path, is a major factor in defining the water flow velocity, as defined by Darcy's law (section 2.10). The gradient is expressed in meters drop of height per kilometer of horizontal flow trajectory, as shown in Fig. 4.11. The drop in water table levels between adjacent wells is computed from their water table data, and the distance between the wells is commonly read from a map on which the wells have been accurately marked. The gradient also may be extracted from a water table map, by reading the drop in water table altitude between selected points, as shown in Fig. 4.12. Water flows mainly along the maximum gradient, as shown in Fig. 4.4b.

4.4 The Need for Complementing Data to Check Deduced Gradients and Flow Directions

Determination of the flow gradient between two wells (Fig. 4.11) is based on the assumption that the two wells are hydraulically interconnected, as shown in Fig. 4.13a. However, wells may be separated by an impermeable rock bed with no hydraulic interconnection (Fig. 4.13b). Thus, the gradient measurement between the two wells in the second case is meaningless. The same holds true for deduced directions of water flow—they are meaningful, provided the involved wells are hydraulically interconnected. Water flows from I to II in Fig. 4.13a, but it does not flow from III to IV in Fig. 4.13b.

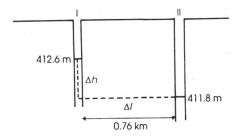

Fig. 4.11 Determining flow gradient. The drop of water table between wells I and II is $\Delta h = 412.6 - 411.8 = 0.8$ m; the distance (read from a map) is 0.76 km. Hence, the hydraulic gradient is 0.8m/0.76km = 1.0m/km.

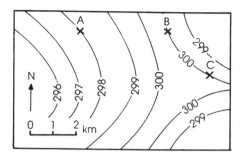

Fig. 4.12 Extracting flow gradients from a water table map (contours in masl). Δh, from B to A, is 300.0m − 298.0m = 2.0 m, and the distance (read from the map) is 3.7 km. Hence, the gradient between B and A is 0.54 m/km. The gradient from B to C is 0 m/km.

Geological information of the rock sequence and tectonic settings may be instructive but not definitive, as it is hard to translate field data into hydraulic conductivity values: a shale bed may be fractured and let water flow through in one case, and a clay bed may be weathered and act as an aquiclude in another. In addition, a variety of processes lower the local water conductance, occasionally preventing lateral flow. An example of such a process is chemical clogging (Goldenberg et al., 1983).

Other parameters are needed to check for hydraulic interconnections between studied wells. Chemical constituents of the water may indicate whether studied wells tap the same water type or belong to distinct water groups. In the latter case, hydraulic interconnections can be ruled out (section 6.5). The temperature of groundwater commonly increases downflow, and hence if colder water is encountered along a suggested flow path, straightforward interconnection

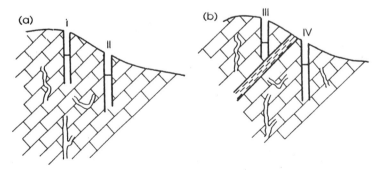

Fig. 4.13 Hydraulic interconnection between wells in fissured limestone terrain: (a) wells I and II are interconnected and water flows down-gradient; (b) similar looking wells, separated by an impermeable rock bed; wells III and IV are not interconnected, in spite of the apparent gradient.

Physical Parameters

between wells is disproved (section 4.8). The isotopic composition of water provides additional clues to hydraulic interconnections (sections 9.10 and 9.11) and so do the radioactive age indicators tritium (section 10.7) and ^{14}C (section 11.9).

4.5 Velocities and Pumping Tests

As stated in section 2.10, the velocity by which groundwater flows is commonly calculated from the water table gradient and the coefficient of permeability (k, or the related parameter of transmissivity). The k value is determined by a pumping test. During such a test a studied well is intensively pumped and the water table is monitored in it as well as in available adjacent observation wells. The change in water table level as a function of the pumping rate serves to compute the aquifer permeability.

Pumping tests call for expertise that is beyond the scope of this book, but as shown in the next section, the incorporation of hydrochemical methods is essential for safe interpretation of pumping test data.

4.6 Chemical and Physical Measurements During Pumping Tests

The common interpretation of pumping test data is based on the assumption that only one aquifer is pumped and tested. However, the intensive pumping during the test causes a significant local pressure drop in the pumped aquifer that may cause water from an adjacent aquifer to breach in (Fig. 4.14). If the pumping test is done in a phreatic aquifer, water of a lower confined aquifer may flow in. Similarly, in pumping tests in confined aquifers, an overlying phreatic aquifer may be drawn in.

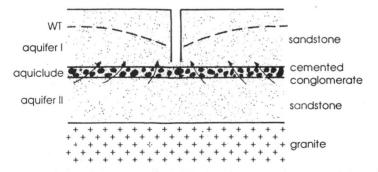

Fig. 4.14 Possible effect of a pumping test. The water table of aquifer I is drawn down near the well, a feature called a depression cone. Occasionally water from a lower confined aquifer may breach in (arrows across the aquiclude).

The breaching in of water from a second aquifer has to be detected so it can be included in the pumping test interpretation. Otherwise, the conclusions based on the test will be erroneous: apparent conductivities that are too large may be deduced, and exaggerated operation pumping rates may be suggested.

To obtain the necessary information on the number of aquifers effectively included in a pumping test, continuous measurements of temperature, conductivity, and other parameters are recommended before, during, and after the test. Temperature is sensitive to aquifer depth (section 4.7) and is most useful in distinguishing different water systems (section 4.8). A constant temperature value, as shown in Fig. 4.15a, is a favorable indication that the pumping test remained restricted to a single aquifer. In contrast, in the example given in Fig. 4.15b, warmer water intruded into the pumped aquifer, and the later part of the pumping test included water from two aquifers.

Electrical conductivity is readily measured in the field and reflects the total amount of salts dissolved in the water (section 8.4). It is highly recommended that electrical conductivity be measured in short intervals throughout a pumping test. The data can be plotted against time, or preferably, against cumulative abstraction (Fig. 4.16). In the example, it is seen that the intruding water is saltier. This result may be of importance for the regular pumping rate to be decided for the well. If the additional salinity is not harmful than a relatively high abstraction rate may be recommended. But if the additional salinity is unacceptable, then a low rate of routine abstraction has to be adapted.

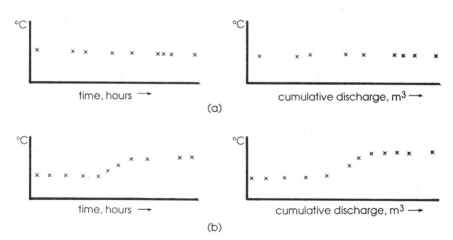

Fig. 4.15 Temperature measurements during pumping tests, expressed as a function of time and cumulative discharge: (a) temperature remained constant, indicating pumping remained constricted to a single aquifer; (b) temperature suddenly increased, indicating water from a warmer (probably lower) aquifer breached in.

Physical Parameters

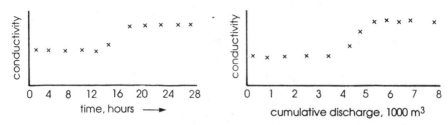

Fig. 4.16 Conductivity measurements during a pumping test: saltier water intruded in the pumped system after 12 hours, during which 3000 m³ were pumped.

In these types of considerations the presentation of measured parameters as a function of both time and cumulative discharge is most useful. In the example of Fig. 4.16 the salty water broke through after pumpage of 3000 m³ in 12 hours. Hence, the routine pumping rate should be significantly lower if only fresh water is wanted, and abstraction may be closer to the break-through value if water quantity is of prime importance and the quality of the mixed waters is acceptable.

In addition to direct measurements, water samples should be collected for detailed chemical and isotopic analysis. As a rule of thumb, a pumping test should be accompanied by one sample taken before the test, representing the nondisturbed water system, and 10 samples collected periodically during the test. It is then suggested that the first and last samples be fully analyzed. If they come out alike, and no temperature and electrical conductivity changes have been noticed, then one may confidently conclude that the pumping test effected only one aquifer. If the first and last samples differ from each other, then the rest of the collected samples should be analyzed to define the nature and mixing percentages of the waters involved. Laboratory analyses are costly, but the information derived is of prime importance. Failure to notice the intrusion of a second water type may turn out to be a failure to predict possible deterioration of water quality in an overpumped well system. A case study of a documented pumping test is shown in Fig. 4.17, taken from a study in the Aravaipa Valley, Arizona. The outcome of a fully documented pumping test is a corner stone of the hydrochemist's report (Chapter 19).

4.7 Temperature Measurements

4.7.1 Temperature Field Measurements

Commercial thermometers are adequate for most hydrological studies, their readings being good to $\pm 0.5°C$. However, they should be calibrated in the laboratory, as they are often off by about $+1°C$. Calibration may be done by reading the thermometer in a freshwater-ice mixture, providing $0°C$, and in

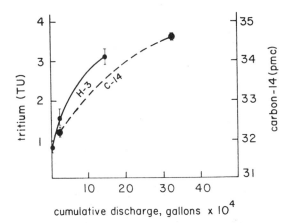

Fig. 4.17 Tritium and ^{14}C measurements during a pumping test conducted in a confined aquifer at the Aravaipa Valley, Arizona (Adar, 1984). Recent water of high tritium and ^{14}C concentrations intruded from the overlying phreatic aquifer.

boiling water, providing the local boiling temperature (determined by the local altitude). Needed corrections should be written on the instrument and applied to the field readings. It is recommended that you always use two thermometers, as a check and as a spare in case of breakage.

Temperatures should be read while the thermometer is dipped in the water, because upon removal the reading will drop rapidly while drying. If reading while dipped in the water is not feasible, a bottle should be filled with the water and the thermometer dipped in it.

The temperature of the water of a spring should be measured as deep as possible. If the water pool of a spring is large, several readings should be taken to reach the extreme value—warmest or coldest—most different from ambient air temperature.

If the temperature measured at a spring is close to the ambient air temperature, temperature reequilibration of the near-surface spring water could have happened. A second measurement, at night or at a different season, may clear this point. If the same value is repeated, in spite of a variation in the ambient air temperature, then the measured spring temperature is the true one, reflecting the temperature at depth.

4.7.2 Depth of Groundwater Circulation Deduced from Temperature Data

Temperature, measured in wells and mines, reveals a general increase with depth, or a geothermal gradient (Fig. 4.18). The value of the geothermal gradient varies from one location to another, an average value being 3°C/100 m.

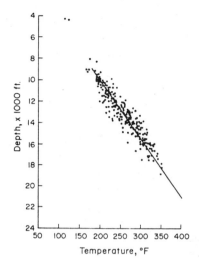

Fig. 4.18 Temperature measurements in 48 deep wells in south Louisiana (from Bebout and Guttierrez, 1981). A gradient of 1.8°F/100 ft, or 3°C/100 m, is observed.

Groundwater is commonly temperature-equilibrated with the aquifer rocks. Temperatures measured in springs or wells reflect the temperature attained at depth, and therefore provide information on the depth of circulation. The calculation is straightforward:

$$\text{depth (m)} = \frac{T_{\text{measured}} - T_{\text{surface}}}{\Delta T/100}$$

where T_{measured} is the measured spring or well temperature, T_{surface} is the local average annual surface temperature obtained from climatological maps or local meteorological institutes, and ΔT is the local geothermal heat gradient established by geophysical studies.

Example: A spring of 62°C emerges in a region with an average annual surface temperature of 22°C. The local heat gradient has not been measured, hence the value of 3°C/100 m may be applied. The depth of circulation is

$$\frac{62 - 22}{3/100} = 1330 \text{ m}$$

Such values are minimum depth values, as the water may cool during ascent. This effect is smaller the larger the water flux is. Calculations reveal that in a spring or well with a discharge of 30 m^3/h or more, cooling is negligible. For cases of lower discharge the calculated circulation depths are to be regarded as minimum values.

Temperature is the best and most reliable tool to establish the depth of groundwater circulation. Knowledge of the depth of circulation may be used to infer the types of rocks in the geological column and the area of recharge, topics to be discussed later on in light of case studies.

4.7.3 Type of Groundwater Flow Traced by Water Temperature Measurements

Recharge water often has a temperature that differs significantly from the aquifer temperature. Continuous temperature measurements may, thus, serve as excellent recharge indicators. Fig. 4.19 shows biweekly temperature measurements, conducted by Shuster and White (1971) in springs of two regions of carbonate rocks in the central Appalachians. The researchers recognized, by independent observations, two types of springs:

A conduit type of rapid karstic flow, recognized by adjacent sinkholes and by the muddy, or turbid, water.

A diffuse type, with slow flow through minute fractures in the rock, recognized by their clear water and lack of adjacent karstic morphology.

The temperature records in Fig. 4.19 correlate well with this classification. The conduit type revealed significant temperature variations, in good agreement with the known seasonal variations of karstic flow, whereas the springs with the diffuse type of flow revealed steady temperatures. Once the temperature pattern was established by calibration against other observations, the tool could be applied to the Paradise Spring (Fig. 4.19), defining it as a conduit type although no supporting field data were available.

4.7.4 Mixing of Groundwaters

Mixing of different water types may occasionally be detected by temperature measurements, as for example, a drop of temperature caused by the arrival of snowmelt recharge seasonally added to a regional base flow. Temperature measurements are most useful for detecting intermixing, especially when coupled with chemical and isotopic measurements, a topic discussed in sections 6.6 and 6.7.

4.8 Tracing Groundwater by Temperature— A Few Case Studies: The Mohawk River

A large-scale temperature study was reported by Winslow et al. (1965), studying recharge from the Mohawk River into gravel deposits underlying the river

Physical Parameters

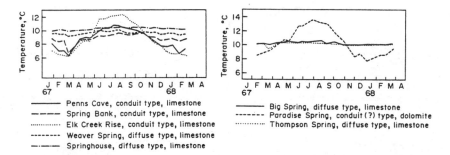

Fig. 4.19 Seasonal temperature observations in springs of the Penns Greek basin, central Appalachians (from Shuster and White, 1971). The authors classified the springs (text) into a conduit type with rapid karstic flow, and a diffuse type with slow flow through minute rock fractures. The former revealed significant seasonal temperature variations, whereas the latter manifested a steady temperature.

floodplain. They monitored temperatures in well fields with a total yield of 7 × 10^4 m^3/day. The temperature data tracing the river recharge has been worked out in a number of modes, all providing superb examples for processing temperature data.

4.8.1 Contours of Equal Temperature Indicating Recharge from a River

Temperature data are occasionally presented in contours of equal groundwater temperature (results of measurements in a large number of wells). Contour maps are shown in Fig. 4.20 for 12 successive months during 1960–1961. The river temperatures (written in the lower left corner of each monthly map) are seen to decrease from 64°F in the autumn to 32°F (freezing) in the winter, and in the spring they warm to reach 77°F in the summer. Thus, twice a year the temperature gradient between the river and the aquifer is reversed. The temperatures in the aquifer are seen to respond to these changes, but with a certain time lag. In October the temperature in the aquifer is about 75°F near the river and 50°F 900 m south, as compared to 64°F in the river. This range drops to a minimum of 40°F near the river and 50°F 900 m away. By that time the river has already started to warm up. The aquifer reaches maximum temperatures of 75°F near the river and 50°F 900 m away in September, when the river is 77°F. Give another glance to the October map to close the cycle. These significant variations in the aquifer temperature, lagging behind the changes in the river, prove that the aquifer is recharged by the river. The changes are greater near the river (35°F–75°F) and are dampened away from it, until at a distance of 900 m to the south the temperature is steady (50°F)

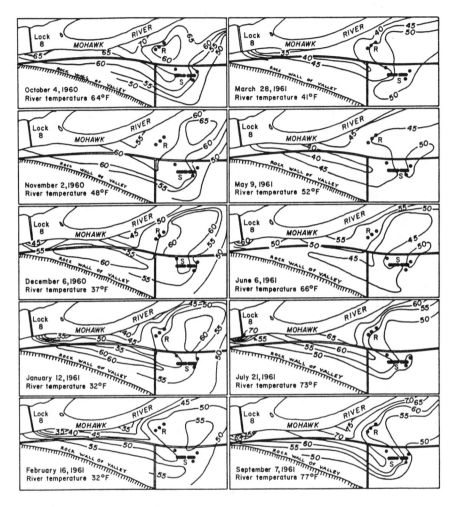

Fig. 4.20 Monthly maps of equal groundwater temperature for a well field bordering the Mohawk River (after Winslow et al., 1965). The river temperature changed over an annual cycle from 77°F to 32°F. The aquifer followed these temperature changes, indicating recharge from the river.

year-round. This lateral temperature gradient points again to recharge from the river.

4.8.2 Seasonally Repeated Depth Profiles in a Well Indicating a Horizon of Preferred Flow

A second way of presenting the data of the temperature survey at the Mohawk River is in the form of seasonally repeated depth profiles in a well (Fig. 4.21).

Physical Parameters

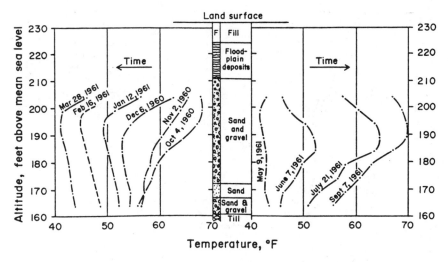

Fig. 4.21 Seasonal temperature profiles in well 61, 100 m away from the Mohawk River (from Winslow et al., 1965). The temperatures are seen to decrease from October–March and to increase from June–September, similar to the trends seen in the temperature maps of the previous figure. The profiles reveal that the largest temperature variations occurred at a depth interval of 180–200 ft above sea level, indicating recharge occurred mainly through this part of the rock section, which in turn must have a higher conductivity.

The temperatures are seen to increase for half a year and then decrease, proving recharge, similar to the mode seen in the temperature maps (Fig. 4.20). The profiles show the vertical dimension of the recharge—temperature fluctuations are accentuated between 180 and 200 ft. Recharge is most efficient in this horizon, indicating highest conductance.

4.8.3 Map of Contours of the Annual Range of Temperature Variations Indicating a Zone of Preferred Flow

Winslow et al (1965) also expressed the results of their temperature study as a map of contours of the annual range of temperature variations (Fig. 4.22). The recharge from the river and the movement southward are seen in the southward damping of the range of annual temperature variations. A detail clearly born out is that of a zone of preferred conductivity, marked A in Fig. 4.22.

4.8.4 Temperature Time-Data Series Indicating Effective Flow Velocities

A fourth way in which temperature data from the Mohawk River were processed is shown in Fig. 4.23. Temperature observations over a whole year are

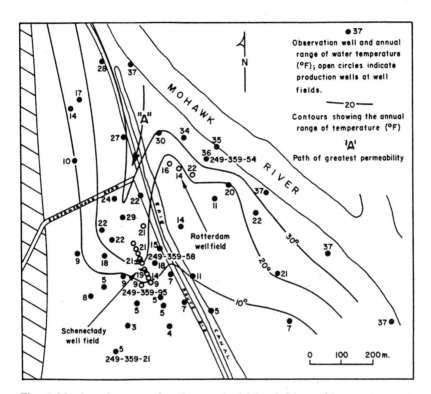

Fig. 4.22 Location map of wells near the Mohawk River with annual groundwater temperatures (from Winslow et al., 1965). Using these values, contours of equal annual temperature variation were drawn. The decrease of these contour values indicates water moves from the river into the aquifer, especially through the zone marked A.

plotted for six wells and for the river. Wells 54 and 59 are seen to follow the pattern of the river, but with a time shift of about 2 months. Well 21, in contrast, reveals no resemblance to the river at all, and the rest of the wells have intermediate patterns. The degree of similarity to the river is correlated with the distance of the well from the river, or as more precisely stated by the researchers, the degree of similarity reflects the hydraulic distance. By this they mean the combined effect of distance, conductivity, amount of water in transit, and temperature equilibration with the aquifer materials.

The temperature graphs in Fig. 4.23 serve not only to demonstrate recharge, but the time lags in the response of each well to the temperature changes in the river may be divided by the distances from the river to provide effective water velocities. A treatment of this kind for all wells showed a zone of high velocities, marked A in Fig. 4.22.

Physical Parameters

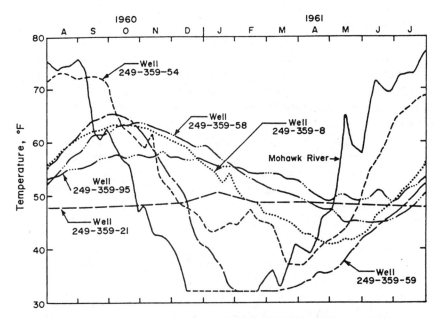

Fig. 4.23 Temperature records over a whole year in wells near the Mohawk River (following Winslow et al., 1965). Wells 54 and 58 follow the river temperature changes with a time lag. Well 21, most distant from the river, revealed a steady temperature over the year, indicating that river recharge is probably not contributing to this well. The rest of the wells showed intermediate degrees of temperature response to the river temperature variations in proportion to their hydraulic distances.

4.9 Cold and Hot Groundwater Systems

Groundwater temperatures vary from a few degrees above freezing to boiling, and in geothermal wells steam at over 300°C is produced. This high range of temperatures is caused by a variety of discharge mechanisms and, to a larger extent, by differences in the depth of circulation and local heat gradient values.

Temperature data closely reflect hydrological conditions, and several groups may accordingly be distinguished:

Groundwater colder then local annual surface temperature occurs at high altitudes and is caused mainly by snowmelt recharge.

Groundwater with a temperature close to the local average annual surface temperature belongs to the shallow active water cycle, circulation being limited to 100 m, and rarely 200 m.

Groundwater that is more than 6°C above local average annual surface temperature circulates to appreciable depths, deducible from the heat gradient (section 4.7).

Warm springs with temperatures up to 65°C are common in tectonically bisected terrains, where groundwater can circulate to appreciable depths, or is stored in deep stagnant (trapped) aquifers. Most rift systems and active orogenic regions host hot springs that are explained in this way. Certain large synclines, or sedimentary basins, contain huge amounts of warm groundwater, exploitable for space heating. The Paris basin is an example.

Geothermal systems host boiling springs and steam. Classical examples can be found in United States (e.g., Yellowstone), Italy (e.g., Pisa), Iceland, New Zealand, and Japan. In the last decades methods of harvesting the energy of geothermal systems have been greatly advanced and a whole field of geothermal prospecting and production has evolved. Geothermal manifestations are closely correlated to recent magmatic activity and movement of crustal plates. The associated waters are concentrated brines rich in CO_2.

The topic of geothermal systems will not be dealt with in this book, but as hydrological systems their methods of study overlap those of nonthermal groundwater.

4.10 Discharge Measurements and Their Interpretative Value

Discharge is the measure of the amount (volume) of water emerging from a spring, or pumped from a well, per unit time. A wide array of discharge units exist in the literature, but cubic meters per hour (m^3/h) is recommended. Discharge may, in certain cases, be measured in the field with the aid of a container of known volume and a stopwatch. The discharge is calculated from the volume (number of times a vessel is filled) and the time involved. Occasionally spring water flows from several directions, the output is too large, or the well discharges into a closed system of pipes. In such cases the discharge information has to be obtained from the well operator or local water authority.

Discharge of a spring or a well is a most informative parameter because it provides insight into the quantitative aspects of groundwater hydrology. Fig. 4.24 includes monthly measurements of discharge, temperature, and chlorine content for a spring. Constant discharge is observed, indicating that water is delayed in the aerated zone, flowing through a porous medium (in conduit-dominated recharge, seasonal recharge fluctuations are reflected in variations in spring discharge). In addition, constant discharge indicates that the water storage capacity of the system seems to be large compared to the annual recharge or discharge. The accompanying temperature and chlorine values are also steady, indicating one type of water is involved in the system. Figure 4.24 shows two useful modes of presentation of recharge data: as a time series and

Physical Parameters

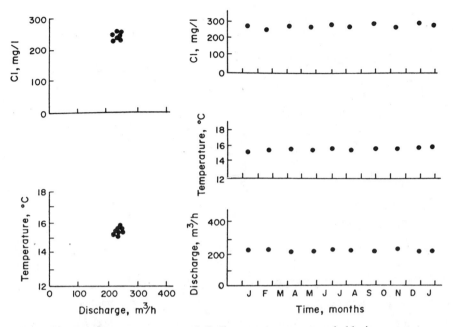

Fig. 4.24 Monthly measurements of discharge, temperature, and chlorine concentration in a (hypothetical) spring. The three parameters are constant over the year, indicating only one type of water is involved, recharged through a porous medium (nonkarstic), and the system's storage capacity is large compared to the annual recharge and discharge.

as a function of other measured parameters. The conclusion that one type of water is involved can be deduced from the horizontal line in the time-series diagram or from the narrow cluster of values in the parametric diagrams.

Data plotted in Fig. 4.25 are from a different case, but resemble the parameters reported in Fig. 4.24. The obtained pattern is different: the discharge varied in an annual cycle, but the temperature and chlorine content remained constant; thus, the interpretation is again of one type of water being involved, but recharge is via conduits of a karstic nature. Temperature is close to the average annual surface temperature and fluctuations in the recharge water temperature are damped by intermixing in the aquifer. Thus, the storage capacity of the system is large as compared to the peak recharge.

A third example, given in Fig 4.26, reveals a case in which recharge varies seasonally in a spring and temperature and chlorine content varied as well,

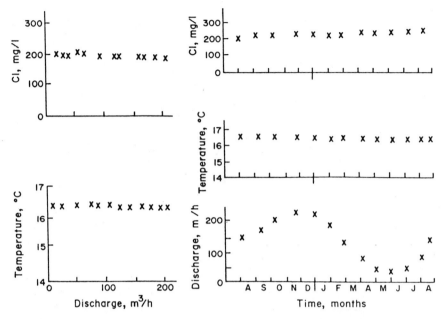

Fig. 4.25 Monthly measurements of discharge, temperature, and chlorine in a spring. Temperature and chlorine are constant, revealing one type of water is involved. The significant discharge variations indicate recharge is fast and of a karstic nature.

but in an opposite pattern (right diagrams n Fig. 4.26). The data plot along straight lines in the left diagrams of Fig. 4.26, revealing a negative correlation between recharge and temperature or chlorine content. Such correlation lines indicate intermixing of two water types, a topic fully addressed in section 6.6. A warmer and more saline type of water intermixes (in varying percentages) with a colder and less saline water. The highest possible temperature of the warmer end member might be deduced by extrapolation of the best-fit line to zero discharge on the temperature-discharge graph. A value of 20.3°C is obtained. In a similar way, the highest possible value of chlorine concentration for the warm end member may be deduced by extrapolation to zero discharge in the chlorine-discharge diagram in Fig. 4.26. A value of 420 mg/Cl/l is obtained. As zero discharge has no meaning in our context (a dry spring has no temperature or chlorine concentration), it is clear that the true intermixing water end members have values that lie between the extrapolated values (20.3°C, 420 mg Cl/l) and the highest observed values (19°C, 400 mg/Cl/l). In this example the two sets of values are very close. The topic of water mix-

Physical Parameters

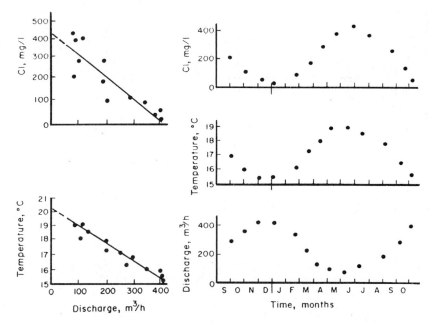

Fig. 4.26 Monthly measurements in a spring. Discharge, temperature, and chlorine are seen to vary considerably, indicating more than one type of water is involved (e.g., base flow and seasonal additions) and recharge is karstic. The data plot along straight lines in the temperature-discharge and chlorine-discharge graphs, indicating two end members are intermixing in the spring system. The temperature and chlorine values of the warm and more saline end member may be deduced by extrapolation of the best-fit lines to zero discharge (text).

tures is discussed in sections 6.6 and 6.7. Discharge measurements often provide negative correlation lines, essential in finding end member properties.

A case study from the Yverdon spring, western Switzerland (altitude 438 masl) is summarized in Fig. 4.27. Maxima in the discharge curve are seen in general to be accompanied by minima in the temperature and electrical conductivity (reflecting salinity) curves, indicating a warm and more saline water intermixes with a colder and fresher water. The discharge and temperature data of the Yverdon spring have been replotted in Fig. 4.28. The best-fit line, extrapolated to zero discharge, intersects the temperature axis at 30°C. This is a maximum possible temperature of the warm end member in the Yverdon complex. The true end member temperature is between this extrapolated value (30°C) and the highest observed temperature (24.3°C).

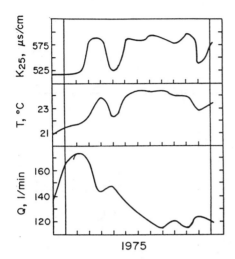

Fig. 4.27 Discharge, temperature, and conductivity (reflecting salinity) in a spring at Yverdon, western Switzerland (Vuatax, 1981). Temperature and conductivity varied in a negative correlation to discharge (Fig. 4.28), indicating mixture of two end members (text).

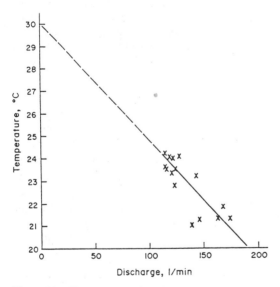

Fig. 4.28 Temperature-discharge data (from Fig. 4.27) of the Yverdon spring, western Switzerland. The negative correlation enables one to deduce an upper limit for the temperature of the warm end member (30°C) by extrapolating to zero discharge (text).

Physical Parameters 89

4.11 Summary Exercises

Exercise 4.1: Draw schematically a vertical selection through a well and indicate (a) the casing and zone of perforation, (b) measuring reference point at the top of the well, and (c) the water table.

Exercise 4.2: The altitude of a reference point of a well is 570 masl (what does this mean?) and the water table was measured to be at a depth of 27 m. What is the altitude of the water table (use the relevant units)?

Exercise 4.3: Study Fig. 4.4. Do the arrows show the only direction of water flow?

Exercise 4.4: Study Fig. 4.6. On what evidence has the fault been plotted? Could a different interpretation be offered?

Exercise 4.5: Is there a good correlation between local precipitation and water level in the case plotted in Fig. 4.8? Can you comment on this correlation?

Exercise 4.6: Observe the high correlation and immediate responses between the river level and the water table in adjacent wells, shown in Figs. 4.9 and 4.10. What is the nature of water flow from the river to the wells?

Exercise 4.7: Why, in the terms discussed above, is there no flow from well C to the direction of well B in Fig. 4.12?

Exercise 4.8: Study Fig. 4.14. What may happen if aquifer II is strongly pumped?

Exercise 4.9: Playing devil's advocate, can you claim that the results in Fig. 4.15a (upper diagrams) do not exclude breaching-in of water from a second aquifer?

Exercise 4.10: Fig. 4.20 is loaded with data. Do you understand the conclusion drawn? Try to go in steps: compare the given river temperature with the equal temperature line for the wells closest to the river in (1) October 1960, (2) January 1961, (3) March 1961, and (4) September 1961. What trend is seen?

Exercise 4.11: The contours in Fig. 4.22 depict _____ (fill in). What is the general trend of these values as a function of distance from the river? What can be concluded?

Exercise 4.12: The lines in Fig. 4.27 make a dramatic impression (check). However, one should also study the figures given on the axes. Doing so, are the changes in the parameters of the Yverdon spring really significant?

5
ELEMENTS, ISOTOPES, IONS, UNITS, AND ERRORS

The outcome of hydrochemical studies is based on chemical data and their interpretation. A clear knowledge of basic terms, units applied, and the nature of analytical errors is essential to avoid confusion and mistakes. The following concepts and definitions have to be mastered as a prerequisite to data processing.

5.1 Elements

To define an element one has to discuss the basic units of nature—the atoms. Atoms are built of a nucleus that consists of two main types of particles: protons, which have a positive charge, and neutrons, which are neutral. Around the nucleus orbit much smaller particles with a negative charge, called electrons. In each atom the number of orbiting electrons matches the number of protons in the nucleus, balancing the negative and positive charges and making the atom neutral. The chemical properties of atoms are defined by the number of electrons, which in turn are defined by the number of protons in the nucleus.

Chemical elements consist of atoms of only one kind. Hydrogen atoms contain one proton, helium atoms contain two protons, oxygen atoms contain eight protons, and uranium atoms contain 92 protons (Table 5.1).

5.2 Isotopes

The number of neutrons in an atomic nucleus either equals the number of protons or is slightly greater. Most elements have several types of atoms, differing in the number of neutrons that accompany the protons, whose number is

Elements, Isotopes, Ions, Units, and Errors

Table 5.1 Atomic weights and valences

Element	Symbols	Protons	Atomic weight	Electrons	Valency
Hydrogen	H	1	1.0	1	+1
Helium	He	2	4.0	2	0
Carbon	C	6	12.0	2,4	+4
Nitrogen	N	7	14.0	2,5	−3 to +5
Oxygen	O	8	16.0	2,6	−2
Sodium	Na	11	23.0	2,8,1	+1
Magnesium	Mg	12	24.3	2,8,2	+2
Silicon	Si	14	28.0	2,8,4	+4
Sulphur	S	16	32.0	2,8,6	−2 to +6
Chlorine	Cl	17	35.5	2,8,7	−1
Potassium	K	14	39.1	2,8,8,1	+1
Calcium	Ca	20	40.0	2,8,8,2	+2
Bromine	Br	35	80.0	2,8,8,17	−1

Compound	Formula	Ionic weight		Valency
Bicarbonate	HCO_3^-	1 + 12 + 48	= 61	−1
Sulphate	SO_4^{2-}	32 + 64	= 96	−2
Nitrate	NO_3^-	14 + 48	= 62	−1

fixed for each element. For example, hydrogen atoms (one proton) are known with no neutron, with one neutron, and with two neutrons. Such varieties in the number of an element's neutrons are called isotopes (Greek: *isos*, equal, and *topos*, place, referring to the place in the Periodic Table). Hydrogen thus has three isotopes, and they have been given special names and symbols:

H—common hydrogen, 1 proton.
D—deuterium, heavy stable hydrogen, 1 proton + 1 neutron.
T—tritium, radioactive hydrogen, 1 proton + 2 neutrons.

Another way to describe an isotope is to add the number of particles in the nucleus (protons + neutrons). This number is placed at the upper left corner of the symbol for the element. Thus, the hydrogen isotopes may be written

^1H—common hydrogen, 1 proton.
^2H—deuterium (also written D), 1 proton + 1 neutron.
^3H—tritium (also written T), 1 proton + 2 neutrons.
  Carbon isotopes are:
^{12}C—common carbon, 6 protons + 6 neutrons.
^{13}C—heavy stable carbon, 6 protons + 7 neutrons.
^{14}C—radiocarbon, 6 protons + 8 neutrons.

Oxygen Isotopes are:

^{16}O—common oxygen, 8 protons + 8 neutrons.
^{17}O—heavy (very rare) oxygen, 8 protons + 9 neutrons.
^{18}O—heavy oxygen, 8 protons 10 neutrons.

The relative amounts of the individual isotope species in each element, expressed in percent, are called the *isotopic abundances*. For example, in seawater the relative abundances of hydrogen isotopes are

^1H—99.984 percent
^2H (or D)—0.016 percent
^3H (or T)—approximately 5×10^{-6} percent

and the isotope abundances of oxygen in seawater are

^{16}O—99.76 percent
^{17}O—0.04 percent
^{18}O—0.20 percent

The isotopic abundances vary slightly in different natural materials. Tables have been compiled giving the average isotopic abundances as observed in terrestrial materials.

Differences in isotopic compositions observed in groundwaters provide the basis for powerful tracing methods, discussed in Chapter 9.

5.3 Atomic Weight

The atoms of different elements and of their isotopes differ from each other in their masses or weights. By international convention common carbon (^{12}C) has been defined as having an atomic weight of 12. On this basis common hydrogen (^1H) has an atomic weight of 1, common oxygen (^{16}O) has an atomic weight of 16, and ^{238}U has an atomic weight of 238.

Isotopes have specific atomic weights, according to their number of neutrons. Thus, the atomic weight of ^{35}Cl is 35, that of ^{37}Cl is 37, and the average atomic mass of chlorine is in between at 35.5. The average atomic weights of the elements have been determined from two sets of data: the atomic weight of each isotope and the relative abundances of those isotopes. From these data the weighted average atomic weight of each element has been calculated. Atomic weights for elements of hydrochemical interest are included in Table 5.1.

5.4 Ions and Valences

The arrangement of electrons orbiting around atomic nuclei obeys strict rules. As the number of electrons increases, from light atoms to heavy atoms, they are arranged in concentric orbits, constituting shells:

Innermost shell—up to 2 electrons.
Second shell—up to 8 electrons.
Third shell—up to 8 electrons.
Fourth shell—up to 18 electrons.
Fifth shell—up to 18 electrons.

In heavier elements the rules are slightly more complicated.

Atoms with complete electron shells are chemically inert. Examples are the noble gases:

He (2 electrons).
Ne (10 electrons, 8 in the outer shell).
Ar (18 electrons, in shells of 2, 8, and 8).
Kr (36 electrons, in shells of 2, 8, 8, and 18).
Xe (54 electrons, in shells of 2, 8, 8, 18, and 18).

All other atoms tend to interact with each other by sharing the electrons in their outer shells. The interacting atoms thus achieve complete outer shells of 2, 8, or 18 electrons. For example,

Na (electron shells of 2,8,1) + Cl (2,8,7) = NaCl.

In this case sodium gives away its single outer electron, being left with its next complete shell of eight electrons. As a result, the sodium atom loses a negative charge and one of its protons is left nonbalanced. In other words, the sodium atom in the NaCl atom combination has one positive, nonbalanced, charge. Thus, it is no longer an atom, which is defined as neutral entity. The new product is called an *ion* and it is marked by a + sign next to the chemical symbol: Na^+. Similarly, the chlorine atom in NaCl gained an electron and completed an outer shell of eight electrons. Thus, the chlorine is now an ion too, but with a negative charge: Cl^-.

Positively charged ions are called *cations* (so called because in electrolysis they move to the cathode) and negatively charged ions are called anions (traveling to the anode). Let us examine another example:

Ca^{2+} + Cl^- + Cl^- = $CaCl_2$.

In this case each of the chlorine atoms received an electron and was turned into an anion, Cl^-. The calcium atom (20 electrons, arranged in shells of 2, 8, 8, and 2) donated its two outermost electrons, remaining with a complete outer shell of eight electrons. Thus, a double-charged cation, Ca^{2+}, was formed.

The number of electrons that an atom gives, or gains, is called its *valency*. In the examples so far discussed, Cl^- has one (negative) valency and Ca^{2+} has a double (positive) valency. In other words, calcium is bivalent. Valences of selected elements are given in Table 5.1.

5.5 Ionic Compounds

Groups of atoms of certain elements combine in a rather stable state, making up ionic compounds. An example is $CaCO_3$, the compositional formula of limestone, composed of $Ca^{2+} + CO_3^{2-}$. A second example is $CaSO_4$, the compositional formula of anhydrite, composed of $Ca^{2+} + SO_4^{2-}$.

In the above examples CO_3^{2-} and SO_4^{2-} are ionic compounds. Their ionic weights are the algebraic sum of the atomic weights of the atoms involved: the ionic weight of CO_3^{2-} is $12 + (3 \times 16) = 60$, and the ionic weight of SO_4^{2-} is $32 + (4 \times 6) = 96$. The valences of ionic compounds are the algebraic sum of the valences of the atoms involved. Thus, the valence of carbonate is

$$C^{4+} + O^{2-} + O^{2-} + O^{2-} = CO_3^{2-}$$

and the valency of sulfate is

$$S^{6+} + O^{2-} + O^{2-} + O^{2-} + O^{2-} = SO_4^{2-}.$$

Ionic compounds commonly dissolved in groundwater include HCO_3^- (bicarbonate), CO_3^{2-} (carbonate), SO_4^{2-} (sulfate), and NO_3^- (nitrate).

5.6 Concentration Units

The concentration of ions dissolved in water is expressed in a variety of ways:

Weight per volume. The units most commonly used are milligram per liter (mg/l) for major ions, and microgram per liter (µg/l) for trace elements. For example, data for groundwater from the Uriya 3 well are given in Table 5.2 in weight per volume units.

Weight per weight. These units are commonly applied to highly saline waters. Units in use include grams per kilogram (g/kg) and milligrams per kilogram (mg/kg). An example for the Dead Sea brines is given in Table 5.3. The conversion from weight per volume to weight per weight units is performed by dividing the former by the density of the brine, for example, 1.23 for the Dead Sea example in Table 5.3.

Ionic equivalence units. The number of anions in solution always equals the number of dissolved cations. Interactions of aqueous solutions with rocks always end up in ionically balanced solutions. Thus, for the discussion of chemical processes it is most meaningful to express chemical data in ionic equivalence units, or equivalents:

$$1 \text{ equivalent} = \frac{\text{number of grams equal to ionic weight}}{\text{valence}}$$

For example, 1 equivalent of Na^+ is $23/1 = 23$ g, and 1 equivalent of SO_4^{2-} is $96/2 = 48$ g.

Table 5.2 Dissolved ions, Uriya 4 well, Israel (temperature 28.6°C, pH 7.05)*

	mg/l (ppm)	meq/l
K	7.0	0.18
Na	253	11.0
Ca	84	4.2
Mg	35	2.9
Sr	1.9	
Cl	405	11.4
SO_4	74	1.5
HCO_3	354	5.8
Si	7	
	µg/l (ppb)	
Al	5	
B	510	
Br	90	
Cl	<0.2	
Cr	4	
Co	3	
Fe	34	
Mn	32	
Ni	5	
P	100	
Total cations		18.3
Total anions		18.7
Total dissolved ions (TDI)		37.0
Reaction error		−1.1%

*From Arad et al. (1984) and Kroitoru (1987). Estimated analytical errors: ±5% for major ions; 10–15% for minor ions.

The data needed to calculate the equivalents of common ions are given in Table 5.1. Equivalent units are too large for convenient use for common waters, and therefore the milliequivalent (meq) unit is used: 1 meq of Na^+ = 23 mg, and 1 meq of SO_4^{2-} = 48 mg. Milliequivalent units are applied per volume units, as in meq/l, or per weight units, as in meq/g.

Table 5.3 Chemical composition of Dead Sea brines (March 1977, density 1.23 g/cm³)*

	K^+	Na^+	Ca^{+2}	Mg^{2+}	Br^-	Cl^-	SO_4^{2-}	TDI
g/l	7.65	40.1	17.2	44.0	5.3	225	0.45	340
g/kg	6.22	32.6	14.0	35.8	4.3	183	0.37	276

*From Kroitoru (1987).

5.7 Reproducibility, Accuracy, Resolution, and Limit of Detection

Data should be reported with a description of their quality. For example, the temperature of a spring measured by one person five times in succession was 16.5°C, 17.2°C, 14.0°C, 15.7°C, and 16.9°C. The same spring measured by a second person five times in succession yielded 16.0°C, 16.1°C, 15.8°C, 16.1°C, and 16.4°C. A quick glance reveals that the second person's measurements are closer to each other, or to use a more technical expression, the second set of data reveals a higher degree of *reproducibility*. This degree of reproducibility may be expressed quantitatively by calculating the mean deviation for the sets of data. The mean value in the first set of temperature measurements is 16.1°C and the mean deviation is ± 1.0°C. Hence, the mean value of the first set of measurements is 16.1°C $\pm$ 1.0°C. The mean value of the second set of measurements is 16.1°C $\pm$ 0.2°C. Hence, the reproducibility in the second set was better and the data of that set were of higher quality. Reproducibility may be best expressed in percentages; for example, the reproducibility of the first person's temperature measurements was $1.0 \times 100/16.1 = \pm 6.2\%$, whereas the second person's reproducibility was $\pm 1.2\%$.

A second property of data quality is their agreement with standard measures. For example, is a thermometer used in the field calibrated by comparison to a standard thermometer? Does it show 0°C in a bath of ice and distilled water, and does it show precisely the local boiling temperature? The agreement with standard measures is called *accuracy*. A good thermometer has an accuracy of ± 0.2°C, but many commercial thermometers have lower accuracies, as they may be off a calibrated instrument by 1.0°C or more. Accuracy of chemical analytical data is tested by the analyst with solutions that are carefully prepared from chemically pure compounds.

A third property of data quality relates to the *resolution*. Thermometers may be long or short. In the first case readings in 0.2°C intervals are possible, whereas in the second case only 1°C intervals are readable. Thus, the resolution of the long thermometer is 0.2°C and that of the short is 1.0°C. Similarly, chemical laboratory data should be reported along with resolution, defined by the instruments involved. Knowledge of the resolution is needed to distinguish differences that are analytically significant. For example, if chlorine is determined with a resolution of 0.5 mg/l, then values of 22.7 mg/l and 23.1 mg/l are nondistinguishable, whereas 22.5 mg/l and 25.8 mg/l reveal a difference that is more than twice the resolution and is therefore analytically significant (this point is further discussed in section 5.8).

A fourth property of significance in data quality assessment is the *limit of detection*, that is, the lowest value detectable. This value has to be known in order to properly process very low concentration values. If the limit of detection for lithium is 0.015 mg/l, then a value of 0.07 mg/l is analytically significant (this point is further discussed in section 5.8).

5.8 Errors and Significant Figures

As seen in the previous section, measured values are not absolute, but are obtained with a certain degree of uncertainty. The uncertainty is caused by the combined effect of several error sources. Four major sources for data uncertainty were described in the previous section: reproducibility, accuracy, resolution, and limit of detection. To these may be added other factors: instability of instrumentation, contamination, accuracy in preparation of standard solutions, etc. The sum of all uncertainties is called the *analytical error*. The analytical error is a cumulative outcome of all errors involved in a measurement. Data included in a laboratory report should always be accompanied by the relevant analytical error, written with a $\pm$ sign to the right of the result; for example, 25.62 $\pm$ 0.50 mg/l. The analytical error is occasionally expressed as a percentage of the obtained value. Thus, 25.62 $\pm$ 0.50 mg/l may also be stated as 25.62 mg/l $\pm$ 2%. In certain cases the analytical error is not computed for each value, but is given in a general mode, for example, in the bottom line of a table, as: analytical errors: Na: ± 0.50 mg/l, Ca: ± 0.70 mg/l, etc.; or Na: $\pm 2\%$, Ca: ± 2.55, and so on.

The analytical error is needed to decide which data differ from each other with analytical significance: only data that differ by more than the relevant analytical error should be regarded as different for purposes of data processing. Accordingly, data should be reported only in significant *figures*. SO_4^{2-} concentrations of 16.273 mg/l or 106.16 mg/l are meaningless if the analytical error is, for example, ± 0.7 mg/l. In such a case the data should be reported using only significant figures, namely 16.3 mg/l and 106.2 mg/l.

5.8.1 Reaction Error

The sum of cations equals the sum of anions in each solution. The deviation from such an equality provides another way to assess data quality. The equation used is

$$\text{Reaction error} = \frac{\Sigma_{cations} - \Sigma_{anions}}{\Sigma_{ions}} \times 100$$

The reaction error is thus expressed as a percentage of the total ion concentration. Positive reaction errors indicate cation excess; negative errors indicate anion excess. Reaction errors are caused by the analytical errors of the individual parameters, and the fact that not all possible ions are commonly measured.

In certain cases it is worthwhile to enlarge the list of ions analyzed in order to lower the reaction error: for example, to include NO_3^-, Fe^{3+}, or PO_4^{4-}.

At the beginning of each study a decision has to be made as to which reaction errors will be acceptable. A cutoff at 2% or 5% is common. Analyses with high reaction errors are omitted in the data processing and, if possible, they should be discussed with the laboratory personnel.

5.9 Checking the Laboratory

Only in rare cases do field hydrochemists themselves measure all the parameters required. In most cases samples are sent to laboratories for part, or all, of the measurements. It is the hydrochemist's duty to discuss data quality with the laboratory and obtain, at least, the analytical error and limit of detection for each parameter measured. In addition, laboratories should be checked by their clients. There are several kinds of laboratory checks.

5.9.1 Duplicate Samples

Each batch of samples sent to the lab should include duplicates of one sample, sent with different names and sample numbers. The results from the duplicate sample give a picture of the quality of the data. If the duplicates fall in the range of the quoted analytical error, the data for the whole batch of samples is acceptable. If, however, the duplicate values differ by more than the stated analytical error, the results should be discussed with the laboratory personnel and the data for the whole sample batch should be regarded as questionable. The differences observed between duplicate samples of several sample batches establish the analytical error of the specific laboratory for each parameter.

5.9.2 Dilution of a Sample

Samples should be diluted with measured amounts of distilled water. The results of the diluted sample are acceptable if they agree with the calculated diluted value within the stated analytical error.

Example: A water sample has been diluted with 1 volume of distilled water. The laboratory results for magnesium were 105 mg/l for the nondiluted sample and 52.9 mg/l for the diluted sample; the analytical error was 0.8 mg/l. Thus, the reported diluted value, 52.9 ± 0.8 mg/l, included in its range the calculated value for a 1:1 diluted sample, 52.5 mg/l, and the magnesium data from the laboratory may be accepted for the whole batch.

5.9.3 Standard Water Sample

A highly recommended procedure is to collect a large sample of groundwater, keep it in a cold dark place (to avoid bacterial decomposition), and add a sample of it with every batch of samples sent to the laboratory. This provides a continuous check, revealing the analytical error and serving as a monitor of laboratory performance.

5.10 Evaluation of Data Quality by Data Processing Techniques

Awareness of data quality information is, unfortunately, limited. A large number of articles are published with no analytical errors or limits of detection, and the results are often stated using nonsignificant figures. A major source for hydrochemical data are archives containing valuable historical data (section 7.5). These, too, are often without quality descriptions. As a result, investigators tend to discard such data as nonreliable, in spite of their high potential value for describing early stages of local water exploration.

Data processing methods provide a substitute for the missing description of analytical data. Repeated measurements of the same water source are occasionally available. In such cases a mean value may be calculated for each parameter (dissolved ions, pH, or temperature). The standard deviations of these values serve as an estimate of the analytical error. In fact, these standard deviations also incorporate the natural fluctuations in the measured water source. Therefore, these mean deviations serve as conservative estimates to the analytical errors. These estimated errors may then be applied to all the data reported from the same laboratory during the same period. Experience shows that old data are often good and acceptable.

5.10.1 Reproducibility Deduced from Clustering of Data

A basic step in the processing hydrochemical data is to plot the various parameters as a function of the total dissolved ions (TDI) or other parameters (sections 6.3 and 6.4). Occasionally a group of hydrologically related samples cluster, or group, around the same values (e.g., wells tapping the same aquifer). In such cases the mean deviation from the mean value, observed for each dissolved ion, serves as an upper limit for the analytical error (it also includes the natural fluctuations).

5.10.2 Reproducibility Deduced from Data Plotting on a Mixing or Dilution Line

As discussed in section 6.6, data occasionally plot along a well-defined line (in most cases a mixing line). The mean deviation from such a line provides an upper limit for the sum of analytical errors for the parameters plotted.

Table 5.4 Dissolved ions, Wisdom Spring, meq/l*

Sample No.	K^+	Na^+	Ca^{2+}	Mg^{2+}	Cl^-	HCO_3^-	SO_4^{2+}
1	0.40	4.52	1.23	0.91	43.2	1.95	0.21
2	0.37	4.93	1.44	0.97	4.75	2.46	0.01
3	0.41	5.24	1.67	1.03	5.11	2.72	0.17
4	0.37	4.92	1.35	1.09	5.01	2.51	0.02
5	0.33	4.67	1.23	0.95	4.82	2.03	0.1

*Perfect Analytics Laboratory, Chemistryland.

5.11 Putting Life into a Dry Table

Mastering the units and data quality concepts discussed in this chapter is absolutely necessary in order to proceed with the remaining chapters in this book, and to enjoy hydrochemistry.

A new hydrochemical study, commenced in Wonderland, included the collection of *five identical* sample bottles from the Wisdom Spring. They were transferred to a laboratory at different dates for determination of major dissolved ions; the data are given in Table 5.4. What can be deduced, or calculated, in light of the data quality concepts and data processing approaches discussed in the previous sections?

Sending several bottles of the same water provides the data needed to calculate the laboratory's reproducibility (section 5.7). This procedure is recommended at the beginning of a hydrochemical study, when using a new laboratory, or as a periodic check of a known laboratory (section 5.9).

From the data of Table 5.4 it can be seen that the chlorine value of sample 1 is significantly higher than the chlorine values of the other four samples. An inquiry to the laboratory may reveal that it was a typographical error, and 43.2 should be corrected to 4.32. The mean sodium value is

$$\frac{4.52 + 4.93 + 5.24 + 4.92 + 4.67}{5} = 4.856$$

or, in *significant figures*, 4.86 meq/l.

The deviation of each sodium measurement from the average is given below:

Sample	Na	Deviation
1	4.52	−0.34
2	4.93	+0.07
3	5.24	+0.48
4	4.92	+0.06
5	4.67	−0.19
mean	4.86	0.23

Thus, the value of 4.52 reported for sodium in sample 1 deviates from the mean value by

4.86 − 4.52 = −0.34 meq/l.

A minus sign is needed to specify that measurement 1 was lower than the mean. The sum of negative deviations (in our example: −0.34 + −0.19 = −0.53) should be close to the sum of the positive deviations (0.07 + 0.48 + 0.06 = 0.61), indicating that the mean value (4.86) and the deviations have been correctly calculated. The two sums may differ slightly (0.53 versus 0.61) due to rounding up of each value to its significant figures.

The *mean deviation* is calculated from all deviations regardless of their positive or negative sign. In the present example

$$\frac{0.34 + 0.07 + 0.48 + 0.06 + 0.19}{5} = 0.228$$

or, in significant figures, 0.23. The mean deviation of the sodium measurements reported in Table 5.4 is ± 0.23 and the mean sodium concentration is 4.86 ± 0.23 meq/l.

The mean deviation and the *standard deviation* (more familiar to some students) are numerically very close. They are often also called sigma (after the Greek letter). Data are commonly reported with one sigma (4.86 ± 0.23 in our sodium example). Statistically, if additional measurements of the same type are done, 67% of the cases will fall in the one-sigma range, 90% of the cases will fall in the two-sigma range (4.86 ± 0.46 in our Na example), and 97% of the cases will fall in the three-sigma range. By convention, results are reported with *one sigma*, unless otherwise specified. Table 5.4 has been reworked in Table 5.5, which includes the average, or mean, concentrations of the various ions and the reproducibilities.

The total concentration of dissolved cations—total cations—is the sum of the concentrations of K, Na, Ca, and Mg:

0.40 + 4.52 + 1.23 + 0.91 = 7.06 meq/l.

The average *total anions* is

4.32 + 1.95 + 0.21 = 6.48 meq/l.

The value of total dissolved ions (TDI) is

7.06 + 6.48 = 13.54 meq/l.

The reaction error (section 5.8) is the difference between total cations and total anions, expressed as a percentage of the TDI. For sample 1 of Table 5.4 this would be

$$\frac{(7.06 - 6.48)}{13.54} \times 100 = 4.3\%$$

Table 5.5 Mean concentrations (meq/l), mean deviations, and reproducibilities for the Wisdom Spring data (Table 5.4)

Sample	K$^+$	deviation	Na$^+$	deviation	Ca^{2+}	deviation	Mg^{2+}	deviation
1	0.40	+0.02	4.52	−0.34	1.23	−0.15	0.91	−0.08
2	0.37	+0.01	4.93	+0.07	1.44	+0.06	0.97	−0.02
3	0.41	+0.03	5.24	+0.48	1.67	+0.29	1.03	+0.04
4	0.37	−0.01	4.92	+0.06	1.35	−0.03	1.09	+0.10
5	0.33	−0.05	4.67	−0.19	1.23	−0.15	0.95	−0.04
Mean	0.38	±0.03	4.86	±0.23	1.38	±0.14	0.99	±0.06
Reproducibility		±7.9%		±4.7%		±10%		±6.1%

Sample	Cl$^-$	deviation	HCO$_3^-$	deviation	SO$_4^{2-}$	deviation
1	4.32	−0.48	1.95	+0.38	0.21	+0.11
2	4.75	−0.05	2.46	+0.13	0.01	=0.09
3	5.11	+0.31	2.72	+0.39	0.17	+0.07
4	5.01	+0.21	2.51	+0.18	0.02	−0.08
5	4.82	+0.02	2.03	−0.30	0.11	+0.01
Mean	4.80	±0.21	2.33	±0.28	0.10	±0.07
Reproducibility		±4.5%		±12%		±70%

The data of Table 5.4 have been reworked again in Table 5.6, which includes the total cations, total anions, total ions, and reaction errors. The data in Table 5.4 have been expressed in meq/l. However, most people are used to the mg/l units for assessment of the degree of salinity of water. The conversion of meq/l data into mg/l was been discussed in section 5.6, and is here demonstrated on the Wisdom Spring data (Table 5.7):

$$\text{Na}^+ : \frac{4.52 \times 23.0}{1} = 104 \text{ mg/l}$$

Table 5.6 Total ions (meq/l) and reaction errors, calculated for the Wisdom Spring data (Table 5.4)

Sample	K$^+$	Na$^+$	Ca^{2+}	Mg^{2+}	Cl$^-$	HCO$_3^-$	SO$_4^{2-}$	Total cations	Total anions	Total ions	Reaction error
1	0.40	4.52	1.23	0.91	4.32	1.95	0.21	7.06	6.48	13.54	4.3%
2	0.37	4.93	1.44	0.97	4.75	2.46	0.01	7.71	7.22	14.93	3.3%
3	0.41	5.24	1.67	1.03	5.11	2.72	0.17	8.35	8.00	16.35	2.1%
4	0.37	4.92	1.35	1.09	5.01	2.51	0.02	7.73	7.54	15.27	1.2%
5	0.33	4.67	1.23	0.95	4.82	2.03	0.11	7.18	6.96	14.14	1.6%

Elements, Isotopes, Ions, Units, and Errors

Table 5.7 Dissolved ions in samples of Wisdom Spring (mg/l)

Sample	K^+	Na^+	Ca^{2+}	Mg^{2+}	Cl^-	HCO_3^-	SO_4^{2-}	TDI
1	15.6	104	24.6	11.0	153	119	10.1	428
2	14.5	113	28.8	11.8	169	150	0.5	488
3	16.0	120	33.4	12.5	181	166	8.2	537
4	14.5	113	27.0	13.2	178	153	1.0	500
5	12.9	107	24.6	11.5	171	124	5.3	456

$$K^-: \frac{0.40 \times 39.1}{1} = 15.6 \text{ mg/l}$$

$$Ca^{2+}: \frac{1.23 \times 40.0}{2} = 24.6 \text{ mg/l}$$

$$Mg^{2+}: \frac{0.91 \times 24.3}{2} = 11.0 \text{ mg/l}$$

$$Cl^-: \frac{4.32 \times 35.5}{1} = 153 \text{ mg/l}$$

$$HCO_3^-: \frac{1.95 \times 61}{1} = 119 \text{ mg/l}$$

$$SO_4^{2-}: \frac{0.21 \times 96}{2} = 10.1 \text{ mg/l}$$

Comparing the raw data in Table 5.4 with Table 5.5, 5.6, and 5.7 and the relevant discussion, the reader may, perhaps, be amazed how much life can be put into a single table of dry data.

5.12 Evaluation of Calculated Reproducibilities and Reaction Errors

Reproducibility of measurements of major dissolved ions can, in theory, be better than $\pm 1\%$. However, in real life poorer reproducibilities are common. In Table 5.5 the reproducibility values of $\pm 4.7\%$ for the sodium data and $\pm 4.5\%$ for the chlorine data are on the limit of acceptance. However, $\pm 10\%$ for calcium and $\pm 12\%$ for HCO_3 are shaky, and the value of $\pm 70\%$ for SO_4 is totally unacceptable.

In the Wisdom Spring example, the investigator should discuss the results with the laboratory staff and see whether they can improve. Otherwise, it will be necessary to shift to another laboratory and repeat the check.

As stated, reaction errors are another way to estimate data quality (section 5.8). The reaction error can (and should) be better than 1 percent, but they are often larger. Many investigators will accept up to 5 percent. Using the latter criterion the reaction errors calculated in Table 5.6 (4.3%, 3.3%, 2.1%, 1.2%, and 1.6%) are acceptable. This example brings out a major shortcoming of the reaction error quality test: the poor reproducibility of the SO_4 data, demonstrated above, is not reflected in the reaction error because SO_4 is a minor compound of the Wisdom Spring water. Low reaction errors indicate acceptable quality of the data for the major ions alone.

The reaction errors in repeated measurements are expected to be random, that is, in part of the measurements the total cations will exceed the total anions and in the rest of the cases the total cations will exceed the total anions. However, occasionally a systematic pattern is seen. In Table 5.6 all the measurements reveal total cations to be higher than total anions (check it). Possible explanations are:

A systematic analytical error in one (or more) of the ions.

The list of measured anions should include other ions that happen to be important, for example, carbonate (CO_3^{2-}), nitrate (NO_3^-), or bromide (Br^-).

5.13 Summary Exercises

Exercise 5.1: Provide the following information for magnesium: atomic weight, valence, and how many mg/l are there in 1 meq/l of magnesium?

Exercise 5.2: Match the symbols in the first column with the items in the second column:

1. Ar   A. Anion of sulfate.
2. HCO_3^- B. The light isotope of oxygen.
3. ^{16}O C. Noble gas argon.
4. NO_3^- D. Deuterium, a heavy isotope of oxygen.
5. SO_4^{2-} E. The element bromine.
6. Ca^{2+} F. The bicarbonate anion.
7. D   G. The heavy and common isotope of helium.
8. Br   H. The anion nitrate.
9. ^3H I. Tritium, a radioactive hydrogen isotope.
10. ^4He J. The bivalent cation of calcium.

Exercise 5.3: What is the unit meq/l? What is its definition? What is 1 meq/l of chloride?

Exercise 5.4: What is the reaction error for the data of sample 73 of the Green Mice Spring reported in Table 6.2?

Exercise 5.5: What is the reproducibility concept? Five bottles of water collected at the same time at a spring were sent to a laboratory and the following values were obtained for the potassium concentration: 12.2 mg/l, 11.9 mg/l, 11.9 mg/l, 12.1 mg/l, and 12.0 mg/l. What is the reproducibility?

Exercise 5.6: The reported resolution of chlorine measurements at a certain laboratory is 2 mg/l. Evaluate the following set of data obtained by sending the same water sample five times: 100.67 mg/l, 99.27 mg/l, 97.31 mg/l, 102.95 mg/l, and 202.02 mg/l.

6
CHEMICAL PARAMETERS: DATA PROCESSING

6.1 Data Tables

Hydrochemical studies generate large amounts of data of different parameters obtained in the field and reported by various laboratories. The first stage in data processing is to organize the data into tables. This stage is important and warrants some thinking. Have a look at Tables 6.1, 6.2, and 6.3. They contain the same data, but differ in their structure. Which of the three tables is "impossible" and which is most handy and most informative?

Table 6.1 has no caption that relates the data to specific wells or springs, the analytical units are not specified, and the data are arranged in increasing order of the sample number, which has no meaning. Table 6.1 is useless.

Table 6.2 has a caption relating the data to Green Mice Springs; the names of the springs from which the samples were collected; the units (meq/l); the name of the laboratory that determined the dissolved ions; the sum of dissolved ions, TDI; and the range of analytical errors. The columns in Table 6.2 are arranged by having the field data first (on the left), followed by the cations, and finally (to the right) the anions. The lines of Table 6.2 are arranged by increasing concentration of TDI.

Table 6.3 is even more organized: the cations are arranged by increasing concentration ($K < Mg < Ca < Na$), and the anions are arranged by increasing concentrations in the most saline sample (spring H), namely $SO_4 < HCO_3 < Cl$. This order is seen in all the samples, except the first (spring A). Table 6.3 reveals that one deals with a complex of springs that have a similar abundance of the dissolved ions, but differ in salinity (TDI) and temperature.

Table 6.1

Sample No.	K	Cl	Mg	Na	SO_4	Ca	HCO_3	Temp. (°C)
71	0.60	9.11	1.78	8.10	1.09	2.72	3.00	21.4
72	1.03	15.5	2.76	11.5	1.36	3.84	4.22	26.1
73	0.02	0.53	0.47	0.91	0.73	1.22	1.36	15.2
74	0.16	2.67	0.80	2.71	0.82	1.60	1.76	16.8
75	0.31	4.82	1.13	4.51	0.91	1.97	2.17	18.3
76	0.75	11.3	2.10	9.90	1.18	3.10	3.41	23.0
77	1.18	17.7	3.08	15.3	1.45	4.21	4.62	27.6
78	0.89	13.3	2.43	11.7	1.27	3.47	3.81	24.5
79	0.46	6.97	1.46	6.30	1.00	2.35	2.59	19.9

Table 6.2 Chemical composition of the Green Mice Springs complex (meq/l)*

Sample No.	Spring	Temp. (°C)	K	Na	Ca	Mg	Cl	HCO_3	SO_4	TDI
73	A	15.2	0.02	0.91	1.22	0.47	0.53	1.36	0.73	5.24
74	C	16.8	0.16	2.71	1.60	0.80	2.67	1.76	0.82	10.5
75	E	18.3	0.31	4.51	1.97	1.13	4.82	2.17	0.91	15.8
79	B	19.9	0.46	6.30	2.35	1.46	6.97	2.59	1.00	21.1
71	D	21.4	0.60	8.10	2.72	1.78	9.11	3.00	1.09	26.4
76	I	23.0	0.75	9.90	3.10	2.10	11.3	3.41	1.18	31.7
78	F	24.5	0.89	11.7	3.47	2.43	13.4	3.81	1.27	37.0
72	G	26.1	1.03	13.5	3.84	2.76	15.5	4.22	1.36	38.8
77	H	27.6	1.18	15.3	4.21	3.08	17.7	4.62	1.45	47.5

*Big Chemistry Laboratory, Dataland. Temperature measurement error: ±0.2°C; analytical errors are ±2% for Na, Ca, Cl, and HCO_3; and ±5% for K, Mg, and SO_4.

Table 6.3 Rearranged chemical composition data of the Green Mice Springs complex (meq/l)*

Sample	Spring	Temp.	K	Mg	Ca	Na	SO_4	HCO_3	Cl	TDI
73	A	15.2	0.02	0.47	1.22	0.91	0.73	1.36	0.53	5.24
74	C	16.8	0.14	0.80	1.60	2.71	0.82	1.76	2.67	10.5
75	E	18.3	0.31	1.13	1.97	4.51	0.91	2.17	4.82	15.8
79	B	19.9	0.46	1.46	2.35	6.30	1.00	2.59	6.97	21.1
71	D	21.4	0.60	1.78	2.72	8.10	1.09	3.00	9.11	26.4
76	I	23.0	0.75	2.10	3.10	9.90	1.18	3.41	11.3	31.7
78	F	24.5	0.89	2.43	3.47	11.7	1.27	3.81	13.4	37.0
72	G	26.1	1.03	2.76	3.84	13.5	1.36	4.22	15.5	38.8
77	H	27.6	1.18	3.08	4.21	13.3	1.45	4.62	17.7	47.5

*Big Chemistry Laboratory, Dataland. Temperature measurement error: ±0.2°C; analytical errors are ±2% for Na, Ca, Cl, and HCO_3; and ±5% for K, Mg, and SO_4.

6.2 Fingerprint Diagrams

Figure 6.1 is a fingerprint diagram of the data in Table 6.3. In this figure each spring is represented by one line that provides a visual description of the relative abundance pattern of the dissolved ions (the shape of each line) and the relative salinity (the position of the line at the upper or lower part of the diagram). Each line is the compositional imprint of a water sample, and various samples can be compared to each other in the way people can be sorted and identified by their fingerprints. It was Schoeller who, in 1954, applied a fingerprint diagram for the first time in relation to groundwater analyses. the fingerprint diagram is a most powerful tool in the hands of the hydrochemist, provided that it is done with good thinking regarding the following points.

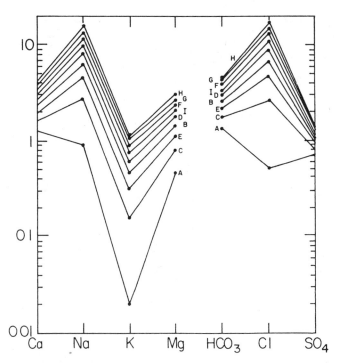

Fig. 6.1 A fingerprint diagram of the data of Table 6.3. Cations are by convention plotted on the left and anions on the right. In Fig. 6.5 the same data have been replotted in increasing order of cation concentration and decreasing order of anion concentration.

Chemical Parameters: Data Processing

Table 6.4 Synthetic data generated to simulate dilution of a saline water by a fresh water (meq/l)

Data set	K	Mg	Ca	Na	SO_4	HCO_3	Cl	TDI
1	0.2	0.8	1.4	2.6	0.3	2.4	2.3	10.0
2	0.4	1.6	2.8	5.2	0.6	4.8	4.6	20.0
3	0.6	2.4	4.2	7.8	0.9	7.2	6.9	30.0
4	0.8	3.2	5.6	10.4	1.2	9.6	9.2	40.0
5	1.0	4.0	7.0	13.0	1.5	12.0	11.5	50.0

6.2.1 The Logarithmic Concentration Axis

The set of data in Table 6.4 was generated so that the concentration of dissolved ions increases from data set 1 to set 5, but the relative abundance of the ions is preserved (check it, for example by comparing the Mg:Ca ratio in data sets 1–5, Table 6.4). This imitates dilution of a saline water by different amounts of a fresh (ideally, distilled) water, a common occurrence in nature. The data of Table 6.4 have been plotted once on a regular millimeter paper with a linear concentration axis (Fig. 6.2) and once on a semilogarithmic paper with a logarithmic concentration axis (Fig. 6.3). The outcome is striking: the same data, plotted with a different concentration axis, reveal intrinsically different patterns. On the regular millimeter plot the lines of the individual water samples differ in their gradients, whereas on the semilogarithmic plot the lines have the

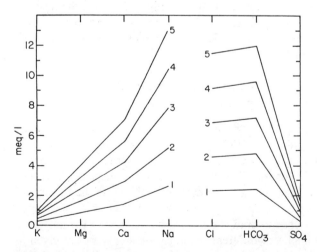

Fig. 6.2 A linear fingerprint diagram of samples that resulted from different degrees of dilution of a saline water (Table 6.4). The data reveal compositional lines of different patterns, although their relative ion abundance is the same.

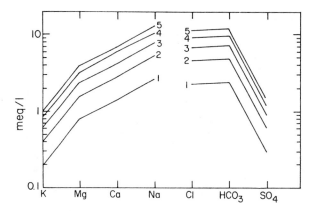

Fig. 6.3 The data of Fig. 6.2 (Table 6.4) replotted on a semilogarithmic paper. Parallel lines of the same pattern are obtained, well reflecting different degrees of dilution of the same saline water.

same gradients. The latter well reflects the dilution of saline water with fresh water, whereas the former diagram is misleading in giving the impression that each of the plotted water samples has its own relative ion abundances. For this reason semilogarithmic paper, or a logarithmic vertical axis in computer programs, should always be used to produce fingerprint diagrams.

The advantage of the logarithmic concentration axis is that it provides equal room for ions of low concentration as for ions of high concentration: the distance between the K points of samples 1 to 5 on Fig. 6.3 equals the distance between the Na points of samples 1 to 5. Not so in the linear concentration diagram (Fig. 6.2), in which the K and SO_4 points are densely packed and the Na and Cl points are widely spread.

6.2.2 Selecting the Right Number of Cycles in the Semilogarithmic Diagram

The number of cycles needed is determined by the lowest concentration value and the highest concentration value in the data set. For example, in Table 6.3 the lowest concentration value is the K concentration in spring A (0.02 meq/l) and the highest is the Cl concentration in spring H (17.7 meq/l). Thus, four logarithmic cycles were needed to plot the data of Table 6.3 in the fingerprint diagram of Fig. 6.1 (check it). Looking at the data of Table 6.4, how many logarithmic cycles are needed to draw the fingerprint diagram? (The answer is given in Fig. 6.3). Examples of different semilogarithmic papers are given in Fig. 6.4 in order to illustrate the selection of the proper number of logarithmic cycles, a task that is quite easy with most graphic computer programs.

Chemical Parameters: Data Processing

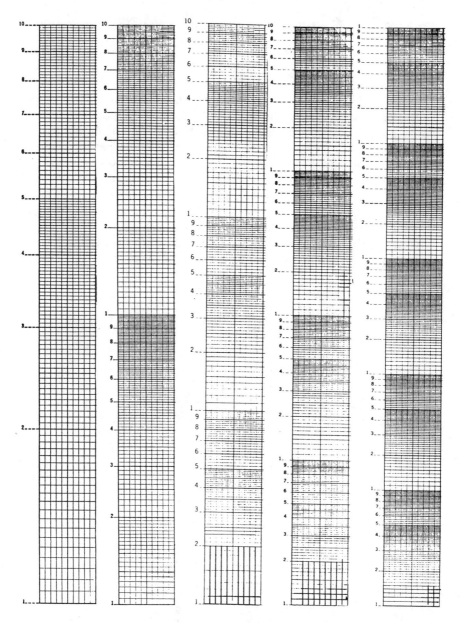

Fig. 6.4 Semilogarithmic papers with 1, 2, 3, 4, and 5 cycles. The 5-cycle paper is most convenient for the majority of hydrochemical data processing tasks.

6.2.3 Selection of the Order of Ions on the Horizontal Axis

Figures 6.1 and 6.5 portray the data of Table 6.3, the difference being the order of the ions on the horizontal axis: in the first case the order is Ca, Na, K, Mg, HCO_3, Cl, SO_4, whereas in the second case the order is K, Mg, Ca, Na, Cl, HCO_3, SO_4. The resulting visual images are completely different. Two considerations lead the decision on the order of ions in the fingerprint diagram: (1) placing ions of a geochemical importance close to each other, for example, K close to Na, or (2) arranging the ions by concentration, as has been done in Fig. 6.5. The advantage of this mode is that simple chemical imprint lines are obtained so that pattern differences between the lines on the diagram are more obvious. Zigzag lines, as in Fig. 6.1, are more confusing.

Another consideration is incorporated in Fig. 6.5: the cations are arranged in increasing order of concentration and the anions are arranged in decreasing order of concentration. In this way the cation and anion lines gain a symmetry and are easy to follow (imagine Fig. 6.5 with the anions arranged in increasing order of concentrations—what would be the outcome?).

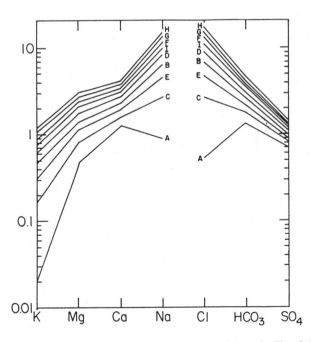

Fig. 6.5 Fingerprint diagram of the same data as in Fig. 6.1 (Table 6.3), but cations are arranged in an increasing order of concentration and anions are arranged in a decreasing order of concentration, resulting in simple lines that can readily be compared.

6.2.4 Separation of Cations and Anions

As a convention it is suggested that the cations be plotted to the left and the anions to the right of the fingerprint diagram. This separation is in accordance with geochemical thinking: waters are described by the concentration order of the cations and anions (section 6.9) and water-rock interactions are discussed by cations and balancing anions. For clarity it is recommended that there be two line segments for each water sample—a line connecting all cations and a separate line connecting all anions, as in Figs. 6.3 and 6.5.

6.2.5 A Methodological Note

The discussed mode of construction of fingerprint diagrams has been found by this writer to be most informative. Other combinations of the fingerprint diagram principles are in use by various investigators. As a result, different types of fingerprint diagrams are included in the discussion of case studies in the following chapters.

6.3 Composition Diagrams

Pairs of measured parameters may be plotted in x-y diagrams, or composition diagrams. These may be closely placed on one page, as shown in Fig. 6.6 for the data given in Table 6.3. The compositional diagram of Fig. 6.6 portrays the Green Mice Springs data in a visual form. The following features can be observed: the springs vary considerably in their concentrations, and the data plot on straight lines, revealing a positive correlation between K, Na, Ca, Mg, Cl, HCO_3, and SO_4 with the TDI. Similarly, the temperature is seen to be positively correlated with the TDI. The composition diagram provides a handy way to visually express large amounts of data, complementing the fingerprint diagram. Each type of diagram has its advantages.

6.4 Major Patterns Seen in Composition Diagrams

Let us plot chlorine against TDI with data obtained from samples collected in regional studies carried out on different water sources. The following patterns are possible:

A cluster (Fig. 6.7). Chloride concentrations in five adjacent springs reveal a cluster when plotted against TDI. This may indicate all five springs are fed by one type of water, or in hydrological terms, the five springs are fed by the same aquifer. One may argue that different waters are involved, each having the same chlorine concentration. Therefore, other parameters are plotted as well, Ca and SO_4 in the example of Fig. 6.7. If the same pattern of a single cluster is obtained, the conclusion of one type of water is confirmed.

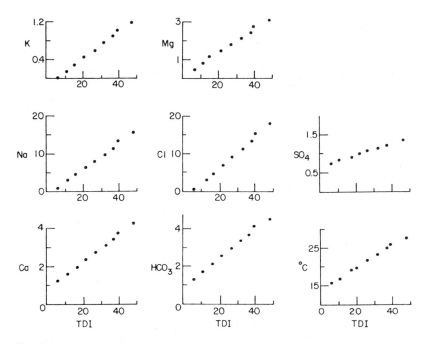

Fig. 6.6 A composition diagram of the Green Mice Springs (Table 6.3). Concentrations are given in meq/l. The compositional interrelations of the nine springs are clearly exhibited: mixing lines indicate a saline warm end member mixes in various proportions with a fresh cold end member.

Two clusters (Fig. 6.8). Two clusters of springs are revealed in Mg, Na and temperature plots versus Cl. Two water groups emerge: a group of five springs is fed by a water that has relatively higher Mg, Na, and Cl concentrations and is relatively cold; and a group of three springs fed by water with relatively lower Mg, Na, and Cl concentrations and a higher temperature. Also, three or more water groups may be reflected in composition diagrams.

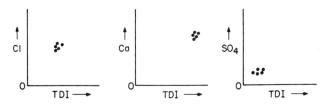

Fig. 6.7 A cluster pattern in a set of composition diagrams of five adjacent springs. The pattern indicates one type of water is involved. The more parameters that are checked, the higher is the confidence of the conclusion.

Chemical Parameters: Data Processing

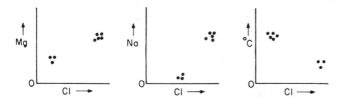

Fig. 6.8 Two clusters in a set of eight adjacent springs. The pattern indicates two distinct types of water occur in the studied region (with no intermixing): a water type of low Cl, Mg, and Na and an elevated temperature, and a water type of high Cl, Mg, and Na and a low temperature.

Data plotting on lines. This type of pattern is shown in Fig. 6.9a–c. Such lines may be of different kinds: Fig 6.9a depicts a line extrapolating to the zero point; Fig. 6.9b shows a line extrapolating to a value on the TDI axis; and Fig. 6.9c extrapolates to a value on the vertical axis. Such lines are formed by mixing fresh and saline water in various percentages, for example, intermixing of water ascending from depth with seasonal rain and snowmelt recharge. The topic is further discussed in sections 6.6 and 6.7.

Triangular distribution. Occasionally data plot in triangles in composition diagrams, as seen in Fig. 6.10. Such a distribution points to intermixing of three distinct water types:

A water type of low TDI, high Na, and low Mg and Cl.
A water type of medium TDI with low Na and high Mg and Cl.
A water type of high TDI, high Na, low Mg, and medium Cl.

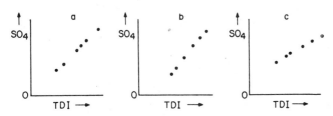

Fig. 6.9 Data from a well field plotting in straight lines in composition diagrams. These are mixing lines, of which three variations are shown: (a) the line extrapolates to the zero points indicating mixing of a saline water with a water that has negligible SO_4 concentrations (dilution); (b) the line extrapolates to a point in the TDI axis, indicating the fresher end member contains significant concentrations of ions other than SO_4; and (c) the line extrapolates to the SO_4 axis, indicating both intermixing waters contain significant concentrations of SO_4.

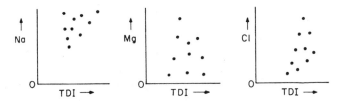

Fig. 6.10 Data of a group of wells falling in triangular areas on compositional diagrams, indicating three distinct water types intermix in varying proportions.

Random distribution. Fig. 6.11a shows data of a studied region that reveal a random distribution on a composition diagram. Random distributions of data may indicate:

The measured samples are from nonrelated water sources of different compositions.

The analytical quality of the data is poor.

The latter case may be established if other pairs of parameters show a distinct pattern. In the example of Fig. 6.11, the SO_4 reveals a random distribution as a function of TDI, but Cl and HCO_3 reveal distinct patterns of mixing between fresh and saline end members. Thus, the SO_4 values are suspected as erroneous and this parameter has to be remeasured to check for a distinct pattern.

6.5 Establishing Hydraulic Interconnections

6.5.1 Direct Proofs for Hydraulic Interconnections

If dyes injected into one well are then found in an adjacent well, this directly proves hydraulic interconnections and the direction of groundwater flow. Fungal

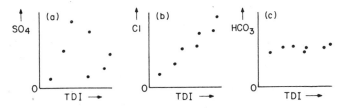

Fig. 6.11 Compositional diagrams of a set of well samples: (a) random SO_4 distribution; (b) a positive Cl-TDI correlation; and (c) a constant HCO_3 value. Possible interpretation: mixing of fresh and saline end members that both have the same HCO_3 concentration; the SO_4 measurements are suspected to be erroneous and should be repeated.

Chemical Parameters: Data Processing

spores, salt, and various radioactive isotopes have been used to trace groundwater flow. A drop in the water table as a result of pumping in an adjacent well is another direct proof of hydraulic interconnections.

The drawback of these methods is that they can be traced over small distances, on the order of tens to hundreds meters, and even that requires substantial effort. For greater distances, other indirect tracing methods are at our disposal. But first let us see why is it important to establish hydraulic interconnections.

6.5.2 Water Level Gradient: A Necessary Condition for Determining Flow Direction, but not a Sufficient One

Water flows from high points to low points, and hence a water level gradient is an essential condition for underground flow. A common practice among hydrologists is to reconstruct groundwater flow directions based on water level gradients (section 4.3). However, a glance at Figs. 6.12–6.14 reveals that a gradient is a necessary, but insufficient condition. Figure 6.12 portrays three wells tapping the same aquifer and the water flows from the region of well I to well II and on to well III, down-gradient. In Fig. 6.13 three wells manifest a relatively high water table at well I, a medium water table at well II, and a relatively low water table at well III. However, the three wells are not interconnected; they are separated by aquicludes. The three wells shown in Fig. 6.14 are separated by a buried anticline disconnecting well I from wells II and III, in spite of the apparent water table gradient. Wells II and III are hydraulically interconnected.

As already mentioned in section 4.4, one can never deduce flow directions from water levels alone.

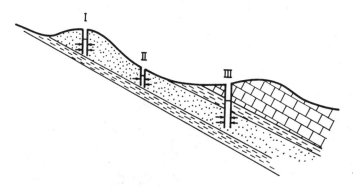

Fig. 6.12 Three wells tapping the same aquifer. Water flows down-gradient from the area of well I to well II and on to well III.

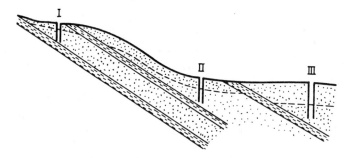

Fig. 6.13 Three wells with water tables similar to those seen in Fig. 6.12 but separated by aquicludes. They have no hydrological connections in spite of the apparent water table gradient.

6.5.3 The Use of Groundwater Parameters to Establish Hydraulic Interconnections

The patterns of data point distribution on composition diagrams, discussed in section 6.4, provide the means to check hydraulic interconnections:

Clustering around a single value, as in Fig. 6.7, indicates that sampled springs have the same type of water and are, therefore, most likely interconnected.

Several clusters, as in Fig. 6.8, indicate separate hydraulic systems are involved, each having distinct water types and isolated from the others.

Data plotting on mixing lines, as in Fig. 6.9, is likely to indicate mixing of two water types in various proportions, a topic further discussed in the next section. Mixing of two (or more) water types indicates that at some point two separated water systems are interconnected, either naturally or by drilling. The same holds true for triangular data patterns on composition diagrams, indicating three distinct water types are interconnected.

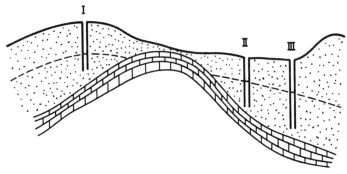

Fig. 6.14 Three wells with an apparent water table gradient of well I > well II > well III. However, a concealed folded structure isolates well I from wells II and III.

6.5.4 A Practical Example of Hydrochemical Identification of Separated Groundwater Systems

The data of Table 6.5 are plotted on a fingerprint diagram in Fig. 6.15. Three water groups emerge: A, B, and C. The groups are homogeneous; the samples of each group having nearly the same ion concentrations. Fig. 6.16 depicts the data of Table 6.5 in composition diagrams. The three distinct composition groups are well seen.

Table 6.5 Dissolved ions in the Hot Fudge well field (meq/l)*

No.	Na	Cl	SO_4	Mg	HCO_3	Ca	K
1	7.2	8.1	0.52	1.5	3.8	3.7	<0.02
2	3.8	4.5	3.1	2.2	5.6	6.0	1.2
3	4.5	5.2	3.7	2.7	5.1	5.5	1.0
4	11.5	11.0	5.4	4.0	9.5	7.2	2.7
5	6.9	8.2	0.84	1.2	3.1	4.0	0.04
6	11.0	12.0	6.0	4.6	8.8	8.3	2.7
7	7.5	8.6	0.60	1.8	4.2	4.1	<0.02
8	12.0	11.5	5.8	4.4	8.3	7.8	2.4

*Good Day Laboratory; analytical errors: ±4% for K, Mg, Ca, SO_4, and HCO_3.

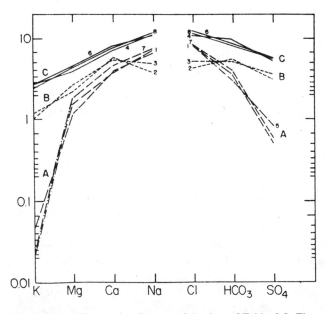

Fig. 6.15 A fingerprint diagram of the data of Table 6.5. Three distinct compositional groups emerge: A, B, and C (seen also in Fig. 6.16).

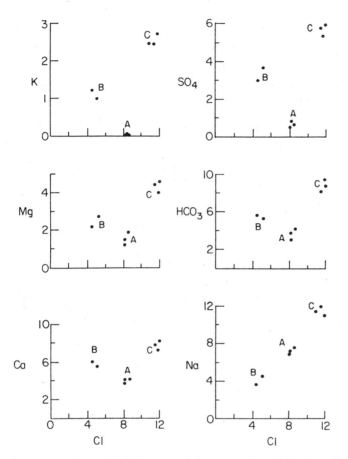

Fig. 6.16 Composition diagrams of the data of Table 6.5. Three distinct compositional water groups emerge: A, B, and C.

Now that a compositional picture has been gained, Table 6.5 may be reorganized by water groups and by increasing concentrations of cations and anions. The outcome (Table 6.6) is informative and reflects the composition pattern of the Hot Fudge wells—they are not fed by one uniform aquifer, but tap three distinct groundwater systems. The wells of group A are most likely hydraulically interconnected, and so are the wells of group B and of group C. But, the water systems A, B, and C are not hydraulically connected.

The next stage deals with a search for the geological, hydrological, or geographical meaning of the observed composition groups, turning them into geochemical groups.

Chemical Parameters: Data Processing

Table 6.6 Reorganized table of the Hot Fudge data (Table 6.5) (meq/l)

Group	No.	K	Mg	Ca	Na	SO_4	HCO_3	Cl	TDI
A	1	<0.02	1.5	3.7	7.2	0.52	3.8	8.1	24.8
	7	<0.02	1.8	4.1	7.5	0.60	4.2	8.6	26.8
	5	0.04	1.2	4.0	6.9	0.84	3.1	8.2	24.3
B	2	1.2	2.2	6.0	3.8	3.1	5.6	4.5	26.4
	3	1.0	2.7	5.5	4.5	3.7	5.1	5.2	27.7
C	8	2.4	4.4	7.8	12.0	5.8	8.3	11.5	52.2
	4	2.7	4.0	7.2	11.5	5.4	9.5	11.0	51.3
	6	2.7	4.6	8.3	11.0	6.0	8.8	12.0	53.4

6.6 Mixing Patterns

The data given in Table 6.4 were generated by mixing the water of set 1 (fresh) with the water of set 5 (more saline). The fraction (x) of type 5 water in the various samples may be calculated by applying each of the parameters of Table 6.4. For example, applying the K values, the fraction of type 5 water in set 3 is $1.0x + 0.2(1 - x) = 0.6$, hence, $x = 0.50$, or set 3 contains 50 percent of type 5 water and 50 percent of type 1 water. (Calculate the fraction of type 5 water in set 3, applying the respective concentrations of Ca, Na, and Cl. Is the value obtained with K confirmed?) The calculation of mixing percentages is further discussed in the next section.

The series of synthetically generated mixtures of two water types given in Table 6.4 is drawn in Fig. 6.3 in a fingerprint diagram on semilogarithmic paper. The data plot on parallel lines, reflecting the dilution of a saline water type with a salt-devoid fresh water. The same data are plotted on a composition diagram in Fig. 6.17. The data plot on straight mixing lines that extrapolate to the zero points. In contrast, the lines in Fig. 6.6 do not extrapolate to

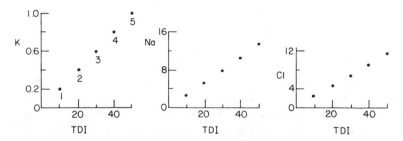

Fig. 6.17 Composition diagrams of the data of Table 6.4 (in meq/l). The data have been computed to represent different mixing ratios between water of type 5 and distilled water. The data plot on straight mixing lines that extrapolate to the zero points.

the zero points, indicating mixing of two water types, each containing a significant concentration of dissolved ions. Thus, a glance at a composition diagram reveals:

The occurrence of mixing (straight lines).
Dilution of a saline water with fresh water (extrapolation to the zero points).
Mixing of two water types, both with a load of dissolved ions (lines extrapolating to one of the axes).

The patterns in a fingerprint diagram provide the same information:

Parallel lines (e.g., Fig. 6.3) indicate dilution occurs.
Lines with a fan shape (e.g., Fig. 6.5), caused by progressive changes in concentrations and relative abundance, indicate mixing of two water types.

6.7 End Member Properties and Mixing Percentages

6.7.1 Hydrologically Deducible End Members

A group of springs, Kaneh-Samar, issues at the Dead Sea shore. Results of a hydrochemical study are given in Table 6.7 and plotted in Fig. 6.18, revealing mixing lines. Contributions from the adjacent Dead Sea were suspected. Thus, the Dead Sea values were entered into the Kaneh-Samar composition diagrams of Fig. 6.18. Since the Dead Sea values fall on the same line as the Kaneh-Samar data, the role of the Dead Sea as the saline end member is established. The procedure has been repeated for four different ions (Fig. 6.18) and the conclusion is therefore reached with a high degree of confidence. The fresh end member, in this case, is recharged in the Judean Mountains, with a salinity that is negligible compared with the Dead Sea brine, so one actually talks of dilution.

Examples of hydrologically deducible end members include dams that contribute to adjacent wells, and seawater intruding into coastal aquifers. Known sources of man-made pollution may be treated as hydrologically suspected end members.

6.7.2 Calculating Mixing Percentages

Once the end members are identified, their mixing ratios can be calculated, as discussed in section 6.6. Dealing with real data of complex natural systems, the calculations of mixing percentages warrant some discussion. Table 6.8 contains calculations of the percentages of Dead Sea water in each Kaneh-Samar spring (Table 6.7). It is seen that the percentages derived via the various parameters somewhat differ from each other due to analytical errors and, possibly, due to superposition of secondary processes. The average value may be taken as best representing the true values (last column in Table 6.8).

Table 6.7 Dissolved ions in the Kaneh-Samar Springs and Dead Sea brine (meq/l) (Mazor, et al., 1973)

No.	Li	Sr	K	Na	Mg	Ca	Br	Cl	SO$_4$	HCO$_3$	Total anions	Total cations	Total ions	Reaction error
1	<0.00014	<0.08	0.2	1.8	3.2	3.2	<0.00001	3.3	0.1	4.92	8.3	8.5	16.8	1.2
2	<0.00014	<0.002	0.2	1.9	3.6	2.8	<0.00001	3.7	0.08	4.49	8.3	8.5	16.8	1.2
3	<0.00014	0.012	0.5	1.9	3.6	3.2	<0.00001	4.3	0.13	4.54	8.9	9.2	18.1	1.6
4	0.001	0.012	0.4	3.3	5.6	4.4	0.02	8.0	0.08	5.23	13.3	13.7	27.0	1.5
5	0.001	0.016	0.8	4.6	6.8	4.0	0.04	10.9	0.55	4.75	16.2	16.2	32.4	0
6	0.01	0.04	1.9	15.2	24.0	8.8	0.32	39.2	1.21	4.82	49.9	49.9	95.4	4.6
7	0.01	0.04	2.0	16.0	22.8	8.8	0.32	39.6	1.02	5.39	46.3	49.6	95.9	3.4
8	0.01	0.04	1.7	16.1	24.8	9.6	0.38	43.4	1.09	4.8	49.7	52.2	101.9	2.5
9	0.02	0.04	2.2	20.0	30.8	12.8	0.52	62.6	1.27	5.28	69.7	65.9	135.6	2.8
10	0.02	0.04	2.7	27.2	34.0	13.6	0.55	64.8	1.47	5.28	72.1	77.6	149.7	3.7
11	0.02	0.04	2.0	19.0	37.2	14.0	0.58	69.7	1.23	5.49	77.0	72.3	149.3	3.1
12	0.03	0.08	3.1	33.9	46.4	14.8	0.79	83.8	1.30	4.52	90.4	98.3	188.7	4.2
13	0.02	0.08	2.4	29.9	45.2	14.8	0.75	87.4	1.82	5.57	95.5	92.4	187.9	1.6
14	0.03	0.08	3.5	32.6	51.6	17.6	0.89	98.3	1.35	5.79	106.3	105.4	211.7	0.4
15	0.04	0.12	3.2	40.7	72.0	22.0	1.29	136.0	1.02	4.66	143.3	138.1	281.4	1.8
16	0.05	0.12	6.4	54.3	94.8	29.6	1.68	181.0	1.73	4.57	189.3	185.3	374.6	1.1
17	0.06	0.16	5.7	65.2	113.2	33.6	2.10	213.0	1.62	4.66	222.1	217.9	440.0	1.0
Dead Sea	2.5	5.9	185.0	1590	3045	687.0	58.0	5486	6.3	3.77	5554.0	5515.0	11069.0	0.4
Dead Sea	2.5	5.9	169.0	1650	3260	756.0	56.0	5814	6.0	3.8	5880	5843	11723	0.3

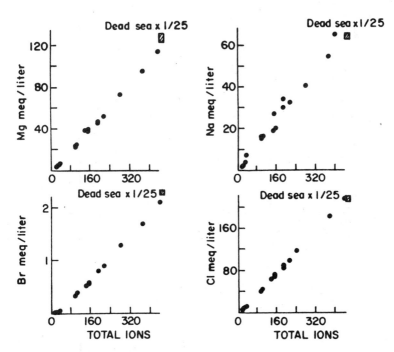

Fig. 6.18 The Kaneh-Samar Springs, on the western shore of the Dead Sea. The composition of the Dead Sea water falls on the continuation of the mixing line obtained by the spring data, supporting the initial hypothesis that Dead Sea water is locally intermixed with fresh water (Mazor, 1973). The ion concentrations in the Dead Sea were divided by 1/25 to accommodate the Dead Sea in the diagram, which had to be designed for the much fresher spring waters.

6.7.3 Extrapolated End Members

Mixing lines can also be of negative correlations, as for example, for salt-poor warm water mixing with a saline cold end member (Fig. 6.19). In such a case the maximum possible temperature of the warm end member can be deduced by extrapolating the best-fit line to zero TDI. A value of 46°C is obtained in the example shown in Fig. 6.19. The temperature of the true warm end member lies between the warmest measured value and the extrapolated value. In the example given in Fig. 6.19, these two values are rather close: 39°C and 46°C, respectively. Negative correlations in mixed groundwater systems are often obtained with tritium and ^{14}C data plotted versus dissolved ions (old saline water diluted by recent fresh water). The value of tritium and ^{14}C in this respect will be demonstrated in sections 10.6 and 11.10.

Chemical Parameters: Data Processing

Table 6.8 Percentages of Dead Sea brine diluted by fresh water in the Kaneh-Samar Springs (based on the data of Table 6.7), calculated from the concentrations of various dissolved ions

No.	Na	Mg	Cl	Br	Average
1	0.11	0.10	0.06	—	0.09 ± 0.02
2	0.12	0.11	0.07	—	0.08 ± 0.03
3	0.12	0.11	0.08	—	0.08 ± 0.02
4	0.20	0.17	0.14	—	0.17 ± 0.02
5	0.28	0.22	0.19	—	0.23 ± 0.03
6	0.94	0.76	0.70	—	0.80 ± 0.09
7	0.99	0.72	0.71	—	0.81 ± 0.12
8	1.0	0.79	0.78	0.67	0.81 ± 0.10
9	1.2	0.98	1.1	0.91	1.0 ± 0.10
10	1.7	1.1	1.2	0.97	1.2 ± 0.2
11	1.2	1.2	1.2	1.0	1.2 ± 0.1
12	2.1	1.5	1.5	1.4	1.6 ± 0.2
13	1.8	1.4	1.6	1.3	1.5 ± 0.2
14	2.0	1.6	1.8	1.6	1.8 ± 0.2
15	2.5	2.3	2.4	2.3	2.4 ± 0.1
16	3.4	3.0	3.2	3.2	3.2 ± 0.1
17	4.0	3.6	3.8	3.6	3.8 ± 0.3

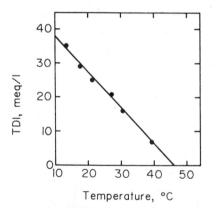

Fig. 6.19 A negative linear correlation between temperature and TDI in a group of springs. By extrapolation, the highest possible temperature value of the warm end member is 46°C. The real temperature of the end member lies between the warmest measured spring and the extrapolated value.

6.7.4 Mixings Deduced from Contradicting Parametric Combinations

Dissolved free oxygen is characteristic of well-aerated groundwater, whereas dissolved H_2S is characteristic of anaerobic conditions. Thus, water contains either dissolved free oxygen or H_2S. Yet, in special cases, both constituents are found in significant concentrations as a result of "last minute" mixing.

A second example of a contradicting combination is occasionally observed in groundwaters with small amounts of ^{14}C and significant amounts of tritium. The first feature, small amounts of ^{14}C, indicates an old age of several thousands years, whereas large amounts of tritium indicate water recharged post-nuclear bomb tests, that is, after 1953. The use of tritium and ^{14}C as age indicators is discussed in Chapters 10 and 11. Water cannot be old and young at the same time, and the combination of small amounts of ^{14}C and large amounts of tritium can result only from mixing of old and young waters.

6.8 Water-Rock Interactions and the Types of Rocks Passed

The bearing of lithology on water composition was discussed in section 3.1 and summed up in Table 3.1. Also the role of soil as a source of CO_2 was mentioned in section 2.9, relating to CO_2-induced water-rock interactions. These points deserve further attention.

6.8.1 Water-Rock Interactions Induced by CO_2

The concentration of CO_2 in the atmosphere is 0.03% by volume. In the soil air, that is, the air between the soil grains, the concentration of CO_2 is 50 to 100 times higher. The source of the extra CO_2 is biogenic: respiration of plant roots and bacterial decomposition of buried plant remains.

Distilled water is fairly nonreactive and practically does not interact with carbonates or silicates, such as limestone, dolomite and marl, or granite, basalt, or sandstone. However, when enriched with CO_2, water turns into carbonic acid and can dissolve rocks by means of the reaction:

$$CaCO_3 + H_2O + CO_2 \leftrightarrow Ca(HCO_3)_2$$
limestone   biocarbonate (soluble)

The double arrows indicate the reaction can go both ways: to the right, dissolution of limestone is described; to the left, precipitation of limestone is described. A high supply of CO_2 drives the reaction to the right (limestone dissolution, as in the forming karstic conduits); depletion of CO_2 drives the reaction to the left (precipitation of calcite and aragonite, as in the formation of stalactites and stalagmites). The result of CO_2-induced limestone dissolution

Chemical Parameters: Data Processing

is groundwater with calcium as a significant dissolved cation and HCO_3 as a significant anion. The maximum concentration of these ions is, however, limited by the saturation value of calcite in aqueous solutions. Therefore, groundwater in limestone terrains is commonly of good quality.

Carbon dioxide also induces interactions of water with silicate rocks:

$$2KAlSi_3O_8 + 6H_2O + CO_2 \leftrightarrow Al_2Si_2O_5(OH)_4 + 4SiO(OH)_2 + K_2CO_3$$
potassium feldspar — clay — soluble — soluble

$$(Ca,Fe,Mg)(SiO_3) + 2H_2O + 2CO_2 \leftrightarrow (Ca,Fe,Mg)(HCO_3)_2 + SiO(OH)_2$$
pyroxene — soluble — soluble

Thus, water containing HCO_3 indicates CO_2-induced interactions with rocks, and the balancing cations indicate the types of rocks passed: calcium comes from interaction with limestone, and calcium and magnesium together come from interaction with dolomite; potassium and, even more often, sodium in bicarbonate water come from silicate rocks rich in potassium or sodium feldspars.

6.8.2 The Diagnostic Value of Dissolved Ions and Cations

The discussion in this chapter and in section 3.1 and Table 3.1 reveals the diagnostic value of anions as indicators for the dominant rock types through which a given groundwater has passed:

Highly saline groundwater, containing chlorine on the order of 180,000 mg/l, and containing sodium as the dominant cation, indicates that halite is present in the aquifer rocks.

Relatively fresh groundwater that contains chlorine in concentrations that indicate undersaturation with respect to halite indicates this mineral is absent in the host rocks.

SO_4-dominated water with concentrations of more than 2,300 mg/l, and containing equivalent concentrations of calcium, indicates that gypsum or anhydrite are present in the aquifer rocks.

Groundwater that is distinctly undersaturated with respect to gypsum indicates this mineral is absent in the aquifer rocks.

Bicarbonate-dominated water with TDI concentrations up to about 600 mg/l has not passed evaporites, and the nature of the rocks interacted with can be deduced from the cations: calcium-dominated HCO_3 water has passed limestone; calcium- and magnesium- dominated HCO_3 water is produced by contact with dolomite; sodium-dominated and potassium-rich HCO_3 waters have interacted with feldspar, plagioclase, and pyroxene contained in igneous or volcanic rocks.

The discussion on the application of dissolved ions as indicators of the rocks passed through by groundwater is of a generalized nature—to show the direction of hydrochemical thinking—and is useful in establishing the constrains needed to formulate phenomenological conceptual hydrological models (section 1.5). The topic of chemical water-rock interactions is discussed by Drever (1982), Erikson (1985), and Hem (1985).

6.9 Water Composition: Modes of Description

Chemical data may be presented in tables and in graphs, but we also need ways to describe the chemical composition of water in written texts. This can be done in various ways:

Dominant cations and anions. Water can be described by its dominant cation and dominant anion. For example, the water of spring A in Table 6.3 has calcium as the dominant cation and HCO_3 as the dominant anion, or the water is of a Ca-HCO_3 type. Similarly, the water of spring H in Table 6.3 is of a Na-Cl type.

Order of cation and anion concentrations. A more detailed description of water composition includes the relative abundance of cations and anions. The example of spring A in Table 6.3 can be described as

$Ca > Na > Mg \gg K$ and $HCO_3 > SO_4 > Cl$

Chlorinity, total dissolved salts (TDS), and total dissolved ions (TDI). The concentrations of chemical compounds can be expressed in various ways:

Chlorinity—the concentration of chlorine
Total dissolved salts or solids (TDS)—the amount of all chemical constituents, determined via electrical conductivity measurements, often called salinity.
Total dissolved ions (TDI)—the sum of dissolved cations and anions.

Mode of composition description: by equivalents or by weight per volume of water. Each description of water composition has to be accompanied by a statement of the units in which the data applied have been expressed, such as meq/l or mg/l. This point is demonstrated by the following set of data:

	K	Na	Mg	Ca	Cl	SO_4	HCO_3	TDI
in meq/l	0.10	5.1	2.5	1.8	5.2	0.3	4.7	19.7
in mg/l	3.9	117	30.2	36.0	185	14.4	225	611

This water can be described in the following two ways:

Na > Mg > Ca >> K ; Cl > HCO$_3$ > SO$_4$ (in equivalents per volume of water)
Na > Ca > Mg >> K ; HCO$_3$ > Cl > SO$_4$ (in weight per volume of water)

It is important to notice that the order of the ions is different in the two modes of expression and therefore the units applied have to be specified. The same holds true for chlorinity, TDS, or TDI. The TDI value of the above example is 19.7 meq/l or 611 mg/l.

Water description by its quality or usefulness.
Examples:

Potable water—up to 600 mg/l TDI.
Slightly saline water, adequate for drinking and irrigation—up to 1000 mg/l TDI.
Medium saline water, potable only in cases of need, can be used for irrigation of special crops, fish raising —up to 2500 mg/l TDI.
Saline water, adequate for fish raising and industrial use—up to 5000 mg/l.
Brackish water—up to TDI of seawater, or 35 g/l.
Brine—most saline water, with TDI higher than seawater.

6.10 Compositional Time Series

The chlorine concentration has been repeatedly measured in a well, as shown in Fig. 6.20. What hydrological conclusions may be reached? Two water types, of different chlorine concentrations, intermix. What is the nature of this mixing? To answer this question, the nature of the time periodicity has to be discussed. A case study of this kind has been reported by Tremblay et al. (1973) from a coastal well on the Prince Edward Island (Fig. 6.21). It was pumped daily from 8 a.m. to 5 p.m., and the chlorine concentration was measured at these hours. The chlorinity increased during the day and dropped until the following morning. This simple series of observations revealed that seawater

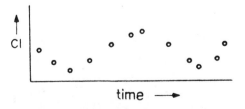

Fig. 6.20 Repeated chlorine measurements in the same water source (well or spring). Mixing of two water types is revealed (in the text).

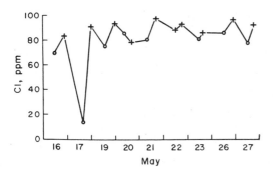

Fig. 6.21 Chloride measurements in a coastal well on Prince Edward Island. The well was operated daily from 8 a.m. to 5 p.m. and it was analyzed at these times: 8 p.m. (circles) and 5 p.m. (+) (following Tremblay, et al., 1973). Encroachment of seawater was concluded, with immediate bearing on management (text).

was drawn into the well due to overpumping, but the inflow of fresh water was sufficient to suppress the seawater intrusion overnight. Hence, if the higher chlorine concentration is not wanted, the pumping rate must be reduced. This practical conclusion could be reached by a simple series of repeated chlorine measurements, whereas a single analysis could not reveal the dynamics of this water system. It would be desirable to check this conceptual model by repeated measurements of additional parameters.

Wilmoth (1972) reported chlorine measurements in a group of wells in Charleston, West Virginia (Fig. 6.22). A gradual chlorine increase occurred from 1920 to the end of 1950, when pumping was ended. A check in 1970

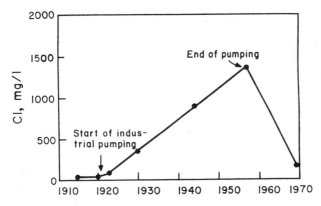

Fig. 6.22 Changes in chlorine concentration in groundwater of an overdeveloped aquifer, Charleston, West Virginia (from Wilmoth, 1972).

Chemical Parameters: Data Processing

revealed that the wells have nearly returned to their original chlorinity. Wilmoth concluded that saline water, underlying the pumped freshwater aquifer, gradually migrated upward due to the pumping. This is another example of the value of repeated measurements, further discussed in section 7.6. This is also a demonstration of the value of historical data, further discussed in section 7.5.

6.11 Some Case Studies

No case study is exactly like another case. Each case study has its own features, defined by the natural setting and the nature of data obtained. The following case studies are an assortment of studies, heavily based on chemical data.

6.11.1 Seawater Encroachment

Cotecchia et al. (1974) studied the salinization of wells on the coast of the Ionian Sea. A fingerprint diagram (Fig. 6.23) served to define a conceptual model. The lowest line (MT) is of a freshwater spring and the uppermost line (I.S.) is of the Ionian Sea water. The lines in between (SR and CH) are of groundwaters with increasing proportions of seawater intrusion. The CH well met the nondiluted seawater at a depth of 170 m. This interpretation seems to

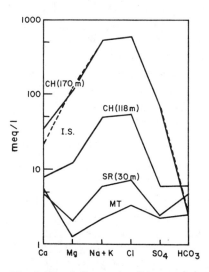

Fig. 6.23 A fingerprint diagram of water in coastal wells of the Ionian Sea (I.S.). MT is a freshwater spring. Well SR has slight contributions of seawater, a feature that is more pronounced in the deeper well (CH), which encountered the seawater at a depth of 170 m (data from Cottechia et al., 1974).

be well founded as it is based on six dissolved ions. The whole story is condensed into one fingerprint diagram.

6.11.2 Classification into Lithologically Controlled Geochemical Water Groups

A major application of fingerprint diagrams is in sorting geochemical data into groups. Figure 6.24 represents the results of an extensive study of mineral waters in Switzerland (Vuataz, 1981). Three compositional groups emerged: Na-SO_4, Ca(Na)-HCO_3, and Ca-SO_4. A search for the geographical, hydrological, or lithological meaning of these compositional groups showed a match with the last factor—lithology: the Na-SO_4 waters issue in crystalline rocks, the Ca(Na)-HCO_3 waters issue in carbonate rocks, and the Ca-SO_4 waters pass gypsiferous sediments. The next step in such a study may be more quantitative, that is, the conceptual model is checked and worked out with water-rock equilibration equations and calculation of saturation indices with regards to various mineral compositions.

6.11.3 Solubility Control of Groundwater Chemistry in an Arid Region

An extensive study of shallow groundwaters in the Kalahari flatland revealed a wide range of concentrations and different compositions. A composition diagram (Fig. 6.25) produced an evolutionary picture. The concentration of HCO_3 increased with increasing TDI, and at 10 meq HCO_3/l the values leveled off and no further systematic increase in HCO_3 concentration was observed, although the TDI increased significantly. Such a pattern indicates that saturation of a relevant salt controls the concentration of the respective ions.

In the Kalahari study saturation with regard to calcium and magnesium carbonates was suggested. This conclusion was supported by

Equilibrium calculations that revealed that the waters with 10 meq HCO_3/l were saturated with regard to limestone and dolomite.
Calcretes ($CaCO_3$ crusts) are common in the investigated area.

The SO_4 concentration is seen in Fig. 6.25 to be low in the range of HCO_3 increase, that is, up to 40 meqTDI/l. At higher TDI values the SO_4 rises, but it levels off at about 20 meq/l which is the saturation value with regard to gypsum. Sodium and chlorine are low in Fig. 6.25, up to 40 meqTDI/l, and then increase linearly with TDI. This was explained by dissolution of NaCl. All of the studied waters, even the most saline ones, were below NaCl saturation.

The observations discussed in light of the composition diagrams lead to the following conceptual model for the Kalahari groundwaters: infiltrating rainwater

Chemical Parameters: Data Processing

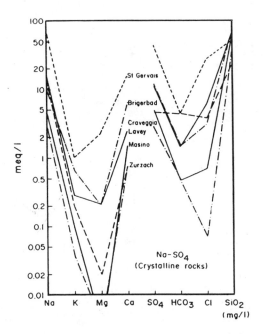
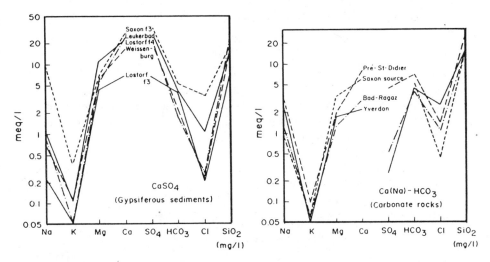

Fig. 6.24 Fingerprint diagrams of data obtained in a study of mineral springs in Switzerland (Vuatax, 1981). Three compositional groups emerged: Na-SO$_4$, Ca(Na)-HCO$_3$, and Ca-SO$_4$. In this case lithology was identified as the major control.

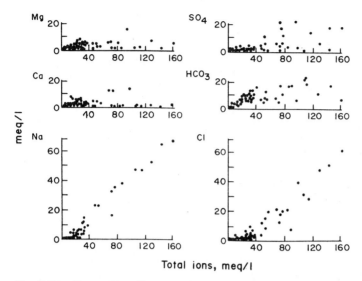

Fig. 6.25 Composition diagrams of an extensive study of shallow groundwaters in the Kalahari flatland (Mazor, 1979). HCO_3, SO_4, Ca, and Mg increase with increasing TDI in the left part of the diagrams, and then level off at the saturation values for Ca and Mg carbonates and gypsum. In contrast, Na and Cl are low at the lower TDI range and then increase with TDI, indicating that the higher salinization is caused by the more soluble NaCl.

is enriched in soil CO_2 and interacts with feldspars in the covering Kalahari sand. As a result, the water is enriched with HCO_3 that is balanced by calcium, magnesium and some sodium. In waters that become more saline, saturation of carbonates is reached and calcretes are formed. Another source of salts is rainborne and windborne sea spray, common over all continents. Substantial evaporation causes these salts to concentrate in the soil, and they are partially washed down with the fraction of rainwater that infiltrates into the saturated groundwater zone. In summary, the evolution of the Kalahari groundwaters is controlled by the solubility of the relevant minerals, which is

$$(Ca, Mg)\, CO_3 \;<\; CaSO_4 \;\ll\; NaCl$$

6.11.4 Mixing of Cold and Warm Waters

In a study of the spring complex of Combiola, southern Switzerland, samples were repeatedly taken and all the data were summarized in a composition diagram (Fig. 6.26). Mixing lines are indicated at Li, Na, K, Mg, SiO_2, TDS, and temperature, plotted as a function of Cl. Thus, a cold fresh water mixes

Chemical Parameters: Data Processing

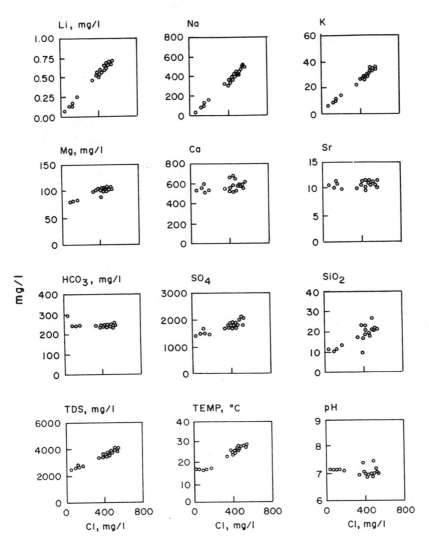

Fig. 6.26 Repeated measurements in a group of springs at Combioula, southern Switzerland (Vuataz, 1981). Positive correlation lines are seen for Li, Na, K, Mg, TDS, and temperature, plotted as a function of Cl. Mixing of a cold fresh water with ascending warm saline water is indicated.

with ascending warm saline water. Horizontal lines are seen for Ca, Sr, and especially for HCO_3, indicating these ions occur in both water end members in the same concentrations. The perfect shape of the Li, Na, K, Mg, HCO_3, and temperature lines in Fig. 6.26 indicates that only one warm water and one cold water are involved in the intermixing. The beauty of this conceptual model lies in its explanation of dozens of different observed water temperatures and ion concentrations as mixings of only two end members.

6.12 Saturation and Undersaturation with Regard to Halite and Gypsum: The Hydrochemical Implications

Water brought in contact with halite readily dissolves it. If the amount of halite is great, as compared to the amount of water, saturation is reached in a matter of a few days, at which time maximum chlorine dissolution takes place, reaching concentrations of more than 180,000 mg/l. Halite is the sole source of chlorine that water has by interaction with rocks. Hence, chlorine concentrations in groundwater of around 180,000 mg/l, balanced mainly by sodium, indicate halite is present in the aquifer rocks. In contrast, groundwater that is significantly undersaturated with respect to halite, indicates that halite is absent from the respective rock system. Most groundwaters contain chlorine in the concentration range of 50–2000 mg/l—two to three orders of magnitude below the halite saturation concentration. Thus, most groundwaters obtain the bulk of their chlorine from outside the rock system—it is brought in with the recharge water.

Saturation with respect to gypsum (or anhydrite) is reached in a few weeks to a few months, attaining an SO_4 concentration of more than 2300 mg/l. Thus, such an SO_4 concentration, mainly balanced by calcium, indicates gypsum or anhydrite are present in the rock system. In contrast, a lower SO_4 concentration indicates undersaturation with respect to gypsum or anhydrite, and thus, absence of these minerals from the rock system. Most groundwaters contain SO_4 in the range of only 20–300 mg/l. The observed SO_4 is in these cases brought in by the recharge water.

6.13 Sea-Derived (Airborne) Ions: A Case Study Demonstrating the Phenomenon

Seawater has a distinct composition, reflected in the concentration of various dissolved ions (Table 6.9). Most groundwaters contain considerably lower ion concentrations than seawater, but the relative abundance of some of the dissolved ions is often similar to the marine abundance. Examples of marine rela-

Table 6.9 Average composition (g/l) of seawater

Na	Mg	Ca	K	Cl	SO_4	HCO_3	Br
10.6	1.3	0.4	0.4	19.0	2.6	0.12	0.07

Chemical Parameters: Data Processing

tive ion abundance in shallow unconfined groundwaters from a small region of the Wheatbelt, eastern Australia, are portrayed in Fig. 6.27. The area is 600 km inland. Linear positive correlations are seen between Cl and Br, Na, Mg, and to some extent K. Hence, it was hypothesized that the source of the major part of the dissolved ions comes not from water-rock interactions, as these would explain neither the observed pattern nor the wide range of concentrations, but that the bulk of the dissolved ions originate from atmospheric salts. The wide range of concentrations was suggested to reflect a wide range of evapotranspiration indices, that in turn reflect local variability in evapotranspiration conditions. Different degrees of retardation of precipitation at or near the surface result in different degrees of salt enrichment in the recharged water fraction. The values of seawater were added to the composition graphs in Fig. 6.27, and they plotted right on the linear correlation lines of Cl and Br, Na, Mg, and to some extent K. The following conclusions could be drawn:

The dissolved Cl, Br, Na, Mg, and K are of a sea-derived origin, whereas observed excesses of HCO_3 and Ca are derived from CO_2-induced water-rock interactions of the type described in section 6.8.

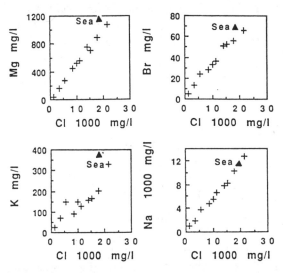

Fig. 6.27 Composition diagrams of shallow observation wells, Merredin region, the Wheatbelt, western Australia (following Mazor and George, 1992). The following observations are made: (1) Cl, Br, Na, Mg, and to some extent K, reveal linear positive correlations; (2) the values spread over a wide range of concentrations, indicating a wide range of evapotranspiration intensitives prior to infiltration into the saturated zone; (3) the seawater value falls on these lines, indicating the marine relative abundance is well preserved. The data provide a detailed insight into the studied hydrological system (text).

The formation of sea spray is accompanied by no ionic fractionation.

The atmospheric transport of sea-derived salts for hundreds of kilometers causes no ionic fractionation.

Evapotranspiration caused ion concentrations to increase up to seawater concentration (Fig. 6.27), maintaining the marine imprint of relative abundance. Thus, salts accumulating on the surface at the end of dry periods are effectively dissolved by the following rains and are eventually carried into the saturated groundwater zone.

The Wheatbelt region, which included study sites in addition to those reported in Fig. 2.27, is hydrochemically mature—the rocks have seen groundwater with the same composition for a long enough time so that all potential ion exchange locations have been used up—and the soil-rock system causes conservative behavior not only for Cl and Br, but also for Na, Mg, and in some cases for K and SO_4.

The distinct variability in ion concentrations, observed in waters tapped in neighboring wells, placed several kilometers apart, indicates the importance of local recharge as compared to the regional base flow.

The preservation of the marine relative abundance of several ions has similarly been observed for large regions of Victoria and other Australian regions, and in other countries. Yet in most setups, water-rock interactions and other ionic inputs change the marine relative abundance, as discussed in the following section.

6.14 Sources of Chlorine in Groundwater

Chloride is perhaps the most informative ion from a hydrochemical perspective, as it is common in groundwater, has a limited number of identifiable sources, and is hydrochemically conservative. These properties turn chlorine into a most useful hydrochemical marker.

Three basic sources of chlorine in groundwater have to be considered in every case study:

Rock salt. Chloride can be contributed by dissolution of halite that is present in the aquifer rocks either as rock salt or as veins crossing other rocks, for example, shale or clay. As stated in section 6.12, this potential source can be ruled out for all groundwaters that are distinctly undersaturated with regard to halite, that is, whenever the chlorine concentration is significantly less than 180,000 mg/l. Thus, this source of chlorine can be ruled out for all fresh groundwater types. This point can be independently checked by the Cl/Br ratio (by weight) in the groundwater: it has to be compared to the marine value of around 293 found in sea spray, in contrast to that of greater than 3000 observed in halite, and thus expected in groundwater that dissolved this mineral.

Seawater. Encroachment of seawater into coastal wells is a phenomenon that is limited to within a few kilometers from the seashore. It is noticeable as rising salinity, and is connected to overpumping that causes significant lowering of the local water table. Seawater encroachment disappears with a decrease in the pumping rate and restoration of the local water table. Seawater stored in aquifer rocks since the last coverage of the terrain by the sea is unlikely in most situations as flushing of aquifer rocks is fast. This source of chlorine can be ruled out for all fresh groundwaters that are located in terrains with a continental history for at least 1 million years.

Sea-derived airborne salts. Also called atmospheric salts, these are the most common source of chlorine in groundwater. The phenomenon was described in the previous section. Atmospheric chlorine is of major importance in all cases where the other sources can be ruled out. Thus, atmospheric chlorine dominates in groundwaters that are fresh and are more than a few kilometers distant from the seashore (and in many cases up to the shore as well). Marine abundance ratios of various ions to chlorine supply direct evidence for their origin from atmospheric sea-derived salts, as discussed in section 6.13.

6.15 Evapotranspiration Index: Calculation Based on Chlorine as a Hydrochemical Marker

Chloride is unique as being most conservative among the common ions. Once it is introduced into the groundwater it stays in it, as there exists no water-rock interaction that can remove chlorine from groundwater. Bromide and lithium are similarly conservative, but their concentrations are low in fresh groundwater, and hence they are seldom analyzed.

Precipitation is split into runoff and into infiltrating water. A major part of the infiltrating water is returned to the atmosphere by the combined action of evaporation and transpiration (section 2.6), and only a small fraction of the water moves deep enough to reach the saturated zone. Evaporation and transpiration are effective on water alone; the contained salts stay behind, dissolved in the fraction of the water remaining in the aerated zone or accumulating on the surface and in the soil at the end of the dry season. The next infiltrating rainwater dissolves these salts and pushes them further down. As a result, groundwater systems are recharged by waters with a wide salinity range, reflecting at each location the relative amount of water returned to the atmosphere by evapotranspiration.

The evapotranspiration index expresses the extent of water loss by evapotranspiration:

ET_{index} = effectively recharged water/infiltrating water, where

> infiltrating water = precipitation − runoff.

Chloride is most useful as a semiquantitative marker for the calculation of the evapotranspiration index, using the chlorine concentration in precipitation and its concentration in a studied groundwater.

***Examples*:**

A chlorine concentration of 5 mg/l is found in the precipitation of many rainy terrains, but the respective groundwaters often contain around 50 mgCl/l. Thus, $ET_{index} = (5/50) \times 100 = 10\%$; that is, 10% of the initially infiltrating water reaches the saturated zone as effective recharge.

A chlorine concentration of 10 mg/l typifies precipitation in semiarid zones, along with groundwaters that often contain 500 mg Cl/l. Thus, in these terrains $ET_{index} = 2\%$.

The complete range of Et_{index} values spans from about 50% in cold rainy regions to 0.5% in warm zones. Large variations are also observed within each climatic zone as a function of topography and the type of local soil or exposed rock, causing different extents of water retardation at or near the surface, resulting in turn in different intensities of evapotranspiration-induced water losses into the atmosphere.

6.16 CO_2-Induced Water-Rock Interactions Revealed by Chlorine Serving as a Hydrochemical Marker

Composition diagrams of water from wells in Tel Aviv, Israel, are given in Fig. 6.28. Each point represents one analysis in a well, and the triangle denotes the composition of seawater (divided by 10 to fit into the diagram). Sodium, magnesium, and even SO_4, plotted as a function of chlorine, are seen to fall on lines on which the marine value also falls. This outcome is interpreted as indicating that the bulk of the Cl, Na, Mg, and SO_4 originate from sea spray, the differences in composition being produced by different evapotranspiration intensities operating on the local discharge water. In contrast, the calcium and HCO_3 values reveal a significant enrichment as compared to the marine relative abundance, indicating enrichment by water-rock interaction. The reported data were collected from 1934–1948, when the young city still had a large number of septic tanks in use. These increased the biogenically produced CO_2 concentration in the aerated zone, making room for the intensified dissolution of carbonate which occurs as the cement of the local sandstone. The observation that magnesium preserved the marine relative abundance (Fig. 6.28), indicates that the dissolved carbonate was a magnesium-poor calcite, and no dolomite is involved.

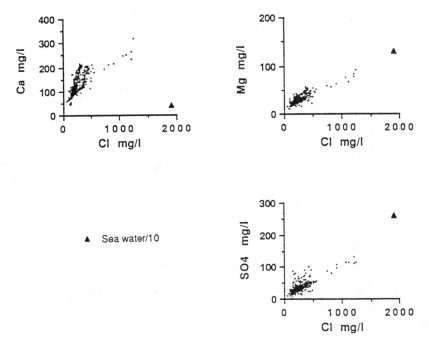

Fig. 6.28 Composition diagrams of 27 wells in the coastal city of Tel Aviv, Israel. Data were repeatedly collected at an early stage of abstraction from 1934–1948 (paper in preparation).

6.17 Exchange of Sodium Versus Calcium and Magnesium Revealed by Chlorine Serving as a Hydrochemical Marker

An extensive survey of the composition of groundwater in a 200-km long section of the Jordan Rift Valley, Israel, revealed in different groundwater subgroups remarkably well-defined linear correlations between the total dissolved ions and Cl, Br, Na, Mg, K, and SO_4 concentrations. The origin of these waters was suggested, based on paleohydrological considerations, to be from seawater that intruded the rift valley during the Pleistocene and got diluted to different degrees by local fresh groundwater (Mazor and Mero, 1969). Compared to the marine abundances, a relative enrichment in calcium and magnesium has been observed in these rift valley groundwaters, balanced by a respective depletion of sodium. Thus, ion exchange of calcium and magnesium for sodium was suggested to have occurred. This ion exchange theory was supported by the observation that coastal waters along the Gulf of Suez revealed similar compositional trends, demonstrating that seawater does undergo ion exchange when

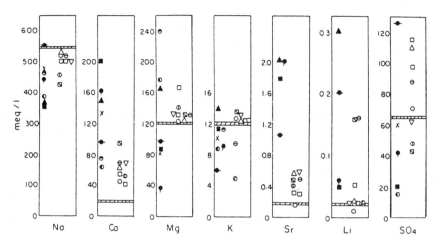

Fig. 6.29 Changes observed in seawater that was shaken for 30 days with pulverized rocks from the rift valley (following Mazor et al., 1973). Open symbols: different rock types (text); closed symbols: rift valley groundwater groups; stippled bar: seawater. The experiments have well imitated the observed trends of the studied groundwaters relative to seawater: Na was taken out of the water in exchange for Ca and Mg, whereas in SO_4 both gains and losses are observable.

intruded into common sedimentary rocks (Mazor, 1968; Mazor et al., 1973). The ion exchange was reproduced in laboratory experiments during which seawater was stirred with crushed chalk, dolomite, basalt, gypsum, bituminous shale, montmorillonite, and phosphorite, as shown in Fig. 6.29.

A similar ion exchange was observed in a large-scale pollution case in which NaCl-rich abattoir effluents were released into a dry river bed (section 16.3.15, Figs. 16.26 and 16.27).

6.18 Summary Exercises

Exercise 6.1: The beginning of a new hydrochemical project includes search for existing data. In many cases such data are scarce and incomplete, but something of value can always be learned. In a real case study of water supply to a mine in Botswana, the initial data available were only a score of chlorine measurements that revealed significant concentration fluctuations at sporadically repeated measurements in the same wells. What would you conclude in such a case? How would this influence your research program?

Exercise 6.2: Types of correlation lines seen in composition diagrams are described in section 6.4 and Fig. 6.9. In this light analyze the data plotted in Fig. 6.26. What can be concluded?

Chemical Parameters: Data Processing

Exercise 6.3: Draw a fingerprint diagram of the data of springs A, B, C, and D given in Table 11.8. Interpret the pattern you will obtain in hydrological terms.

Exercise 6.4: Table 6.7 presents data of springs located on the shore of the Dead Sea and of the Dead Sea brine. Draw a fingerprint diagram of these data, but omit the first three samples because they include nonspecific numbers (e.g., < 0.00014) and theses cannot be drawn, and apply only the second Dead Sea set of values. How many logarithmic cycles are needed? Interpret the data in hydrological terms.

Exercise 6.5: Draw a new fingerprint diagram of the data given is Table 6.7 (without the first three samples and using only the second Dead Sea set of values), but this time divide the Dead Sea values by a factor of 10. Compare this diagram with that you have obtained in the last exercise. Is there an improvement? What is it?

Exercise 6.6: Fig. 6.22 describes repeated chlorine measurements in a well taping an overpumped aquifer, in Virginia, USA. Which operational conclusions can be drawn on the basis of this simple set of data?

Exercise 6.7: Working with archive data, it is often necessary to draw conclusions from incomplete graphic material. Fig. 6.30 summarizes chemical data of six samples collected from a spring over 5 years. The figure is contained in a report and no units are given (mg/l or meq/l). Try to retrieve as much hydrological information as possible.

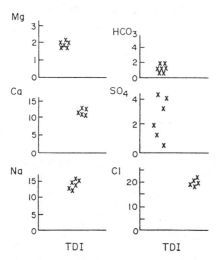

Fig. 6.30 Data (hypothetical) from six samples collected from a spring over 5 years (no units were mentioned in the original report).

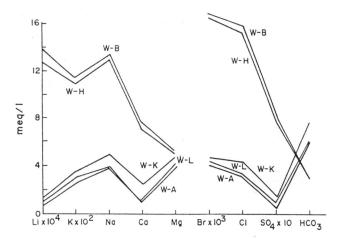

Fig. 6.31 Fingerprint diagram of old municipal wells of which no records of depth and casing construction are available.

Exercise 6.8: Study the fingerprint diagram of Fig. 6.31 and draw all possible hydrochemical and hydrological conclusions. The data are from known municipal wells, but with no records on depth or casing conditions.

Exercise 6.9: A well provided water from a sandstone aquifer underlain by clay, and the water contained (in mg/l): Na:20; Ca:30; Mg:3, Cl:15; SO_4:19; and HCO_3:150. The well was deepened in order to increase the pumping yield and the new water contained (in mg/l): Na:25; Ca:60; Mg:4; SO_4:15; and HCO_3:210. Is the increase in ion concentration, for example, doubling of the calcium concentration, alarming in any way?

Exercise 6.10: A spring emerges in alluvium, and its water composition is (in mg/l): K:4; Na:10; Ca:42; Mg:22; Cl:2; SO_4:20; and HCO_3:220. Which rocks can be excluded as aquifer components?

Exercise 6.11: Shallow water in a crystalline terrain has 200 mg Na/l and 450 mg Cl/l. What may be the origin of the water?

Exercise 6.12: A well was terminated at a depth of 250 m, in sedimentary rocks. The water abstracted had poor taste and was found to contain 450 mg SO_4/l, balanced mainly by calcium. What does the composition indicate? What can be done to improve the water quality?

7
PLANNING HYDROCHEMICAL STUDIES

A good hydrochemical study is well planned in advance, and as little as possible is left to luck or fate. The first stage is a clear definition of the goals or purpose of the study, its extent, and the means available. The answers will differ from one case to another, but there is a common thread: a thorough understanding of the water system under study has to be reached and that in turn has to be based on high-quality data collected in the field or obtained in laboratories. There is always a limit to the number of water sources, wells, and springs, that can be included in a study, there is a limit to the number of periodically repeated measurements possible, and there is a limit to the type and number of laboratory analyses performed. Thus, planning of hydrochemical studies is firmly anchored in optimization decisions. This chapter is devoted to these topics.

7.1 Representative Samples

The discussion may be opened with a problem. Temperature measurements conducted at a spring revealed 16°C at the edge of the water body, 18°C at the center at a depth of 5 cm, and 21°C at the center at the maximal depth, which was 1.3 m. Has the spring three temperatures? Is one value more representative than another? Or should we apply the average measured temperature? The measured temperatures differed substantially, and the cause of these differences has to be understood in order to select the right value. In this example it is clear that the spring water emerged at a certain temperature, but reequilibrium with the ambient air temperature occurred near the surface. This leads to the conclusion that the temperature measured at the maximal depth is closest to the indigenous value. One may generalize: the temperature most different from the ambient air temperature is most representative, not the average value.

One may go on to ask, where in this spring should a sample be collected for laboratory analyses. The temperature can be used as a guide for the location of spring water that has had minimal communication with the surface, that is, minimum evaporation, oxidation, or incorporation of surface materials. Thus, the most representative sample of the spring is as deep as possible and as far as possible from the spring pool edges. Temperature measurements are a sensitive guide to the most representative spring sample, and they should be conducted first, followed by sample collections and other *in situ* measurements (e.g., pH, dissolved oxygen, conductivity).

Wells also have many "faces"—a nonpumped well contains a column of water that has possibly interacted with its surroundings and reveals certain temperature values and a certain composition. Upon pumping, water that was shielded in the aquifer enters the well and the properties of the water in the well change. For this reason, water from pumped wells is in many cases regarded to be most representative of its aquifer water. Pumping should be continued until temperature, conductivity, and other parameters reach a constant value.

By systematically lowering a temperature sensor into the water standing in a nonpumped well, a temperature profile can be obtained. Commonly the temperature near the water table is closer to the local ambient air temperature; deeper, a constant temperature is obtained that is representative of the aquifer. Occasionally, temperature increases with depth and, in rare cases, temperature reversals may be observed, that is, a zone of lower temperature water is overlain by a zone of higher temperature water. The latter case indicates lateral flow of the groundwater and differences in rock conductivity. If such a well is pumped in order to collect samples and conduct field measurements, the values obtained will be a sort of average for the groundwater system being studied. In practice, various modes of sampling are applied: profiles in nonpumped wells, as well as samples in pumped wells. The mode of sample collection has to be taken into consideration at the data processing stage.

Overpumping of a well may change the local pressure distribution in the water system to the extent that water from an adjacent aquifer may breach in and change the water properties. Thus, over intensive pumping can introduce new complications. An obvious example is encroachment of seawater into coastal wells.

The question of what represents water samples collected at wells has also to be addressed in light of the casing perforations. Well casings may be perforated along the entire section of the saturated zone, or they may be perforated only at certain intervals. Thus, a knowledge of the perforation geometry of a well is needed to understand what a collected sample represents.

Planning Hydrochemical Studies

The discussion so far may create the impression that representative water samples are out of reach and nothing can be done about it. In fact, simple precautions provide reasonable solutions:

The mode of data and sample collection in the field has to be written down and taken into account at the data interpretation stage.
Samples protected from surface interaction are preferable.
In nonpumped wells, measuring of profiles is recommended.
In pumped wells, samples should be collected only after stabilization of properties has been achieved. The relevant pumping history has to be documented.
Periodically repeated data collection is desirable.

Confirmation of previous data indicates good representation of all data sets. Variations in the measurements call for careful sorting of the most representative values (which necessitates understanding of the causes of the observed variations).

7.2 Data Collection During Drilling

The fingerprint diagram shown in Fig. 7.1 depicts gradual salinization with depth at the Amiaz 1 well, west of the Dead Sea (Mazor et al., 1969). Fresh water was encountered at a depth of 32 m, whereas saline water of the local Tverya-Noit group was found at a depth of 85 m. A practical consequence of this is that some fresh water may be abstracted from a depth of 30–40 m. The example of the Amiaz 1 well can be generalized: in each area several water bodies may be passed by a drill. All of them should be documented in the driller's records, all should be measured in situ, and samples should be collected for laboratory measurements.

The study of water encountered during drilling causes technical difficulties: most important is the possible interference of water introduced by the drilling procedure, difficulties in noticing natural water horizons that are passed by the drill, and the cost of stopping the drilling operation for measurements. These difficulties can be overcome in the following ways:

Use of every interruption in drilling (e.g., weekends or mechanical breakdowns) to measure the water table, temperature, and conductivity. Repetition of these measurements, during pauses in the drilling operation, may reveal progressive changes, indicating restoration of the water system. At each drilling pause, water samples should be collected for laboratory analyses. A sample of the water applied for the drilling should be analyzed as well. Differences between the latter and the well samples will indicate the water bodies encountered and their properties.

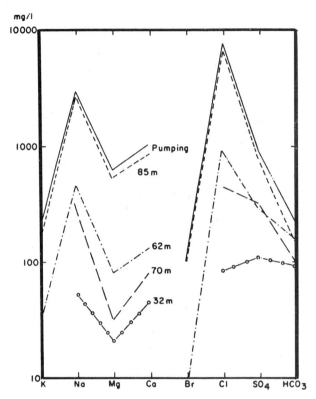

Fig. 7.1 A fingerprint diagram of water from various depths collected during drilling of the Amiaz 1 well, west of the Dead Sea (Mazor *et al.*, 1969). Several water horizons with different water qualities were encountered.

Water in the drill hole should be removed with a bailer, that is, a water sampling rod, lowered and lifted with the drilling equipment. The amount of water removed should be several times the volume of the water-filled section of the drill hole. The bailer also serves to collect the samples. If the water is quickly transferred into bottles, the temperature may even be measured in them. However, lowering a temperature logger, a pump, and a special sampling device into the drill hole is preferable.

Whenever possible, dry drilling (with compressed air) is preferable.

The budget allocated for drilling a well should include the costs of several stops required to measure water bodies that might be passed.

Information obtained during drilling is most valuable. Yet in many cases this information is not collected.

Measurements during drilling provide the hydrological information in the depth axis. Therefore, it is essential that a hydrochemist performs detailed work during each drilling operation. Thus, the planning of hydrochemical studies should routinely include follow-up to drilling operations.

7.3 Depth Profiles

Figure 4.21 depicts temperature profiles in a well close to the Mohawk River, New York. In May a temperature of 43°F is observed at the measured depth interval of 205–165 ft above sea level. The July profile looked totally different: it showed 57°F at the top, 64°F at the center, and 52°F at the bottom. The January profile looks different again: 56°F at the top, 68°F at the center, and 52°F at the bottom. The interpretation of these temperatures profiles is discussed in section 4.8. For our present topic, the planning of hydrochemical studies, it is important to keep in mind that water strata are not necessarily uniform—if they vary in their temperature profiles, they may vary in other parameters as well.

A sample collected from a pumping well provides average water properties and the detailed profile provides the fine structure, which is most important to the understanding of the dynamics of groundwater movement, as discussed in section 4.8.

Depth profiles in observation wells, or nonpumped wells, are conducted by lowering sensors into the well. The depth is usually read from markings on the electrical cable. The lowering of sensors into the well agitates the water and mixes the water column in the well. For this reason:

Measurements were taken downward so the sensor enters undisturbed parts of the water column in the well.
Several parameters are measured together, for example, temperature, conductivity, and pH.
Water level measurements, which may slightly agitate the water column, should be done last.

Samples can be collected from a depth profile by carefully lowering a cylindrical sampling bottle (commercially available). It should be gradually lowered to deeper sections so that each bottle collects undisturbed water. The depths of perforated sections, and other details of the well construction, must be known for proper interpretation of the data obtained in each profile.

Planning of a hydrochemical study has to take into account the need to measure depth profiles and to select the proper time to do them.

7.4 Data Collection During Pumping Tests

Section 4.6 is devoted to chemical and physical measurements during pumping tests. The conclusion reached is that physical and chemical follow-up of pumping tests is essential for their proper interpretation. Pumping of one water body (aquifer) is commonly assumed, but physical and chemical measurements can establish the number of water bodies affected by the pumping. Thus, the hydrochemist should participate in pumping tests and in the interpretation of their data. The equipment needed includes probes for measuring conductivity and temperature, to be placed at the pump outlet to ensure minimum temperature reequilibration. Samples for dissolved ions, stable hydrogen and oxygen isotopes, and tritium measurements should be taken before pumping begins, or immediately at the beginning, and then at intervals up to the end of the pumping. About 10 samples should be taken: the intervals should be planned in light of the pumping test program. The first and last samples should be sent to the laboratory first. If they reveal the same concentrations, and provided the temperature and conductivity remained constant during pumping, then pumping of one water system is demonstrated. But if conductivity and/or temperature changed during pumping, or the first and last sample were found to differ in their properties, then the rest of the samples should be sent for analysis. Modes of data interpretation are discussed in section 4.6.

7.5 Importance of Historical Data

The data collected during the drilling of the Amiaz 1 well (section 7.2 and Fig. 7.1) were put aside once the drill entered saline water and the well was abandoned. Several years later a water source of low output but good quality was needed in the area; on that occasion all historical data were scanned and studied. The existence of the required water body was spotted from the historical data, at no extra cost. A study of groundwater systems along a transect through the Judean Mountains, central Israel, was recently performed (Kroitoru, 1987; Kroitoru et al., 1987). Besides many new measurements, all available data from published reports and unpublished archives were incorporated. One measurement turned out to be of special value: postbomb tritium was observed in the Elisha Spring, situated on the eastern end of the Judean Desert, 22 km east of the recharge area, which is near Jerusalem. The point of importance was that the tritium was observed as early as 1968, at most 15 years after the onset of nuclear bomb tests, held between 1953 and 1963. Thus, it turned out that groundwater flow to the Elisha Spring was at least $22:15 = 1.5$ km/year, which in turn indicated flow in karstic (old) conduits. The historical data provided much additional information in this study and turned out to be most useful (Kroitoru, 1987).

Many researchers tend to disregard data obtained by previous workers and prefer to concentrate on their own new measurements. Explanations given are that the old data are incomplete, the laboratories were not as good in the old days, or the old data are not available. These arguments may easily be countered:

Planning Hydrochemical Studies

Many of the previous researchers produced excellent data, and the quality of the old data can be assessed by use of the processing techniques discussed in section 5.10.

Historical data are available from local water authorities, data banks, and professional publications.

Historical data have the advantage of being available free of charge.

The old data are available at the early stages of a new study and can help in planning the new research campaign.

It is assumed that the reader is now convinced that historical data are useful. But the most important argument has not yet been spelled out: the old data were collected when the water systems were less disturbed by human activity. Thus, historical data record information that cannot now be collected, and serve as a database needed to understand the evolution and dynamics of studied water systems. Historical data also enlarge the time span of repeated observations, the topic of the next section.

7.6 Repeated Observations or Time-Data Series

Repetition of field measurements and laboratory analyses is an essential part of hydrochemical studies. Sources of recharge vary over the year—summer rains, winter rains, and snowmelt. These are distinguishable by their isotopic (and occasionally also chemical) composition (Chapter 9), and their seasonal inputs can be established provided the relevant water systems are monitored over at least 1 year. Water table and discharge responses to precipitation or flood events provide information on hydraulic interconnections. Such responses can be detected only by comparing the results of repeated observations. Chemical variations and temperature fluctuations are indicative of mixing processes, thus justifying the effort invested in repeated observations. Last, but not least, man-induced changes in water systems can be understood by comparing data from repeated measurements. These provide indications of deterioration caused by harmful human impacts, or of improvements achieved by proper management.

Having praised repeated measurements, we now discuss their incorporation into the planning of hydrochemical studies. A common rule of the thumb is that each water source should be studied four times, to cover at least one hydrological cycle, including the four major seasons. This may be regarded as a minimum, because time-data series with only four points suffer from poor resolution.

Many data points—many repetitions—were needed in the hydrographs of the wells depicted in Figs. 4.7–4.10, the temperature observations plotted in Figs. 4.19–4.23, or the discharge, temperature, and chlorine concentration values discussed in the time series of Figs. 4.24–4.27. Thus, in many studies, monthly measurements are needed for at least 14 consecutive months in order to fully cover a hydrological year. Repeated studies are especially necessary in areas that have few accessible wells or springs. When working on single water sources, multisampling provides an insight into the underground plumbing, establishing the existence of a single-water aquifer, several waters that intermix (sections 6.4, 6.5, and 6.6), or conduit-dominated versus porous medium recharge intake (sections 2.2 and 2.9). Thus, measurements may have to be reported once in every few years (Fig. 6.22), seasonally, monthly, daily (Fig. 6.21), or hourly (pumping tests). In the absence of repeated measurements the dynamics of a system may be revealed by measurements performed on many sources (wells or springs) in the studied area.

7.7 Search for Meaningful Parameters

Let us have a look at the compositional diagrams of the Combiola Spring study (Fig. 6.26). Suppose a new study is planned to establish the impact of new wells on the warm mineral springs, and suppose that for budgetary reasons the new study has to be based on repeated measurements of only a few parameters. The detailed study portrayed in Fig. 6.26 revealed that the dominant process in the Combiola Spring complex is intermixing of warm saline water with cold relatively fresh water. The parameters to be included in the new study should be reliable tracers of the mixing. Thus, they have to be selected from the parameters that provide good mixing lines in Fig. 6.26, namely Li, Na, K, Cl, and temperature. Parameters that have equal values in both end members are redundant (these include in our example Ca, Sr, HCO_3, and SO_4). Parameters revealing large data spreads are useful, as with SiO_2. By this method of sorting of parameter efficiencies we are left in the Combiola case with five parameters: Li, Na, K, Cl, and temperature. Temperature, being a physical parameter (with short memory, as it readily reequilibrates), should be included among the most meaningful parameters. The remaining parameters should include at least one anion—chlorine—and one cation, perhaps sodium, which is dominant and easy to measure.

The selection of meaningful parameters is essential for economic reasons, focusing efforts and means on a smaller number of selected, more promising, parameters. The search for meaningful parameters necessitates that each study has a preliminary stage, at which a large number of parameters are measured. From the preliminary results the most informative parameters are then selected. This first stage of measuring a large number of parameters is essential, as no case study resembles previous ones, and the list of meaningful parameters varies from case to case: temperature measurements are essential in warm water systems and less informative in cold systems with constant temperatures; chlorine is less important in many freshwater systems, but essential to trace brines; in contrast, HCO_3 is a sensitive tracer in fresh waters. Stable isotope measurements are meaningful in aspects that are different from the meaning of dissolved ions (e.g., altitude of recharge, identifying evaporation brines, or indicating a thermal history). Hence, if money is tight, the list of dissolved ions measured may be cut in order to save the stable isotopes. Tritium is an essential age indicator for post-1953 water, but it can be dropped if preliminary work reveals that the study deals with older water. Carbon-14 measurements necessitate handling of hundreds of liters of each sample, or payment of a higher price for accelerator determinations. Therefore, although they are a must in the preliminary study stage, they may be reduced once the basic conceptual model is reached.

The last sentence provides a guideline to the timing at which the number of parameters measured in a study can be cut: when an initial conceptual model is reached. Fewer parameters have to be measured to check the validity of the model, or to follow the evolution of a system due to exploitation.

7.8 Sampling for Contour Maps

Extensive hydrological studies, extending over large areas and with large numbers of accessible wells, pose severe difficulties in presenting and processing the wealth of data obtained. One mode is to produce maps with contour lines of the various parameters.

Planning of a hydrochemical study should include the question, will the data be expressed in contour maps? If so, data have to be obtained with an adequate geographical coverage and adequate density of measured points. These requirements are necessary to ensure adequate resolution and to avoid "white holes" in the maps.

Hydrochemical contour maps should be prepared from data collected at the same time. Mixing of data from different seasons, or even different years, reduces the meaning of the map. On the other hand, comparison of contour maps of data obtained at different data may be most informative. The need for many sampling points may result in a need to cut other efforts, for example, to reduce the number of measured parameters, as discussed in the previous section.

7.9 Sampling Along Transects

Sampling at wells and springs lying along a selected transect facilitates data processing and presentation. An example from a recent study in the Judean Mountains is given in Fig. 11.17.

A hydrochemical transect has the advantage of graphically linking parametric variations with geographical locations (topographic transect), geological features (geological transect), and hydrological setting (water table transect), as seen in Figs. 4.5 and 4.6.

Planning a hydrochemical study should include the question of whether the transect approach is desired. If so, the locations of the transect should be decided. Such a decision should take into account:

The geological structures (a transect crossing major structures is desired).
Availability of accessible wells and springs.
Priority to wells and springs of large discharge (best representing the studied water system).
Even distribution of sampling points along the transect.

7.10 Reconnaissance Studies

The term *reconnaissance study* relates to studies conducted for the first time, covering extended areas, to be completed fast and with minimal costs. The purpose of reconnaissance studies is to get a general picture of the water systems involved, to locate promising sites where large amounts of high-quality water can be found, to scan the area for features of special interest, to locate possible pollution processes, and to arrive at a conceptual model. The results of reconnaissance studies are the basis for the planning of detailed studies in selected areas. The major points characterizing reconnaissance studies are the inclusion of the maximum available number of wells and springs and the measurement of as many parameters as possible. The first point is needed for representative samples; the second point is necessary in order to find those parameters that are most informative in the studied system. In planning a hydrochemical study, a clear distinction has to be made between a reconnaissance study and a detailed study.

7.11 Detailed Studies

A detailed study is characterized by:

Relatively small study areas.
Existence of former data—historical or of a preliminary study.
Requirements for thorough research that should provide answers to specific questions.
A quantitative approach, whenever possible.
Inclusion of observations needed to check a proposed conceptual model.

Detailed studies are conducted over a period of a year or more, and include adequate repetitions of field measurements and sample collections for laboratory analyses. These studies are based on a variety of sampling strategies, for example, repeated sampling in springs and wells, depth profiles in nonpumped wells, search for water occurrences during drilling, and measurements during pumping tests.

The number of parameters measured will be large at the beginning of a detailed study, but the number will gradually shrink as the system is better understood and the most-informative parameters are spotted.

7.12 Pollution-Related Studies

Pollution is part of contemporary hydrology, a topic discussed in Chapter 16. Let us stress here only that pollution-related studies may address a specific environmental problem, and/or apply contaminants as tracers to understand flow paths in the natural system. The pollution-related aspect has a bearing on the planning stages of a study. The following information has to be collected first:

An inventory of *potential* pollution sources that may affect groundwater in the studied region (e.g., factories, sewage facilities, agricultural activity).
For each potential pollution source specific information is desired, for example, date of beginning of the activity, type of contaminants, mode of release, periodicity of release, and quantities.
Historical data of water composition in the region in general and in selected study wells in particular, in order to assess the base data pertaining to pre-pollution times.
Other information needed for any hydrochemical study, discussed in the previous sections.
Complaints by local groundwater consumers.

The next stages are to:

Formulate the research problems to be addressed—scientific and applied.
Decide on the nature of the study—reconnaissance or detailed.
Select monitoring points—wells or springs situated near all potential pollution sources (in the downflow direction), surface water bodies (rivers, lakes, operational ponds, etc.), and water supplying wells.
Decide upon parameters to be measured, bearing in mind the specific pollutants already noticed in the groundwater, or expected from known potential polluters.
Plan the monitoring frequency of the various parameters.
Locate the laboratories that will analyze the dissolved ions, gases, and specific pollutants, and get from them detailed water sampling instructions.

7.13 Summary: A Planning List

It is said that a well-formulated question contains half the answer. By the same token, a well-planned study is the key to the success of the study. The following list of planning stages is suggestive and is given as a summary of the present chapter:

1. Define the purpose of the study—practical and scientific.
2. Define the study area.
3. Collect all available data (geographical, hydrological, geological, temperatures, chemical and isotopic compositions, time variations, and so on).
4. Develop a conceptual model, or several possible models, suggested by previous and new workers, based on the available data.
5. Define the nature of the study—reconnaissance or detailed.
6. Develop a sampling strategy—sampling for a contour map or sampling along a transect.
7. Prepare a list of wells, springs, and surface water bodies to be included in the study (in light of points 1-6). The total number of sampling points should be estimated.
8. List the locations for special measurements and sample collections (e.g., depth profiles, sampling during drilling, and during pumping tests).
9. List the parameters to be measured in the field and the type of instruments needed.
10. List the parameters to be analyzed in laboratories, including the names of the laboratories and the types of sampling vessels needed.
11. Decide on the frequency of planned repetitions of field measurements and laboratory analyses.
12. Consider potential bottlenecks that need special care, for example, access to wells on private property, long waiting time in a special laboratory, or cooperation of drilling authorities.
13. Develop a schedule and timetable.
14. Consider budget requirements, based on the above points and ordered by nature of expenditure: salaries, means of transportation, equipment, sample collection vessels, laboratory fees, maps and photos, miscellaneous items.

7.14 Summary Exercises

Exercise 7.1: Acquaint yourself with the research reported from the Saratoga National Historic Park, New York, as presented in Fig. 4.8 and the accompanying text. Discuss it in terms of planning such a study: What was the aim of the study? What is specific about the selected well site? Which type of parameters had to be measured?

Exercise 7.2: You are asked to conduct a study of a spa. The manager brings you bottles with samples he has collected at three of his bathing installations. What will you do first?

Planning Hydrochemical Studies 157

Exercise 7.3: You are planning a study in an area that has a large number of wells marked on the map. How will you select the best wells for your study?

Exercise 7.4: What do you expect to find in the drillers' reports? Is it worthwhile to make the effort to get them for all wells you are going to investigate?

Exercise 7.5: Does it matter whether a well is operated at the time of your arrival for its monitoring, or do you prefer to monitor a well that is constantly pumped before your arrival?

Exercise 7.6: How will the planning of a study be influenced by the fact that available wells and springs are very scarce?

8
CHEMICAL PARAMETERS: FIELD WORK

8.1 Field Measurements

Certain chemical and related physical parameters have to be measured at the well or spring site. The list of parameters for field measurements includes those that change in stored samples and cannot be measured in the laboratory in a meaningful way. Most conspicuous in this respect is temperature, but pH, alkalinity, and dissolved O_2 and H_2S may also change between sampling and analyses in the laboratory.

There are many high-quality instruments available for measuring various parameters. These are accompanied by instruction manuals describing the operation of the sensors and calibration and measurement procedures. Portable versions are available for measuring parameters of water from springs or pumped from wells. Instruments for well logging (depth profiles) are available with long cables, in versions suitable for surveys and in fixed modes appropriate for continuous monitoring. Instrument sets are available that measure several parameters simultaneously, a procedure that is time saving and convenient. A comprehensive and easy to follow text on sampling procedures and biological, chemical, and physical measurements in the field has been prepared by Hutton (1983).

Field measurement instruments should be checked for their limits of detection, accuracy, and reproducibility, and their measurement errors should be established. This can be done with standards, duplicate samples, and dilution of samples, similar to the checks discussed for laboratory measurements (section 5.9). Such checks should be performed in preparation for field work, and

Chemical Parameters: Field Work

the results should be recorded and reported with the data (sections 5.7 and 5.8). Checking the field equipment in advance is essential to secure proper operation in the field.

Field measurements are needed for a number of purposes:

Measurement of parameters that change, or may change, after removal of the water from the sampling point.

Providing immediate information on water quality, needed to decide the extent of sample collection.

Providing checks on the laboratory data. For example, disagreement between electrical conductance measured in the field and TDI determined in the laboratory indicates erroneous field measurements, bad handling of the samples on the way to the laboratory, erroneous laboratory results, or mislabeling of samples. Agreement between field and laboratory data raises the confidence in the data and indicates that no secondary processes have occurred between sampling and laboratory measurement.

8.2 Smell and Taste

On approaching a water source one occasionally senses an unpleasant smell. In most cases this is caused by sulfur compounds, mainly H_2S. This information should be recorded in the field notes and taken into account during data interpretation, because the presence of H_2S and related compounds indicates reducing conditions, bacterial activity, or possible occurrence of sewage or other pollutants.

Human ability to smell H_2S is limited to the ppm range. At higher concentrations the odor is not noticed. Hence, if an H_2S smell is noticed when approaching the site, but it disappears at the well or spring itself, a high H_2S concentration should be suspected and relevant sampling is recommended.

Smell is a crucial factor in the ranking of water quality. In some cases aeration of water is practiced in order to get rid of compounds causing unpleasant smells.

Tasting of water provides immediate quality information, for example:

Estimate of the total salinity.
Identification of major ions—NaCl tastes salty, $MgSO_4$ tastes bitter, and iron "catches" the mouth in an unpleasant way.
Identification of pollutants such as sewage or industrial waste.

Tasting ability can be developed by tasting waters of known compositions. The information gained by tasting is of immediate use in the field: different water types can be recognized, and more detailed sampling may be decided on for special cases.

A word of warning is needed at this point: do not taste water if you suspect it to be contaminated.

8.3 Temperature

Potential uses of temperature data have been discussed in sections 4.7–4.9. The ease with which temperature is measured, and the benefits to the understanding of groundwater systems, make the measurement of this parameter a must in every study. This is also true for repeated measurements, depth profiles, and pumping tests.

Temperature can be measured with mercury-filled thermometers or by thermistors. Resolution and accuracy of $\pm 0.1°C$ are desirable. Thermometers should be calibrated by comparison with standard thermometers, or by immersing in fresh water and ice mixtures (0°C) and in fresh boiling water (the local boiling point is altitude dependent).

Temperature data are needed for water-rock equilibrium calculations, as well as for the identification of water groups, the determination of water end member properties (section 6.7), and to deduce the depth of water circulation (section 4.7).

8.4 Electrical Conductance

An electrical current can be carried through water by the dissolved ions. Electrical conductance is expressed in units of microohm/cm, also called microsiemens (μS). A positive correlation is observed between the electrical conductance and the TDI, expressed in meq/l. This correlation is linear up to 50 meq/l of TDI. At higher concentrations of dissolved ions the line in a conductance-TDI diagram level off.

The electrical conductance of a given water solution increases with temperature. Field probes of electrical conductance are therefore temperature compensated. Conductivity values obtained in the field should be plotted against the corresponding TDI concentrations measured in the laboratory. A good linear correlation confirms the high quality of the data. Outstanding values should be suspected as erroneous, and should be discussed with the laboratory staff for possible detection of errors or repetition of measurements. Conductivity measurements are of special use in the following cases:

Monitoring pumping tests for the detection of possible intrusion of different types of water (section 4.6).
Measuring depth profiles in nonpumped wells to search for different water strata (section 7.3).

Monitoring water flowing from a well opened just before sample collection: samples can be collected once a constant conductivity value is obtained.

Continuous monitoring in observation wells to detect the arrival of rain and flood recharge.

To substitute for TDI concentrations that are needed for data processing but are occasionally missing in cases of incomplete laboratory analyses. The needed TDI value may be read from the conductivity-TDI curve of the studied system.

8.5 pH

The pH of a solution indicates the effective concentration of hydrogen ions, H^+. The units of pH are the negative logarithm of hydrogen ion concentration, expressed in moles per liter:

$$pH = - \log H^+$$

The pH describes the composition of water: pH 7 indicates neutral water, lower values indicate acid water, and higher values indicate basic water. The pH is determined by means of a glass hydrogen ion electrode of known potential. Calibration is carried out with special solutions of known pH.

The pH is controlled by various reactions and the presence of different compounds. In freshwater systems the carbonate system CO_2-HCO_3-CO_3 plays a primary role in determining the pH. In other cases the presence of H_2S or its oxidized form, sulfuric acid, determines low pH. The pH value is an important parameter in water quality assessment in relation to corrosion problems and taste.

A knowledge of the pH, up to ± 0.1 unit, is essential for water-rock equilibrium calculations. Experts may be consulted for such calculations, but accurate pH data may be supplied by the field hydrochemist.

The pH measured in the field should be compared to the pH measured in the laboratory. Agreement between the two values indicates that no related secondary reactions have occurred in the sample bottle, for example, loss of CO_2 or biological activity. Whenever the two pH measurements agree, the water chemistry measured in the laboratory is relevant to the water in situ. However, deviations of the laboratory pH from the pH measured in the field indicate either that one of the measurements is wrong, or that secondary reactions have occurred in the time elapsed since the samples were collected. In such cases the preservation of the sample has to be checked and improved (section 8.10), and the period between sampling and laboratory measurements has to be shortened.

Disagreement between field and laboratory pH measurements places a question mark over the reported concentration of biologically involved ions such as

HCO$_3$, NH$_4$, SO$_4$, H$_2$, and Ca. On the other hand, the reported concentrations of more conservative ions—Cl, Br, Na, K, or Li—may be correct.

8.6 Dissolved Oxygen

Rain and surface waters equilibrate with air, becoming saturated with dissolved oxygen. Equilibration with soil air goes on until water reaches the saturated zone, where it is isolated from further contact with air. The concentration of dissolved oxygen in air-saturated water depends on:

Pressure, which is controlled by the altitude.
Temperature—increasing the temperature decreases the oxygen solubility.
Salinity—increasing the salinity decreases the oxygen solubility (negligible for fresh waters).

The initial concentration of dissolved oxygen in recharged water can be computed for each study area. The concentration at sea level (1 atmosphere), as a function of the average annual local temperature, can be read from Fig. 8.1. At locations with altitudes above or below sea level, a correction factor has to be applied to the oxygen concentration read from Fig. 8.1. The altitude correction factors are given in Fig. 13.2. For example, air-saturated water at sea level at 17°C contains 9.0 mgO$_2$/l and at an altitude of 330 m it contains 9.0 × 0.96 = 8.6 mgO$_2$/l. Comparison of the calculated initial O$_2$ concentration with the value measured in the field reveals the fraction retained. The missing O$_2$ has been consumed by oxidation of rocks and biological activity. The rate of oxygen consumption depends on the aquifer lithology—the occur-

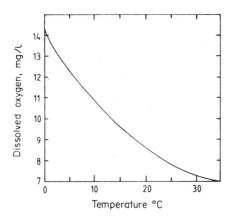

Fig. 8.1 Dissolved oxygen (ml/l) in fresh water as a function of the ambient temperature.

rence of pyrite and other oxygen-consuming minerals, and the availability of organic compounds and nutrients needed for O_2-consuming biological activity. The consumption of dissolved O_2 serves also as a semiquantitative age indicator: older water tends to have lost most or all of its dissolved oxygen. Generally speaking, in phreatic aquifers water retains a significant portion of the initial O_2, whereas in confined aquifers little or no O_2 is retained. Pollution by sewage and other organic compounds results in consumption of the dissolved oxygen.

Determining the amount of dissolved O_2 is, thus, informative in several respects:

Defining the aquifer environment in terms of degree of anaerobic conditions.
Detecting changes occurring in a water system: decrease in dissolved O_2 may indicate an increase in biological activity connected to the arrival of sewage or other polluting fluids.
Identifying systems with dominant conduit-controlled recharge and underground flow. These have limited water-rock contact and fast flow, and retain their initial dissolved O_2 even at long distances from the intake zone.
Identifying confined systems of old water, which are likely to have lost their O_2.
Identifying intrusion of water from adjacent wells as a result of pumping tests and in intensively exploited well fields.
Identifying different water strata in depth profiles.
Identifying mixing of different water types by the occurrence of "forbidden" ion combinations, for example, dissolved O_2 and H_2S.

The concentration of dissolved oxygen is determined by two electrodes immersed in an electrolyte contained in a vessel with an oxygen-permeable membrane wall. The cell is placed in a container through which the measured water flows. Oxygen diffuses into the cell in direct proportion to its concentration in the measured water. The diffused oxygen is consumed by the cathode, creating a measurable current that is proportional to the diffused oxygen and, in turn, to the O_2 concentration in the measured water.

Instruments measuring dissolved O_2 are calibrated by immersion in aerated water samples, for which the dissolved O_2 is computed from given curves that take into account the water temperature and salinity, as well as the altitude.

8.7 Alkalinity

Alkalinity is related to the sum of the carbon-containing species dissolved in the water: CO_2, HCO_3, and CO_3. Bicarbonate is commonly the dominant ion.

Alkalinity has to be measured in the field because CO_2 is often pressurized in groundwater (due to addition of CO_2 from soil and other underground sources). Upon exposure to the atmosphere some CO_2 may leave the water, causing part of the HCO_3 to break down. For these reasons it is highly recommended that alkalinity be determined in the field. Field-measured alkalinity values are needed for water-rock saturation calculations. Various setups are available for alkalinity measurements in the field by titration of the sample with an acid and pH coloring indicator.

8.8 Sampling for Dissolved Ion Analyses

Sampling procedures for determining the amounts of dissolved ions should be discussed with the relevant laboratory. Commonly, half-liter plastic bottles are adequate. Occasionally a separate bottle is needed to which a few drops of pure acid are added to secure low pH, in order to prevent precipitation of some ions, such as carbonates. For SiO_2 analyses, addition of known amounts of distilled water is recommended to avoid precipitation of silica as a result of cooling of the water.

Separate samples are commonly collected for H_2S determination. To these a fixative agent is added, each laboratory preferring its own material.

The details of a sampling expedition have to be worked out with the laboratory staff in advance, and the proper (clean) bottles have to be obtained. In addition to being clean, bottles should be rinsed in the sampled water before taking samples.

Always label the samples in the field—never rely on your memory—and make sure the labels are complete.

Filtering

Water samples contain dispersed soil particles and colloids. These can be removed by filtration, either in the field or in the laboratory. Some workers carefully filter every sample in the field, others argue that the water emerges with particles that are part of the hydrological system under study and therefore their filtration can be postponed until arrival at the laboratory, or skipped altogether.

Filtering requires experience: nonexperienced personnel may introduce contamination and cause fractionation. The topic should be discussed with the laboratories concerned.

8.9 Sampling for Isotopic Measurements

The nature of isotopic data is discussed in later chapters, but sample collection techniques are discussed here.

8.9.1 δD and $\delta^{18}O$ Measurements

Fifty-milliliter samples are adequate. Evaporation of samples has to be avoided by prompt closing of the bottles and storage in the shade. Normally each laboratory prefers its own bottles.

8.9.2 Tritium

One-liter bottles are desirable to allow for repeated measurements and pre-concentration (in the case of low tritium concentrations). Glass bottles are preferred, as tritium can diffuse through PVC bottles and equilibrate with the air.

8.9.3 ^{14}C

Fifty- to two-hundred-liter samples are needed (according to the HCO_3 concentration and the laboratory requirements). These may be collected in large containers with minimum exposure to air. The detailed procedure should be discussed with the laboratory. Samples have to be delivered to the laboratory within days, to avoid equilibration with air.

Precipitation of the dissolved carbon species and their extraction from 50–200 l is preferably done in the field. This procedure requires experience (guided by the laboratory staff or, preferably, carried out by a trained hydrochemist). Precipitation in the field is time consuming, but it reduces the volume and weight of samples that have to be transported to the laboratory.

For ^{14}C accelerator-mass spectrometer analyses, 1 l samples are usually enough, but details should be discussed with the laboratory staff.

8.9.4 ^{13}C

Three-liter samples may be needed, or precipitation and extraction may be done in the field, similar to the treatment of the ^{14}C samples. Separate bottles are needed for ^{13}C, as the treatment of the large samples for ^{14}C may introduce significant fractionation in the ^{13}C value.

8.9.5 Noble Gases

Samples are collected in glass tubes with special stopcocks or in copper tubes with special clamps. The main concern is absolute avoidance of air contamination. This aspect is further discussed in Chapter 13, and should always be coordinated with the laboratory personnel.

8.10 Preservation of Samples

The storage of samples during transport from the field to the laboratory, and on the laboratory shelf, has to be done with care and expertise. Major points are:

Keep the samples cold. Samples left for several hours in the sun after sampling, or samples left in a hot car for several days, may be spoiled by secondary reactions.

Keep the samples in the dark. Place them in boxes or store them in dark rooms. Exposure to light enhances biological activity.

In order to avoid biological activity, samples are occasionally poisoned by adding chemicals containing ions that will not be measured in the laboratory. Mercury iodide is a convenient preservation substance. It is available as small orange crystals, and a single small crystal added to a bottle of water is adequate. Addition of a preservative to a sample should be clearly marked on the bottle. Normally, each laboratory advocates a specific preservative.

8.11 Efflorescences

Small crystals of salts are occasionally seen to cover the soil and rocks near springs and seepages. In arid regions such efflorescences, or salt crusts, are also seen on soils in areas detached from springs or other types of surface water.

Efflorescences are formed by evaporation of rain or spring water leaving behind dissolved salts. Infiltrating rain water redissolves such efflorescences, salinizing local groundwater. The composition of efflorescences is therefore of interest with regard to the hydrochemistry of the water they precipitated from, and with regard to the groundwater they salinize.

Tasting of efflorescences can provide immediate clues as to their composition: white material with a salty taste is halite ($NaCl$); white insoluble material with no taste is most likely gypsum ($CaSO_4 \cdot 2H_2O$); white material that readily dissolves with a bitter taste may be epsomite ($MgSO_4 \cdot 7H_2O$), and white material that readily dissolves with no taste but a sense of cooling is most likely nitrate (KNO_3 or $NaNO_3$).

The occurrence of efflorescences should be documented in the field notebook. Samples, in plastic or glass containers, should be sent to the laboratory for chemical analysis. Conceptual models developed for water systems should incorporate the occurrence of efflorescences of observed compositions.

8.12 Equipment List for Field Work

A great deal of equipment is needed for hydrochemical fieldwork. The details vary according to the specific nature of each expedition, but experience shows that going over an equipment list is always beneficial. The following items of equipment will be required:

1. Sampling schedule (Chapter 7), maps, and writing material.
2. Field measuring equipment for temperature, conductivity, pH, dissolved

Chemical Parameters: Field Work

oxygen, and alkalinity. Specific materials are needed for these instruments for calibration and rinsing. Also carry extra batteries and other parts.

3. Sample collection bottles. These have to be prepared carefully so that (a) all the required varieties are included, (b) all bottles are clean (even new ones have to be cleaned), and (c) the bottles are all well packed in boxes—these will be needed for the filled sample bottles.
4. Filtering equipment (if filtering is to be carried out).
5. Precipitation equipment for ^{14}C and ^{13}C (if precipitation is to be done in the field).
6. Labeling material (labels or inks that have been tested to remain on the relevant bottles).

Note: Proper field measurements and well-collected samples, rapidly delivered to the laboratories, are the key to high-quality data. These are, in turn, the key to high-quality hydrochemical studies.

9
STABLE HYDROGEN AND OXYGEN ISOTOPES

9.1 Isotopic Composition of Water Molecules

The definition of elements and isotopes was presented in sections 5.1 and 5.2. Elements are defined by the number of protons in the nucleus of their atoms. Hydrogen has one proton and oxygen has eight protons. Isotopes are defined as variations of a given element, differing from each other by the number of neutrons. As presented in section 5.2, the hydrogen isotopes are:

^{1}H—common hydrogen; 1 proton
^{2}H—deuterium (also written D); 1 proton + 1 neutron
^{3}H—tritium (also written T); 1 proton + 2 neutrons

Tritium is radioactive and will be discussed in Chapter 10; common hydrogen and deuterium are stable. Oxygen has the following isotopes:

^{16}O—common oxygen; 8 protons + 8 neutrons
^{17}O—heavy (very rare) oxygen; 8 protons + 9 neutrons
^{18}O—heavy oxygen; 8 protons + 10 neutrons

Water is composed of hydrogen and oxygen, so it occurs with different isotopic combinations in its molecules. Most common and of interest to hydrochemists are ^{1}H$_2$ ^{16}O (common), ^{1}HD ^{16}O (rare), and ^{1}H$_2$ ^{18}O (rare). The water molecules may be divided into light molecules (^{1}H$_2$ ^{16}O) and heavy water molecules (^{1}HD ^{16}O and ^{1}H$_2$ ^{18}O).

9.2 Units of Isotopic Composition of Water

The isotopic composition of water is expressed in comparison to the isotopic composition of ocean water. For this purpose an internationally agreed upon sample of ocean water has been selected, called Standard Mean Ocean Water (SMOW) (Craig, 1961a, 1961b).

The isotopic composition of water, determined by mass spectrometry, is expressed in per mil ‰ deviations from the SMOW standard. These deviations are written δD for the deuterium, and $\delta^{18}O$ for ^{18}O:

$$\delta D \text{\textperthousand} = \frac{(D/H)_{sample} - (D/H)_{SMOW}}{(D/H)_{SMOW}} \times 1000$$

and

$$\delta^{18}O \text{\textperthousand} = \frac{(^{18}O/^{16}O)_{sample} - (^{18}O/^{16}O)_{SMOW}}{(^{18}O/^{16}O)_{SMOW}} \times 1000$$

Water with less deuterium than SMOW has a negative δD; water with more deuterium than SMOW has a positive δD. The same is true for $\delta^{18}O$.

9.3 Isotopic Fractionation During Evaporation and Some Hydrological Applications

Evaporation is a physical process in which energy-loaded water molecules move from the water phase into the vapor phase. Isotopically light water molecules evaporate more efficiently than the heavy ones. As a result, an isotopic fractionation occurs at partial evaporation of water: the vapor is enriched in light water molecules, reflected in relatively negative δD and $\delta^{18}O$ values. In contrast, the residual water phase becomes relatively enriched in the heavy isotopes, reflected in more positive δD and $\delta^{18}O$ values. The isotopic separation, or fractionation, is more efficient if the produced vapor is constantly removed, as for example, by wind blowing away vapors produced above an evaporating pond. Figure 9.1 presents the composition of water samples successively collected at a pond in Qatar (Yurtsever and Payne, 1979). The composition of the samples is plotted on a δD-$\delta^{18}O$ diagram. A progressive enrichment in the heavy isotopes is noticed, indicating that the residual water in the pond was progressively enriched in δD and $\delta^{18}O$ as the isotopically light vapor was removed. The original water had a composition of $\delta D = 0$‰ and $\delta^{18}O = -1.4$‰, and the last sample reached values of $\delta D = +33$‰ and $\delta^{18}O = 5.6$‰. The line con-

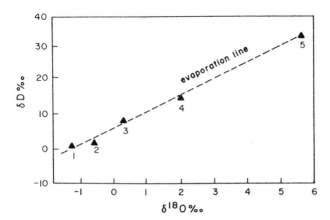

Fig. 9.1 Isotopic composition of water samples successively collected at an evaporating pond at Qatar (Yurtsever and Payne, 1979). The water in the pond became progressively heavier in its composition.

necting the sample points is called the evaporation line, and its slope is determined primarily by the prevailing temperature and air humidity.

In surface water bodies that evaporate to a very great extent, the isotopic enrichment ceases or is even reversed. However, except in very rare cases, natural water bodies, such as lakes or rivers, are only partially evaporated in the range in which isotopic enrichment of the residual water occurs.

Isotopic fractionation during evaporation causes fractionation during cloud formation: the vapor in the clouds has a lighter isotopic composition than the ocean that supplied the water. Upon condensation from the cloud, during rain formation, the reverse is true: the heavy water molecules condense more efficiently, leaving the cloud residual vapor depleted of deuterium and ^{18}O.

Residual evaporation water is thus tagged by high δD and $\delta D^{18}O$ values, an observation used to trace mixing of evaporation brines with local fresh water. An example of two mineral springs located on the Dead Sea shore is given in Fig. 9.2. The Hamei Zohar and Hamei Yesha springs are seen to lie on a freshwater–Dead Sea mixing line, indicating that these springs contain recycled Dead Sea water brought to the surface with the emerging fresh water recharged at the Judean Mountains (Gat et al., 1969). From the information included in Fig. 9.2, one can calculate the percentage of Dead Sea brine intermixed in the Hamei Yesha and the Hamei Zohar springs.

At Orapa, northern Botswana, a new well field of confined water was developed and two hypotheses were proposed to explain the origin of recharge: either underground flow of recharge water from lakes and rivers 45-km distant, or local rain. The consultants of the local diamond mine ruled out local

Stable Hydrogen and Oxygen Isotopes

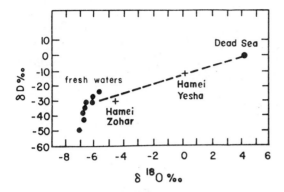

Fig. 9.2 Stable hydrogen and oxygen isotopic composition of the Hamei Zohar and Hamei Yesha mineral springs, Dead Sea shores. The linear correlation indicates that the springs' water is formed by intermixing of Dead Sea brine with local fresh water (Gat et al., 1969).

recharge and favored replenishment from the lakes. Results of an isotopic composition survey, depicted in Fig. 9.3, show the δD and $\delta D^{18}O$ values of the groundwater in the wells, the average annual rain composition in neighboring meteorological stations, and the lakes and rivers mentioned. The δD and $\delta^{18}O$ values of the groundwater in the wells were observed to be significantly lighter (more negative) than in the lake and river waters, which were enriched in the heavy isotopes due to intensive evaporation losses. Thus, the hypothesis of

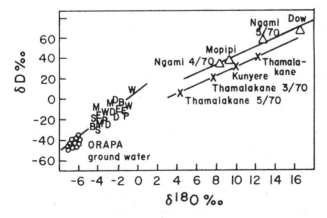

Fig. 9.3 Isotopic data of confined groundwater at Orapa, Botswana (o), average annual rain in neighboring meteorological stations (letters), and 45-km distant lakes (∆) and rivers (x) (Mazor et al., 1977).

recharge from the lakes could be ruled out, and recharge by local rain was supported. However, a slight but analytically significant difference can be seen between the composition of average annual rain and local groundwater. This was explained by an observation that only intensive rains are effectively recharging the groundwater, and these were observed by Vogel and Van Urk (1975) to have an isotopic composition that is lighter than average annual rain composition (the amount effect, section 9.6).

9.4 The Meteoric Isotope Line

Harmon Craig published (1961a) a δD and $\delta^{18}O$ diagram, based on about 400 water samples of rivers, lakes, and precipitation from various countries (Fig. 9.4). An impressive lining of the data along the best-fit line of

$$\delta D = \delta^{18}O + 10$$

has been obtained. Outside this line plot data from East African lakes that undergo significant isotopic fractionation due to intensive evaporation losses.

The data in Fig. 9.4 lie on a straight line in spite of the very wide range of values: δD of $-300‰ - +50‰$, and $\delta^{18}O$ of $46‰ - +6‰$. This line, called the *meteoric line*, has been found, with some local variations, to be valid over large parts of the world.

The meteoric line is a convenient reference line for the understanding and tracing of local groundwater origins and movements. Hence, in each hydro-

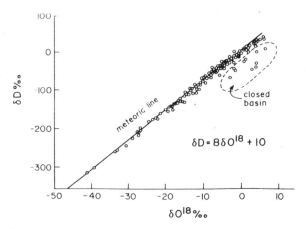

Fig. 9.4 Isotopic data of about 400 samples of rivers, lakes, and precipitation from various parts of the world. The best-fit line was termed the *meteoric line*. Its equation, as found by Craig (1961a), is $\delta D = 8\delta^{18}O + 10$. The data in the encircled zone of "closed basins" is for East African lakes with intensive evaporation.

chemical investigation the *local meteoric line* has to be established from samples of individual rain events or monthly means of precipitation. A specific example of a local meteoric line, from northeastern Brazil, is given in Fig. 9.5. A local meteoric line is obtained from the equation $\delta D = 6.4\ \delta^{18}O + 5.5$ (Salati et al., 1980). Examples of equations of local meteoric lines reported from various parts of the world are given in Table 9.1.

The composition of precipitation is reflected, directly or modified, in the composition of groundwater. Common practice is to plot groundwater data on δD-$\delta^{18}O$ diagrams, along with the meteoric line of local precipitation as a reference line. Examples are given in Figs. 9.6–9.8. In Fig. 9.6 the composition of the groundwater plots are close to the local meteoric line, ruling out secondary processes, such as evaporation prior to infiltration, or isotope exchange with aquifer rocks.

In Fig. 9.7 the groundwater data fall distinctly below the relevant meteoric line, indicating that secondary fractionation has occurred, or that the waters are ancient and were recharged in a different climatic regime that was characterized by a different local meteoric line. What should be checked in order to decide between these two possible explanations? The age. Groundwater data plotted in Fig. 9.8 fall along the meteoric line but reveal a large spread of values. Separation into shallow and deep waters revealed the latter have distinctly lighter isotopic compositions. A possible origin of ancient recharge at a different climatic regime was suggested in this case too. The necessary check has been done: the deeper waters, of light isotopic composition, were indeed

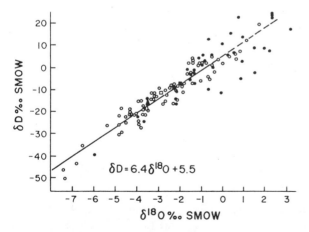

Fig. 9.5 Isotopic composition of precipitation in the Pajeu River basin, Brazil: o, months with rain over 50 mm/month; ●, months with lower precipitation amounts. A local meteoric line is obtained with the equation $\delta D = 6.4 \delta^{18}O + 5.5$ (Salati et al., 1980).

Table 9.1 Examples of regional meteoric lines

Region	Meteoric line (‰)	Reference
'Global' meteoric line	$\delta D = 8\delta^{18}O + 10$	Craig (1961)
Northern hemisphere, continental	$\delta D = (8.1 \pm 1)\delta^{18}O + (11 \pm 1)$	Dansgaard (1964)
Mediterranean or Middle East	$\delta D = 8\delta^{18}O + 22$	Gat (1971)
Maritime Alps (April 1976)	$\delta D = (8.0 \pm 0.1)\delta^{18}O + (12.1 \pm 1.3)$	Bortolami et al. (1979)
Maritime Alps (October 1976)	$\delta D = (7.9 \pm 0.2)\delta^{18}O + (13.4 \pm 2.6)$	
Northeastern Brazil	$\delta D = 6.4\delta^{18}O + 5.5$	Salati et al. (1980)
Northern Chile	$\delta D = 7.9\delta^{18}O + 9.5$	Fritz et al. (1979)
Tropical islands	$\delta D = (4.6 \pm 0.4)\delta^{18}O + (0.1 \pm 1.6)$	Dansgaard (1964)

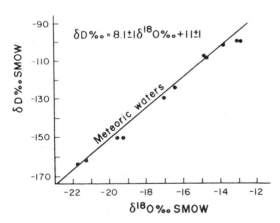

Fig. 9.6 Isotopic composition of water sampled from wells in central Manitoba, Canada. The values fall close to the local meteoric line ($\delta D = 8.1\delta^{18}O + 11$). The researchers (Fritz et al., 1974) concluded that evaporation during recharge and isotopic exchange with aquifer rocks are insignificant.

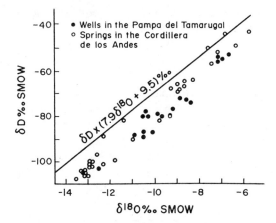

Fig. 9.7 Isotopic composition of groundwaters of northern Chile. The values lie below the meteoric line of local precipitation, explained by the investigators (Fritz, et al., 1979) as reflecting secondary fractionation by evaporation prior to infiltration, or the presence of ancient waters that originated in a different climatic regime. The large variations in the groundwater compositions are useful in local groundwater tracing.

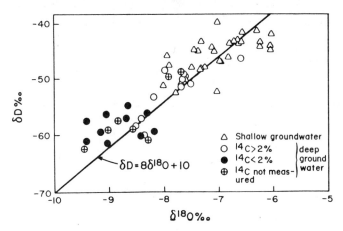

Fig. 9.8 Isotopic composition of groundwaters near Chatt-el-Honda, Algeria (Gonfiantini et al., 1974). The values scatter along the meteoric line but reveal an internal order: values of deep groundwaters are isotopically lighter (more negative) than shallow groundwaters. This was taken as an indication that the deep groundwaters were ancient and originated from rains of a different climatic regime. This is supported by the large number of ancient deep groundwater water samples, as borne out by low ^{14}C concentrations (sections 11.4 and 11.8).

9.5 Temperature Effect

Dansgaard (1964) analyzed a large body of isotopic data gathered by the International Atomic Energy Agency and showed that temperature is the major parameter that determines the isotopic values of precipitation (Fig. 9.9). In his extensive discussion, Dansgaard summed up the knowledge gained in laboratory experiments and field observations. The composition of precipitation depends on the temperature at which the oceanic water is evaporated into the air and, even more important, the temperature of condensation at which clouds and

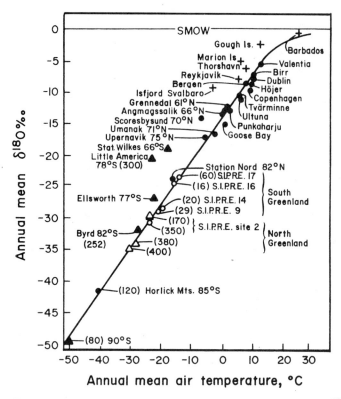

Fig. 9.9 Temperature effect. Correlation between annual mean $\delta^{18}O$ values observed in precipitation and the annual mean temperature of local air: polar ice (circles and triangles; figures in parenthesis indicate total thickness in cm); continental precipitation (●) and island precipitation (+) (Dansgaard, 1964).

rain or snow are formed. The net effect is expressed in the following empirical function (from Fig. 9.9):

$$\delta^{18}O = 0.7T_a - 13‰, \text{ or } 0.7‰/°C$$

and, in a similar way:

$$\delta D = 5.6T_a - 1000‰, \text{ or } 5.6‰/°C$$

A local study at Heidelberg, Germany, revealed the following empirical function (Schoch-Fischer et al., 1983):

$$\delta D = (3.1 \pm 0.2) \, T_a - (172 \pm 3)‰, \text{ or } (2.8 \pm 0.2)‰/°C$$

(T_a is the local mean annual air temperature.) The meteoric line (Fig. 9.4) is thus the result of the combined δD and $\delta^{18}O$ dependencies on temperature. The temperature effect is well seen in seasonal variations in regions with rains during cold and warm seasons. An elegant example from Switzerland is given in Fig. 9.10: monthly measurements in three stations revealed that the annual temperature cycle is followed by corresponding changes of $\delta^{18}O$.

The temperature dependence of the isotopic composition of precipitation, or *temperature effect*, is to a large extent responsible for the large variation in the isotopic composition of groundwaters, thus equipping the hydrologist with a powerful tool. In regions with summer and winter precipitation the isotopic differences in the composition of the precipitation are traceable as winter recharge and summer recharge fronts, important in establishing groundwater velocities and identifying piston flows.

9.6 Amount Effect

Figures 9.11–9.13 show the dependence of the isotopic composition on the amount of rain: heavier rain events, or greater monthly precipitation amounts, result in more negative δD and $\delta^{18}O$ values. Dansgaard (1964) proposed two major explanations for this *amount effect*:

Lower ambient temperatures cause the formation of clouds with lighter isotopic composition (temperature effect, Fig. 9.9); lower temperatures also cause heavier rains.
Falling raindrops undergo evaporation, enriching the falling rain in the heavy isotopes. This effect is less severe both when ambient temperatures are low and when the amount of rain is large (as the air gets more saturated).

The amount of monthly rain varies during the year, causing a seasonal variation in the isotopic composition. This point is demonstrated in a case study from northeastern Brazil (Fig. 9.13).

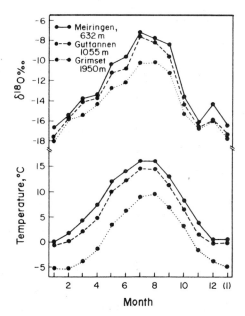

Fig. 9.10 Monthly mean $\delta^{18}O$ values in precipitation and monthly mean air temperatures from 1971-1978 for Swiss stations. The value of January (1) is shown twice to complete the cycle (Siegenthaler and Oeschger, 1980). The $\delta^{18}O$ values are seen to covary with the temperature, reflecting a pronounced temperature effect of 0.35-0.5‰ $\delta^{18}O/°C$. The measurements, carried out at three stations of different altitudes, revealed an altitude effect, precipitation at higher altitudes having isotopically lighter compositions.

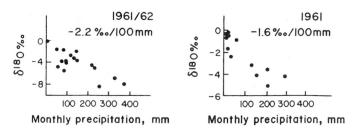

Fig. 9.11 Two cases of isotopic compositions varying with the amount of precipitation (the amount effect), reported by Dansgaard (1964). Left, Binza (Leopoldville), Congo; right, Wake Island. The $\delta^{18}O$ precipitation values are for individual months. The amount effects were -2.2‰ $\delta^{18}O/100$ mm rain for Binza and -16‰ $\delta^{18}O/100$ mm rain for Wake Island.

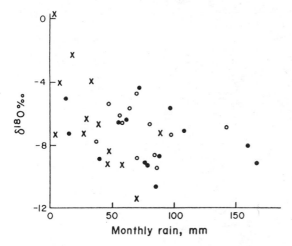

Fig. 9.12 Amount effect. Monthly rain and $\delta^{18}O$ values measured for three years at Rowen Boos, Haute Normandie, France. Dots, October 1974–December 1975; x, 1976 and 1977 (after Conrad et al., 1979). Rainier months reveal isotopically lighter rain.

9.7 Continental Effect

Several workers noticed that the average isotopic composition of precipitation tends to have more negative values further away from the ocean coast. This *continental effect* is well reflected in European groundwaters (Fig. 9.14). The explanation lies in the history of the precipitating air masses. As they travel

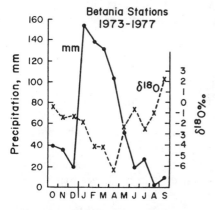

Fig. 9.13 Amount effect. Monthly rain and $\delta^{18}O$ values of northeastern Brazil plotted as a function of sampling date. The curves are mirror images, revealing low $\delta^{18}O$ values at the high rain months (Salati et al., 1980).

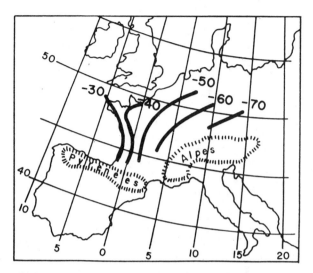

Fig. 9.14 Lines of equal δD values for Europe, based on over 300 samples of groundwater. A trend of lighter isotopic composition is seen as a function of the distance from the ocean, reflecting the continental effect in the precipitation (after Sonntag et al., 1979).

inland, rain is gradually precipitated by condensation, accompanied by more efficient condensation of water molecules with heavier isotopes (opposite to evaporation). The *residual moisture* in the air masses thus becomes gradually lighter in its isotopic composition, and lighter rain is progressively formed. The continental effect is often masked by other effects, for example, temperature (seasonal) effect and altitude effect, as discussed in the next section.

9.8 Altitude Effect

The altitude effect is seen in Fig. 9.15. The $\delta^{18}O$ values in Swiss precipitation are lighter with higher altitudes. The gradient, or *altitude effect*, is $-0.26‰$ $\delta^{18}O/100$ m altitude. The altitude effect is observed in this case to be the same in the precipitation and in the derived surface and groundwaters. This effect is not masked by the seasonal temperature effect (Fig. 9.10).

The altitude effect has to be established in each study area. Leontiadis et al. (1983) reported a value of $-0.44‰/100$ m for the part of Greece that borders Bulgaria.

As clouds rise up the mountains, the heavy isotopes are depleted and the residual precipitation gets isotopically lighter. This effect turns out to be an effective tool in tracing groundwater recharge. Payne and Yurtsever (1974) reported a case study in which isotope hydrology was recruited to find out to

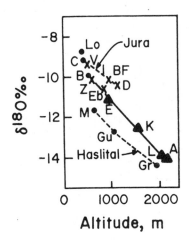

Fig. 9.15 Variations of mean $\delta^{18}O$ as a function of altitude for Swiss precipitation (●), groundwaters (x), and rivers (▲). The lines are parallel, all revealing an altitude effect of -0.26‰ $\delta^{18}O/100$ m (Siegenthaler and Oeschger, 1980).

what extent groundwaters in the Chinahdega Plain, Nicaragua (1100 km², 200 masl), were recharged by local precipitation on the plain or by recharge from the higher slopes of the Cordillera Mountains (up to 1745 m). They worked in three stages:

They analyzed $\delta^{18}O$ in weighted mean precipitation samples from different altitudes and defined an altitude effect of -0.26‰/100 m (Fig. 9.16).
They analyzed groundwaters with known recharge altitudes and found the same altitude effect (Fig. 9.17).
They measured the $\delta^{18}O$ of the studied waters in the plain and established the ratios of local plain recharge (around 25%) to mountain recharge.

A nice control was tritium, found to be higher in wells with higher plain-recharge contributions, revealing shorter travel times, as compared to low tritium contents in wells with higher contributions of mountain recharge (longer travel times).

The altitude effect is applied in an ever-growing number of studies as a tool to calculate the recharge altitude from the deuterium and $\delta^{18}O$ data of springs and wells. Hence, much interest lies in a study of Bortolami et al. (1978) who checked the accuracy of such calculations in a case study in the Italian Maritime Alps. They conducted the following measurements: δD and $\delta^{18}O$ in precipitation in stations of various altitudes during one winter month (April 1976) and one summer month (October 1974). The data (Fig. 9.18) revealed differ-

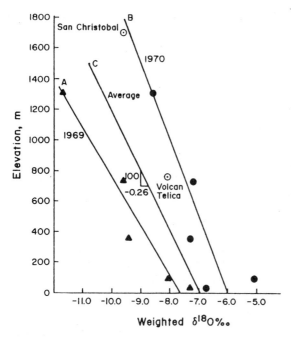

Fig. 9.16 $\delta^{18}O$ values in mean weighted samples of precipitation from different altitudes in Nicaragua. The lines for 1969 (A) and for 1970 (B) are quite parallel, the average (line C) revealing an average altitude effect gradient of $-0.26‰$ $\delta^{18}O/100$ m (Payne and Yurtsever, 1974).

ent lines for the winter and summer $\delta^{18}O$ values, and only one line for δD. The corresponding altitude-isotope equations were found from Fig. 9.18 to be

$$\delta^{18}O = -3.12 \times 10^{-3}h - 8.03$$

and

$$\delta^{18}O = -24.9 \times 10^{-3}h - 51.1$$

where h is the altitude in meters above sea level (masl). Accordingly, two altitude equations can be written:

$$h = -320\ \delta^{18}O - 2564$$

and

$$h = -40.2\ \delta - 2052$$

In the second stage, samples were collected for isotopic analysis in the karstic Bossea Cave, at an altitude of 810 masl. The data, given in Table 9.2, were

Stable Hydrogen and Oxygen Isotopes

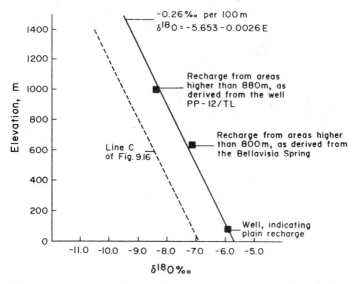

Fig. 9.17 Altitude effect, reflected in groundwater of known recharge areas, Nicaragua. An effect of $-0.26‰$ $\delta^{18}O/100$ m is obtained, the same as observed for the regional precipitation (see Fig. 9.16) (Payne and Yurtsever, 1974).

used individually to calculate recharge altitudes. The agreement between hD and $h^{18}O$ is good in the various samples, but the agreement between repeated samples is rather poor. Values of calculated recharge altitude vary from 1180–1871 m. This is a highly discouraging range of values, showing that calculations based on single water samples are unreliable. The average hD and $h^{18}O$ values, 1565 m and 1539 m, seem however to be very reasonable as average recharge altitudes for the water sampled in the Bossea Cave. The calculations based on the average δD and $\delta^{18}O$ values are practically identical to the earlier ones: 1563 and 1538 m.

The research of Bortolami and his coworkers (1979) provides an excellent example of a study aimed at a local calibration of an isotopic tracing technique. The variability observed between results of single samplings is probably extreme, since they worked on a karstic system which has a rapid response to individual storm events.

9.9 Tracing Groundwater with Deuterium and ¹⁸O: Local Studies

In a work on South African warm springs, the question arose whether their temperature and dissolved ions represent the water at depth, or did intermix-

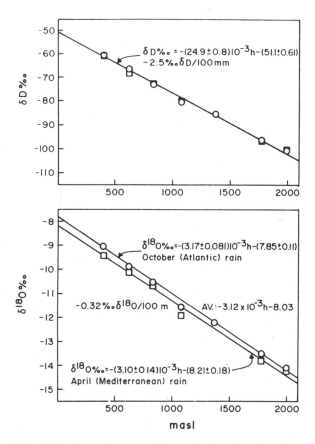

Fig. 9.18 Isotopic compositions, as a function of altitude, for precipitation samples collected in a summer month (April 1976) and a winter month (October 1974), Maritime Alps. The data were used to establish local altitude effect equations (see text). Winter and summer $\delta^{18}O$ values fall on separate lines, reflecting the difference in the origin of the precipitating air masses from the Atlantic and Mediterranean, respectively (after Bortolami et al., 1978).

ing with shallow cold water take place, especially from adjacent rivers. Hence, δD and $\delta^{18}O$ were measured in the thermal springs and the nearest rivers (Mazor and Verhagen, 1983). The results are shown in Fig. 9.19. It is seen that the thermal springs have significantly more negative values than the nearby rivers, indicating that no intermixing took place except in one case. This con-

Table 9.2 Recharge altitude calculated from δD and δ^{18}O in repeatedly collected samples in the Bossea Cave, Maritime Alps, Italy (810 masl)[a]

Date	δD‰	δ^{18}O‰	hD(m)	h^{18}O(m)
13-10-74	−81.4	−11.70	1220	1180
7-3-76	−88.4	−12.86	1502	1551
9-4-76	−94.4	−12.86	1743	1551
18-4-76	−97.3	−13.83	1859	1861
25-4-76	−97.6	−13.85	1871	1868
1-5-76	−90.2	−12.91	1574	1567
14-5-76	−85.8	−12.34	1397	1385
30-5-76	−87.6	−12.54	1470	1449
6-6-76	−87.5	−12.51	1465	1440
Average	−89.93	−12.82	1565[b]	1539[b]
			1563[c]	1538[c]

[a]Data from Bortolami et al. (1979) applying their average equation for the local precipitation: $h = 320\ \delta^{18}O - 2564$ and $h = -40.2\ \delta D - 2052$ (text).
[b]Average of recharge altitudes calculated via δD and δ^{18}O.
[c]Calculation based on average δD = −89.9‰ and average δ^{18}O = −12.82‰.

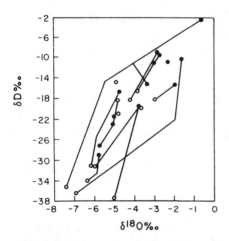

Fig. 9.19 Isotopic composition of thermal waters (o) and adjacent rivers (●) in South Africa. The lines connect thermal springs with the nearest river. It is seen that the spring waters are significantly lighter than the nearby rivers, indicating that no (or negligible) intermixing occurred (Mazor and Verhagen, 1983).

clusion has been supported by the practical absence of measurable tritium in the thermal springs, indicating long travel times.

What might be the reason for the lighter isotopic composition of the South African warm springs, as compared to the river waters, as seen in Fig. 9.19? Three explanations can be offered:

An altitude effect, caused by spring recharge in higher elevations. This would agree with the increased temperatures, indicating relatively deep circulation.
The waters might be ancient, belonging to a different climatic regime (paleowaters).
The river waters were heavy at the time of sampling, in the context of the seasonal variations.

The uncertainty regarding the causes of the isotopic differences did not intervene in the application of this difference to a tracing problem (nonintermixing of thermal and surface waters).

Rivers and shallow groundwater are often saline in arid or semiarid regions. Examples are common in northeastern Brazil in crystalline rocks, remote from the sea, and in regions where no marine transgression has occurred for 100 million years. Hence, evaporation was for a long time suspected to be the cause of the rise in salinity. The problem was studied by examining Cl and $\delta^{18}O$ (Fig. 9.20). In general, the Cl and $\delta^{18}O$ curves covaried during the months of observation, proving that evaporation caused the salinity (Salati et al., 1980). A variance is seen in the details of the peaks of the two curves in Fig. 9.20, such as the peak defined by the last three points. This might be explained by special rain events, for example, a summer rain that underwent much evaporation during descent to the ground (resulting in more positive $\delta^{18}O$), which lowered for a short while the chlorine concentration in the river water (at the low-flow season).

In shallow groundwaters, at a depth of about 1 m, in the Chott-el-Honda salt plain in inland Algeria, a positive correlation was observed between dry residue (salinity indicator) and $\delta^{18}O$ (Fig. 9.21). Evaporation through the thin soil cover was deduced (Gonfiantini et al., 1974). This conclusion has been confirmed by the more saline water lying on an evaporation line in a δD–$\delta^{18}O$ diagram (Fig. 9.22).

The stable isotopes of water are most useful in tracing seawater intrusions, so troublesome in coastal urban areas. One example is seen in Fig. 9.23 for wells west of Hermosillo, Gulf of California.

Table 9.3 reports data on depth profiles in coastal wells of the Salentine Peninsula, Italy. Which trends can be seen, and what is their meaning in terms of groundwater movement? In general, temperature, Cl, δD, $\delta^{18}O$, and $\delta^{13}C$ values are high in the deeper parts of the wells, reflecting seawater encroach-

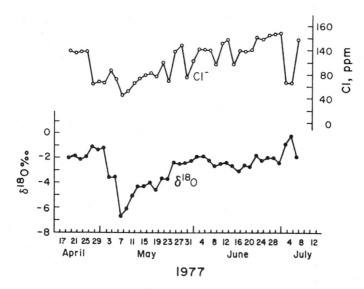

Fig. 9.20 Chloride and $\delta^{18}O$ values in repeatedly collected samples of the Pajeu River, northeastern Brazil. A general correlation is seen, revealing the role of evaporation, most important during July (rise in ambient temperatures and low river flow) (Salati et al., 1980).

ment. The temperature gradient is caused by the regional geothermal gradient. Practical hydrological conclusions can be reached with regard to desired pumping management of these wells, that is, filling up lower parts of wells, and/or limited pumping rates of fresh water only.

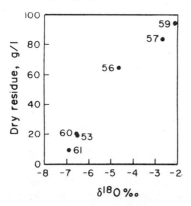

Fig. 9.21 Isotopic composition and salinity (dry residue) in shallow wells south of the Chott-el-Honda salt plain, Algeria. Evaporation is evident, but a further check can be seen in the next figure (Gonfiantini et al., 1974).

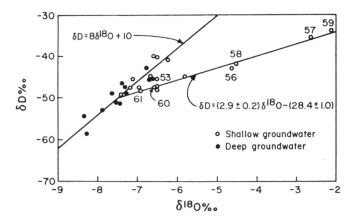

Fig. 9.22 Isotopic composition of groundwaters south of Chott-el-Honda, Algeria. The shallow groundwaters lie on an evaporation line, and the degree of heavy isotope enrichment agrees with the salinity (the sample numbers are the same as in the previous figure) (Gonfiantini et al., 1974).

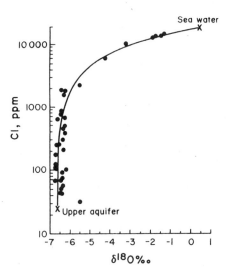

Fig. 9.23 Chlorinity and isotopic composition for coastal wells, Hermosillo, Gulf of California. The percentage seawater infiltration can be calculated for each well (Payne et al., 1980). A mixing line is seen, its curvature being caused by the use of a logarithmic chloride axis.

Table 9.3 Water depth profiles in coastal wells in the Salentine peninsula, Italy (data extracted from Cotecchia et al., 1974)

No.	Source	Date	°C	Cl⁻ (meq/l)	HCO₃ (meq/l)	$\delta^{18}O$(‰)	δD(‰)	$\delta^{13}O$(‰)	^{14}C(pmc)
1	Boraco Spring	9-2-71	17.4	29.2	5.94	−5.5	−34.5	−10.70	41.1
2	Chidro Spring	3-2-71	18.4	59.8	6.15	−5.3	−33.5	−9.50	24.3
3	Iduse Spring	14-7-71	18.2	114.8	4.29	−4.2	−26.5	−9.54	52.7
4	CH well, 40 m	2-2-71	17.4	39.5	6.23	−5.9	−33.5	−10.80	35.1
5	69 m	2-2-71	17.5	42.0	6.24	−5.7	−33.0	−11.30	49.8
6	86 m	3-2-71	17.5	54.9	6.20	−5.7	−32.0	−10.70	35.7
7	118 m	2-2-71	17.5	55.2	6.12	−5.6	−32.0	−12.10	30.4
8	147 m	30-7-71	17.4	594.9	2.40	+1.8	+10.0	−7.09	—
9	170 m	30-7-71	18.0	594.3	2.53	+1.7	+11.0	−5.17	5.5
10	S-1 well, 19 m	9-7-71	17.0	31.9	5.24	−5.4	−31.5	−11.84	64.0
11	30 m	9-7-71	17.0	44.0	5.63	−5.3	−31.0	−13.40	65.4
12	40 m	13-7-71	17.1	581.9	2.36	+1.3	+8.5	−5.00	46.7
13	S3 well, 9 m	13-7-71	17.2	55.4	5.48	−5.1	−29.5	−13.93	66.9
14	19 m	13-7-71	17.4	580.2	2.40	+1.3	+8.5	−3.70	92.6
15	C-S well, 56 m	23-2-71	16.8	2.3	4.37	−5.1	−31.0	−14.37	77.2
16	108 m	24-2-71	16.8	37.8	6.00	−4.9	−30.5	−11.30	—
17	175 m	25-2-71	17.1	562.3	2.68	+1.0	+6.3	−8.87	1.4
18	184 m	9-12-72	17.1	565.2	—	+0.8	+6.1	−7.60	—
19	30 m	17-2-71	17.2	7.3	4.93	−6.10	−36.1	−14.40	64.2
20	70 m	18-2-71	17.5	12.3	4.79	−5.9	−34.0	−13.32	55.8
21	SR well, 146 m	19-2-71	19.4	587.3	2.65	+1.2	+7.5	−5.61	1.8
22	154 m	5-12-72	19.4	594.3	—	+1.4	+9.2	−5.29	—
23	C-S well, 33 m	14-7-71	16.6	15.4	4.59	−5.0	−33.5	−10.90	51.4
24	70 m	14-7-71	18.4	581.5	2.46	+1.3	+9.0	−4.09	31.7
28	Ionian Sea	13-7-71	—	593.2	2.42	+1.3	+9.0	−3.30	111.9
29	Adriatic Sea	14-10-73	—	583.4	2.25	+0.8	+7.5	−2.60	81.0

9.10 Tracing Groundwater with Deuterium and ^{18}O: A Regional Study

Stahl et al. (1974) provided an excellent example of isotopic tracing of groundwater, in the Sperkhios Valley, Greece (insert in Fig. 9.24). The research followed the following logical steps:

1. Artesian wells were grouped according to their δD and $δ^{18}O$ values and geographical distribution. Three such groups emerged (Table 9.4 and Fig. 9.25).
2. Springs for which the recharge altitude could be deduced from field data were analyzed and an isotope-altitude graph was established, defining the local altitude effect (Fig. 9.26).
3. The recharge altitudes of the three artesian well groups were determined with the aid of the altitude-isotopic composition graph (Fig. 9.26), calibrated by the springs of known intake altitudes. The values are given in Table 9.4.
4. Once the three well groups and their corresponding recharge altitudes were obtained, the details of local flow were worked out by additional chemical, radioisotope, and field data (Fig. 9.24).
5. A group of saline (2.5–14.5 g Cl/l) and warm (28°C–40°C) springs and wells in the eastern part of the study area were suspected by earlier workers to contain a seawater component. This was tested and confirmed by Cl-$δ^{18}O$ relations, and the percentage of seawater intermixed in each case could be calculated (Fig. 9.27).

9.11 The Need for Multisampling

In the previous sections the many factors that determine the isotopic composition of groundwater have been described. Many of these, especially the temperature, vary seasonably, while others, such as the intensity of rain, vary from one rain event to the other. As a result, single samples may be nonrepresentative. Table 9.5 gives examples for the variations between daily precipitation events for a single station.

Significant variations in the isotopic composition were seen in water samples collected in a pond (Fig. 9.1), rain samples collected at different dates (Fig. 9.5), water samples from adjacent wells (Fig. 9.6), samples from wells of different depths (Fig. 9.8 and Table 9.3), precipitation samples of various months (Fig. 9.10), rains of different intensities (Fig. 9.11), summer and winter precipitation (Fig. 9.18), or river samples collected during a whole season (Fig. 9.20). The basic conclusion from these variations is the necessity of *multisampling*. The hydrochemist has to plan the sample collection in the right way

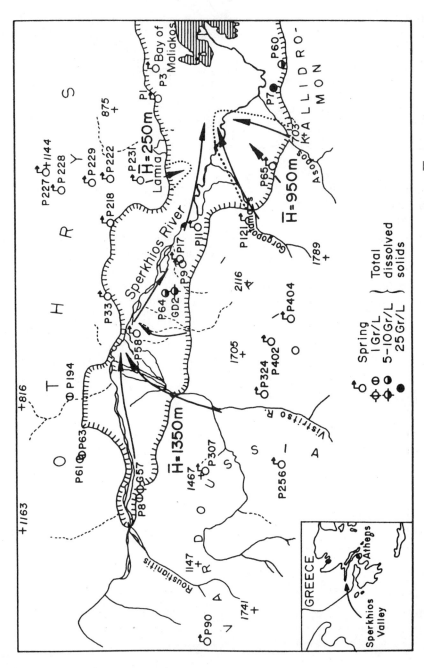

Fig. 9.24 General map of study area in the Sperkhios Valley, Greece. The recharge altitudes ($\overline{H}$) were calculated from deuterium and ^{18}O data (Stahl et al., 1974).

Table 9.4 Isotopically deduced average recharge altitudes for artesian well groups in the Sperkhios Valley (following data by Stahl et al., 1974)

Well group	Av. $\delta^{18}O$(‰)	Av. recharge altitude (masl)
I	-9.36 ± 0.04	1350
II	-8.81 ± 0.07	950
III	-7.80 ± 0.15	250

in order to get enough data for the calculation of meaningful average values and to gain an insight into the fine structure of the investigated system in the lateral, vertical, and time dimensions.

Note: An excellent collection of basic isotope data and discussion of case studies is provided by publications of the International Atomic Energy Agency, included at the end of the reference list. Especially useful is the collection of papers is *Stable Isotope Hydrology: Deuterium and Oxygen-18 in the Water Cycle*, IAEA, Vienna, 1981. Fritz and Fontes (1980 and 1986) are the editors

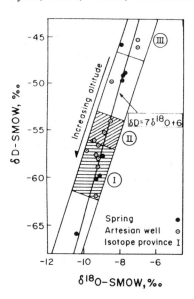

Fig. 9.25 Isotopic data of artesian wells and springs in the Sperkhios Valley. Three well groups, or isotopic provinces, were recognized. A local meteoric line, $\delta D = 7\delta^{18}O + 6$, was established from the spring data (Stahl et al., 1974).

Stable Hydrogen and Oxygen Isotopes

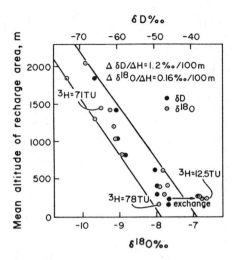

Fig. 9.26 Isotopic composition and recharge altitude of springs in the Sperkhios Valley (Stahl et al., 1974).

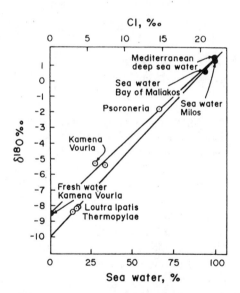

Fig. 9.27 Isotopic composition and chlorinity of warm springs at the Bay of Maliakos (see Fig. 9.24). Values lie on a mixing line with seawater, proving that the last one is intermixing in the mineral springs (Stahl et al., 1974).

Table 9.5 Variations in $\delta^{18}O$ values in daily rain at Ashdod, Israel (Gat and Dansgaard, 1972)

Date (Jan 1968)	Amount (mm)	$\delta^{18}O$ (‰ SMOW)
6	5.3	-5.01 ± 0.06
9	6.5	-2.09 ± 0.06
10	1.0	-0.14 ± 0.07
14	12.5	-6.13 ± 0.07
15	10.1	-6.43 ± 0.13
16	5.2	-2.05 ± 0.13
17	3.3	-2.3 ± 0.11
21	3.4	-4.70 ± 0.11
23	2.4	-2.70 ± 0.10
24	13.0	-5.27 ± 0.11
28	7.1	-3.09 ± 0.11
30	10.7	-4.15 ± 0.11
31	7.6	-6.90 ± 0.11

of the *Handbook of Environmental Isotope Geochemistry* which contains valuable information on the stable hydrogen and oxygen isotopes.

9.12 Sample Collection for Stable Hydrogen and Oxygen Isotopes and Contact with Relevant Laboratories

In principle, sampling is easy. A 100-ml sample is adequate for stable isotope analyses. The main precaution to be taken is to keep the bottle well closed in order to avoid any evaporation. It is highly recommended that sample collection be discussed with the respective laboratory experts. The hydrochemist has to be familiar with analytical difficulties, for example, the presence of H_2S, traces may interfere with the mass-spectrometric measurements.

It is always recommended to take double samples and keep one as a backup sample. These can be used in case of doubt concerning the obtained results.

Samples have to be collected properly. In a spring the sample should be collected as deep as possible from the surface of the water body of the spring eye, a topic also discussed in section 7.1.

9.13 Summary Exercises

Exercise 9.1: From the information included in Fig. 9.2, calculate the percentage of Dead Sea brine intermixed in the Hamei Yesha mineral spring. Do

Stable Hydrogen and Oxygen Isotopes

the calculation using the redundant information available, that is, use the δD and the $\delta^{18}O$ values.

Exercise 9.2: Check the hD and $h^{18}O$ calculations in Table 9.2. Applying the same altitude-isotopic composition equations, calculate the average recharge altitude for (a) a well with a mean annual value of $\delta D = -85.5‰$, and (b) a spring with a steady discharge and a value of $\delta^{18}O = -12.4‰$.

Exercise 9.3: What other measurements could you recommend to further check the evaporation process concluded to operate in the Pajeu River (Fig. 9.20)?

Exercise 9.4: Examine Fig. 9.8. The points with light isotopic composition are postulated to indicate water that was recharged under different climatic conditions? Which type of climate could it be?

Exercise 9.5: Examine Fig. 9.18. It represents a study aimed to calibrate which parameter? How will you select sampling points for a similar calibration in your case study areas?

Exercise 9.6: Examine Fig. 9.22. Which boundary conditions can be deduced for groundwater samples 56-59?

Exercise 9.7: The solubility of HCO_3 decreases or increases as a function of salinity? Retrieve the answer from a close examination of the Salentine Peninsula data given in Table 9.3. From this set of data, is the HCO_3 confirming the conclusion discussed in the text concerning the observed composition changes in the depth profiles of the reported wells?

Exercise 9.8: The seawater dilution (mixing) curve, seen in Fig. 9.23, is curved. How can the same data be plotted so they will fall on a straight mixing line?

10
TRITIUM

10.1 The Radioactive Heavy Hydrogen Isotope

Tritium is the heavy isotope of hydrogen (section 5.2). Its symbol is ^{3}H, or T. Tritium atoms are unstable and disintegrate radioactively, forming stable ^{3}He atoms. The radioactive decay is accompanied by the emission of β^- particles, measurable in specific laboratories:

$$T \xrightarrow{12.3 \text{ y}} \beta^- + {}^3He$$

The rate of radioactive decay is by convention expressed as the half-life, $T_{1/2}$, defined as the time span during which a given concentration of the radioactive element atoms decays to half their initial value. $T_{1/2}$ of tritium is 12.3 years. Thus, after 12.3 years one-half the initial concentration of tritium atoms is left, after 24.6 years only one-quarter is left, and so on. A radioactive decay curve of tritium is given in Fig. 10.1. Using the decay curve it is possible to determine, for example, how many years it takes for a given amount of tritium to decay to 20% of the initial amount. The answer, obtained from Fig. 10.1, is 29 years. Similarly, one can determine what percentage of an initial amount of tritium will be left after 20 years. The answer is 32% (read from Fig. 10.1).

The concentration of tritium in water is expressed by the ratio of T atoms to H atoms:

T/H = 10^{-18} is defined as 1 tritium unit (1 TU)

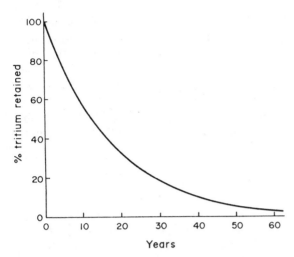

Fig. 10.1 Radioactive decay curve of tritium. After 12.3 years 50% of an initial concentration is left, after 24.6 years 25% is left, etc.

10.2 Natural Tritium Production

Cosmic ray neutrons interact in the upper atmosphere with nitrogen, producing ^{15}N, which is radioactive and disintegrates into common carbon (^{12}C) and tritium:

$$^{14}N + n \rightarrow {}^{15}N \rightarrow {}^{12}C + {}^{3}H$$

The tritium atoms are oxidized to water and become mixed with precipitation, and so enter the groundwater. The natural production of tritium introduces about 5 TU to precipitation and surface water.

In the saturated zone, water is isolated from the atmosphere and the tritium concentration drops due to radioactive decay: the original tritium concentration of 5 TU drops to 2.5 TU after 12.3 years, only 1.2 TU are left after another 12.3 years, and so on.

Provided the natural production of 5 TU is the only source of tritium in the groundwater, we would have at hand a water dating tool. For example, water pumped from a well with 3 TU has preserved $3 \times 100/5 = 60\%$ of its natural tritium content, equivalent to an age of 9 years (readable from Fig. 10.1). However, a hydrologist's life is not that simple. Most aquifers have a capacity equal to several annual recharges, or in other words, water accumulates in the aquifer for many years and the age we have just calculated from the tritium concentration is not a real age, but rather an average or *effective age*. This is still of high value in hydrology. Tritium measurements have a practical limit of 0.2 TU. Assuming an initial input of 5 TU, an amount of 0.2 TU represents $0.2 \times 100/5 = 4\%$, or an age of 55 years (Fig. 10.1). Thus, the tritium method of effective hydrological age seems valid in the range of a few decades.

Low tritium concentrations are determined on samples that have been preconcentrated electrolytically, necessitating samples of about 300 ml. To allow for repeated measurements, samples of 1 l should be collected for tritium measurements.

10.3 Man-Made Tritium Inputs

Nuclear bomb tests, which began in 1952 in the northern hemisphere, added large amounts of tritium to the atmosphere. They reached a peak in 1963, with up to 10,000 TU in a single monthly rain in the United States. An international treaty stopped surface nuclear bomb tests in 1963, and tritium concentrations in precipitation decreased steadily. Since nuclear testing began, tritium (and δD and $\delta^{18}O$) has been measured in a worldwide net of stations (Fig. 10.2), coordinated by the International Atomic Energy Agency (IAEA) in Vienna. The results are published in annual reports: *Environmental Isotope Data: World Survey of Isotope Concentration in Precipitation,* IAEA, Vienna. Figure 10.3 represents annual tritium concentrations in precipitation at various stations from 1961–1975. Figure 10.4 reveals monthly values for several stations for the period 1961–1965, when the bomb-tritium impact was especially high. The following patterns are seen in the tritium curves:

Values rose much higher in the northern hemisphere, where the bomb testing took place.

The maximum peak was reached in 1963, and values decreased thereafter.

A summer peak and winter low were seen each year (Figs. 10.4 and 10.5), reflecting the fact that the tritium was displaced in large amounts in the higher parts of the atmosphere and leaked in the spring and summer into the lower parts.

The man-made tritium, which reached several thousand TU in precipitation during 1963, completely masked the natural tritium production discussed in the previous section. The awareness of the potential importance of tritium to hydrology arose only after the nuclear tests began. By that time the natural tritium content in precipitation could no longer be measured, but a unique solution was found—measurements in stored and dated wine bottles, reflecting the relevant annual rains (Table 10.1). A common value for prebomb tritium is 5 TU.

The described observations may be applied to practical problems. For example, the weighted average tritium concentration in Vienna was 3280 TU in

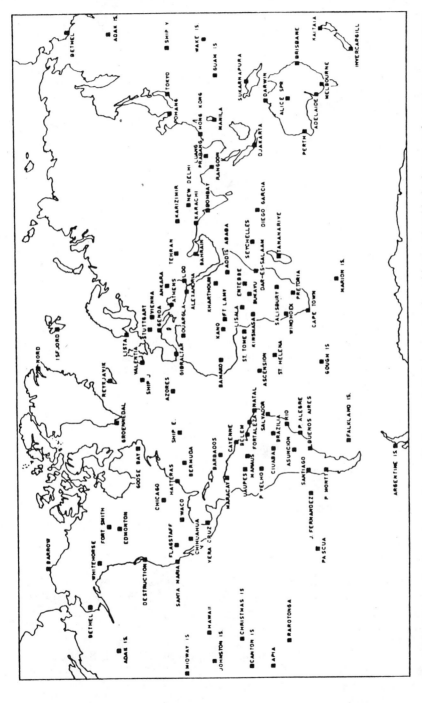

Fig. 10.2 Network of weather stations collecting precipitation samples for tritium and stable hydrogen and oxygen measurements, coordinated by the International Atomic Energy Agency, Vienna.

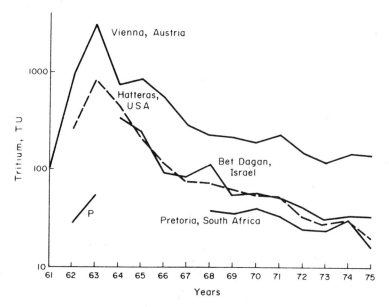

Fig. 10.3 Annual weighted average tritium concentrations of representative weather stations. Values increased dramatically in 1963 and have been steadily decreasing ever since. Values are much higher in the northern hemisphere (where the nuclear tests took place). At present the tritium concentrations are almost back to natural values. (From *Environmental Isotope Data: World Survey of Isotope Concentrations in Precipitation*, vol. 1 (1969) and vol. 7 (1983).)

1963 (Fig. 10.3). How much tritium would be left in water that was recharged in 1963 and pumped in a well in 1988? The answer can be found by using the following steps:

1988 - 1963 = 25 years passed.
From Fig. 10.1 one can read that after 25 years only 25% of the initial tritium is left.
3280 × 25/100 = 820 TU would be left in 1988.

Before 1952, natural tritium was about 5 TU in precipitation. How much would be left in groundwater recharged in 1951 and emerging in a spring in 1988? The answer can be found by using the following steps:

37 years passed.
Radioactive decay would have decreased the concentration to 11% of the initial value (Fig. 10.1).
5 × 11/100 = 0.5 TU would be left in 1988.

Tritium

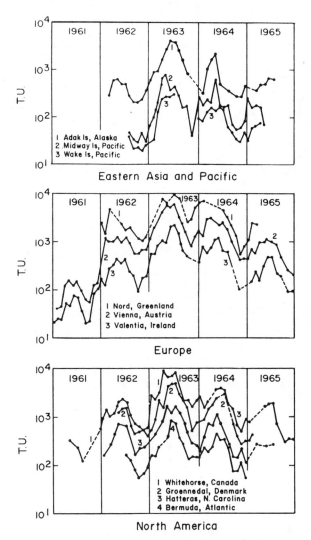

Fig. 10.4 Monthly tritium concentrations in rain at representative stations. Summer peaks indicated that bomb tritium was stored in the high parts of the atmosphere and leaked in the spring into the lower atmosphere. (From IAEA Technical Report No. 73, 1967.)

10.4 Tritium as a Short-Term Age Indicator

When the anthropogenic tritium was noticed, hopes arose that the specific tritium "pulses," contributed by the individual tests, would provide a way to accurately date groundwater. However, it turned out that the input values in precipitation varied considerably from one location to another and from one

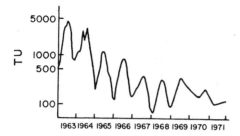

Fig. 10.5 Summer peaks and winter lows of tritium observed in precipitation in Sweden. (From Pearson, 1974.)

season and year to the next. In addition, complicated mixing takes place in each aquifer, and the mode and extent of mixing of each year's recharge with that of previous years' recharge is unknown. Hence, age determinations accurate to the year are impossible and of no meaning to groundwater studies. However, semiquantitative dating is possible and very informative:

Water with zero tritium (in practice <0.5 TU) has a pre-1952 age.
Water with significant tritium concentrations (in practice >10 TU) is of a post-1952 age.

Table 10.1 Estimates of natural, pre-bomb, tritium concentrations in precipitation and wine (text) (Vogel et al., 1974)

Locality	Sample	TU	Researchers[a]
Rhone valley, France	wine	3.4	Kaufman and Libby (1954) Von Buttlar and Libby (1955)
Bordeaux, France	wine	4.3	Kaufman and Libby (1954) Von Buttlar and Libby (1955)
Rhine valley, Germany	wine	5.5	Roether (1967)
New York State, USA	wine	5.8	Kaufman and Libby (1954) Von Buttlar and Libby (1955)
Chicago, USA	wine	7.5	Kaufman and Libby (1954) Von Buttlar and Libby (1955)
Ottawa, Canada	rain	15.3	Brown and Gummit (1956) Brown (1961)
Greenland	ice	12.6	Begemann (1961)

[a]References given in Vogel et al. (1974).

Water with little, but measurable, tritium (0.5–10 TU) seems to be a mixture of pre-1952 and post-1952 water. The topic of water mixtures is further discussed in section 11.10, along with ^{14}C data.

10.5 Tritium as a Tracer of Recharge and Piston Flow: Observations in Wells

10.5.1 Tritium Concentrations and Time Series Providing Insight Into the Local Recharge Regime

Seasonal variations in the tritium concentration in two shallow tribal wells in the Kalahari are seen in Fig. 10.6. Variations in tritium were noticed to follow variations in the water table. This observation provided insight into the local recharge mechanism. The tritium-rich recharge of the rainy season formed an upper layer in the aquifer. As abstraction, mainly during the dry season, advanced lower water layers were encountered, having been stored in the aquifer for several years. The seasonal tritium peak was observed to arrive at the wells several months after the rain peak. Thus, the recharge water moved in the aerated zone not through conduits, but in a porous medium, in a piston flow mode (section 2.1, Fig. 2.2). The time lag provided an idea of the storage capacity of the aerated zone.

10.5.2 Tritium in Soil Profiles Applied to Assess Infiltration Velocity

Several workers measured tritium concentrations in profiles of soil moisture. The samples were collected in most cases by means of a hand drill; the profile samples were carefully wrapped to avoid drying or exchange with tritium

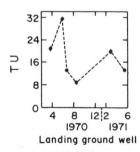
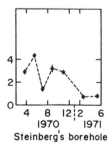

Fig. 10.6 Seasonal variations in tritium concentration in shallow wells in the Kalahari, indicating recharge in a piston-flow mode. (From Mazor et al., 1973.)

in the air. In the laboratory the soil moisture was extracted by distillation, weighted (to get a moisture profile), and measured for the tritium concentration.

Figure 10.7 deals with such a profile from South Africa, along with the local precipitation-tritium curve. The profile was taken in December 1971 and the precipitation-tritium curve has been corrected (solid line, top of Fig. 10.7) for radioactive decay until that date. This corrected line indicates 7 TU for 1958 precipitation, an increase to over 20 TU in the rainy season of 1962, and over 30 TU for 1964. These values were applied to identify the time points marked on the tritium-depth line in Fig. 10.7. The moisture with the 1958 tritium concentration was observed in 1971 at a depth of about 2 m. The infiltrating water moved 2 m in 13 years (1958 to 1971), indicating an average infiltration velocity of 0.15 m/year. The 1962 tritium concentration was observed at a depth of 1.4 m after 1971-1962 = 9 years (1962 to 1971), indicating infiltration

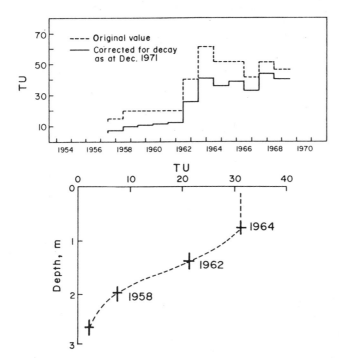

Fig. 10.7 Tritium in a soil profile from the Transvaal, South Africa, and in precipitation at Pretoria (following Bredenkamp et al., 1974). The identification of the 1958, 1962, and 1964 moisture fronts is discussed in the text, along with the applications to calculate recharge percentage.

velocity of 0.16 m/year; the 1964 tritium front was observed at 0.8 m, indicating an infiltration velocity of 0.11 m/year. Variations in the calculated infiltration velocity may occur due to differences in the soil properties or due to variations in the annual rain regimes. One may therefore retrieve from soil profiles average recharge velocities, relative lateral and vertical conductivity (by comparing different profiles), and recharge efficiencies.

10.5.3 Tritium Profiles Applied to Estimate Recharge Amounts

The applicability of the tritium profile data can be demonstrated by means of an example: a tritium front in a soil profile was identified as having descended 3 m during 10 years and the weighted average moisture content was found to be 12% (by volume). These data can be used to calculate the average annual recharge: out of the 3 m profile, $3 \times 0.12 = 0.36$ m, the equivalent column of recharged water. The average annual recharge was $0.36/10 = 0.036$ m, or 36 mm/year. Continuing with this example, if the average annual precipitation was 650 mm/year, then the average recharge percentage was $36 \times 100/650 = 5.5\%$.

A second example: a tritium concentration signifying the 1962 peak was observed at a depth of 4.8 m in a soil profile with a weighted average moisture content of 12% by volume. The observation was made in 1978. The local average precipitation was 720 mm/year. What was the average infiltration velocity and percentage? The average infiltration velocity was $4.8/16 = 0.3$ m/year, the equivalent recharge annual water column was $0.3 \times 0.12 = 0.036$ m, or 36 mm, and the recharge percentage was $36 \times 100/720 = 5\%$.

10.5.4 Active Rain Recharge in a Desert Indicated in a Tritium Soil Profile

Figure 10.8 shows tritium profiles in soil moisture near Grootfontein, South Africa (Bredenkamp et al., 1974). A layered structure is seen, indicating a piston flow type of recharge in the soil. Another set of tritium soil profiles from the southern Kalahari, South Africa, is given in Fig. 10.9 (Verhagen et al., 1979). The value of 10 TU was adopted to indicate the arrival of the 1963 tritium front. The results of this study are of special interest in the light of a long-standing controversy about whether recharge is effective at all in the sand-covered Kalahari dry land (Mazor, 1982). The 1963 recharge has penetrated 7–23 m, thus demonstrating active recharge. Profile B.H. 3 was taken below a large thorn tree. The depth of recharge penetration was only 8 m, about half the depth in the adjacent profiles. This was attributed to the water consumption of the tree.

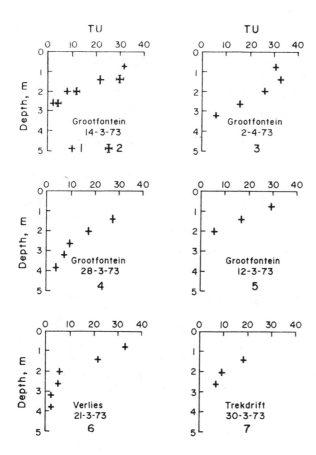

Fig. 10.8 Variations of tritium concentration with depth in soil profiles near Grootfontein, South Africa. Such tritium profiles were an excellent means of demonstrating piston flow of infiltrating recharge water. (From Bredenkamp et al., 1974.)

10.5.5 A Methodological Note: Soil Profiles Overlook Water Flow in Conduits

Tritium profiles are commonly interpreted with the basic premise that the soil is a uniform medium, with a sponge-like structure, through which water infiltrates. However, short-circuit flow through conduits such as cracks, bioturbations, or decayed root channels must also be considered. Cracks, or rodent holes, are efficient intake points of runoff, and occasionally they become filled

Tritium

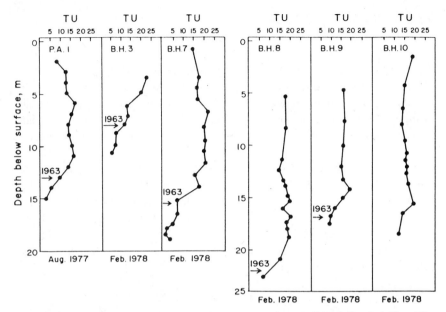

Fig. 10.9 Tritium in soil moisture profiles in the southern Kalahari, South Africa. The measurements were done in 1978, and 10 TU was taken to mark the 1963 recharge. B.H. 3 was profiled under a large camel thorn tree. (From Verhagen et al., 1979.)

with coarse particles, providing routes of relatively high conductivity. When a soil profile builds up, these coarsely repacked channels of higher conductivity add up, forming a network of higher-conductivity conduits. The soil cores do not sample water flowing in cracks or burrowed holes. Thus, the infiltration velocities and percentages, calculated by the above discussed tritium-soil profiles, should be regarded as minimum values.

10.5.6 Tritium Applied to Identify Flow in Conduits

The importance of short-circuit infiltration may vary significantly from one area to another and its intensity may be measured by tritium measurements, e.g. in dripping water in shallow mines or in groundwater in flat terrains: if, in such studies, tritium values encountered are higher than in the overlying soil profiles, then short-circuit recharge has occurred. Such studies are meaningful in cases where lateral recharge can be excluded on grounds of local geology, or topography (e.g., flat terrain or hill tops).

10.5.7 Two Case Studies

Figure 10.10 provides data on repeated tritium measurements in a well and in the adjacent Mohawk river. What hydrological conclusions can be drawn? The variations in tritium concentrations in the well followed variations in the river, as shown in Fig. 10.10. Hence, the river is recharging the well. The time lag in the well's response can be used to calculate recharge velocities. The data reveal piston flow of the recharge water, with little smoothing by dispersion (Fig. 10.10).

Tritium can also be used for regional recharge studies. An example from the Qatar Peninsula has been reported by Yurtsever and Payne (1979). They measured tritium in a large number of wells in a shallow aquifer, producing a map with isotritium contours (Fig. 10.11). Recharge occurred in areas of high tritium flowing to areas of low tritium. The reader is referred to an article by Allison and Hughes (1975) discussing the use of tritium measurements to estimate the amount of recharge.

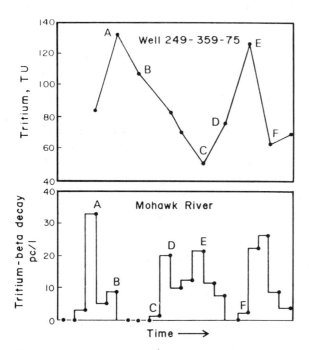

Fig. 10.10 Repeated tritium measurements in a well and in the adjacent Mohawk River, New York, USA. The river was monitored for tritium by beta decay counting and the results are expressed in pc/l (10^{-12} curie/l). (From Winslow et al., 1965.)

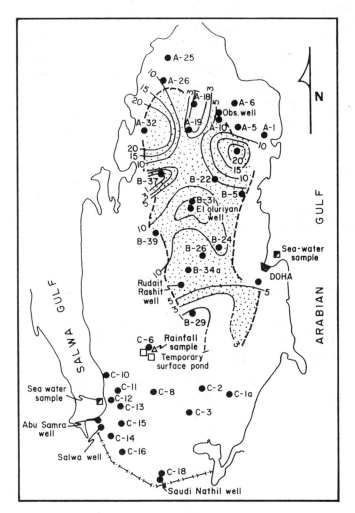

Fig. 10.11 Isotritium lines (in TU) for wells in a shallow aquifer on the Qatar Peninsula. Recharge was from areas of high tritium (stippled on the map) to areas of low tritium. (From Yurtsever and Payne, 1979.)

10.6 The Special Role of Tritium in Tracing Intermixing of Old and Recent Waters and Calculation of the Extrapolated Properties of the End Members

In sections 6.4 and 6.6 mixing of two water sources was discussed in terms of composition diagrams (Figs. 6.4, 6.11, 6.17, 6.18, and 6.19). The parameters of mixed waters plot on straight lines, but in most cases these lines are of positive correlations: for example, in the mixing of a deep warm water with a shallow cold water. In most cases the deep water also has a higher concentration of dissolved ions, and thus the deep water has higher values in all parameters. However, to obtain a negative correlation, we need a parameter that is high in one water and low, or zero, in the other water. Tritium is often such a parameter. In the example given in Fig. 10.12, the data indicate mixing of two water types. An upper limit of 39°C is obtained by extrapolating the best-fit line to zero tritium.

Tritium serves as a sensitive check for possible intermixing of deep water with shallow water. For example, in a study of warm springs in Zimbabwe it was important to establish whether the sampled waters were indigenous to the deeper reservoirs or were intermixed with shallow water of the adjacent rivers (in several cases the spring emerges a few meters from the riverbed or in it). The tritium level in six warm springs (54°C–100°C) was 0.1–0.6 TU, as compared to 6–38 TU in nine river samples (Mazor and Verhagen, 1976). By comparing respective spring and river values it could be concluded that contamination by river water was less than 2%. The springs could be regarded, in this case, as single-component systems.

10.7 Tritium, Dissolved Ions, and Stable Isotopes as Tracers for Rapid Discharge Along Fractures: The Mont Blanc Tunnel Case Study

A unique study of the rock-water interrelation has been reported from the Mont Blanc tunnel by Fontes et al. (1979). The 11.6-km long tunnel begins in a

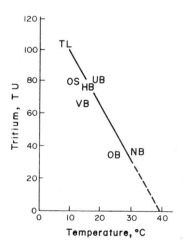

Fig. 10.12 Tritium values from various springs and wells (letters) sampled on the same day in the spa of Vals, Switzerland (Vuataz, 1982). An extrapolation of the best-fit line to zero tritium indicates an uppermost value of 39°C for the warm end member (text).

carbonate rock section in the Italian end, continues through a long section of granitic rocks, and ends in schist at the French end. The geological cross section and chemical data are given in Fig. 10.13 and Table 10.2. A close correlation is seen in Fig. 10.13 between the lithological and chemical transects:

High HCO_3 and pH and low Cl below the carbonate rock section, along with high Ca and Mg (samples I-36 and I-41 in Table 10.2).

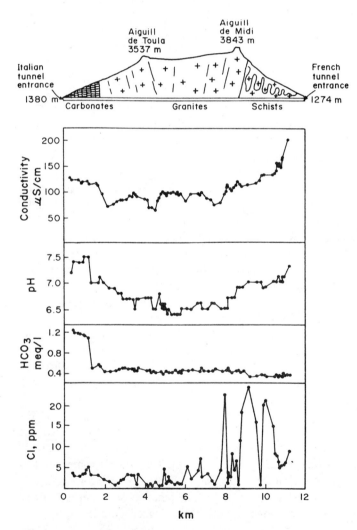

Fig. 10.13 A geological cross section and composition of waters sampled along the Mont Blanc tunnel (after Fontes et al., 1979). More data are given in Table 10.2 and in Figure 10.14. The good correlation with the lithology is discussed in the text.

Table 10.2 Composition (ppm) of representative water samples from the Mont Blanc tunnel (from Fontes et al., 1979)

Sample	Lithology	pH	Ca^{2+}	Mg^{2+}	Na^+	K^+	Cl^-	SO_4^{2-}	HCO_3^-	SiO_2
F-1	schist	6.9	27.6	0.73	18.4	2.97	8.90	72.4	22.0	n.d.
F-27	granite	6.5	23.1	0.24	10.7	1.17	3.54	46.0	29.3	20.9
F-30	granite	6.5	9.22	0.49	13.3	1.17	4.62	26.9	25.6	14.5
F-38	granite	6.5	9.20	0.24	21.3	2.35	5.33	36.8	28.7	15.7
I-21	granite	6.7	9.20	0.49	5.04	2.74	3.55	14.1	30.5	5.6
I-36	carbonates	7.5	26.8	2.41	8.70	0.78	5.32	36.5	67.1	6.9
I-41	carbonates	7.2	32.8	2.19	9.39	2.35	3.55	31.7	75.6	n.d.

Low salt content in the granitic section (Fig. 10.13 and Table 10.2).
Low HCO_3 but relatively high and varying Cl in the schist section.

The last observation has been interpreted by the researchers as indicating the existence of evaporitic rocks in the schist complex, a conclusion that is supported by high Ca, Na, and SO_4 concentrations in the water (e.g., sample F-1 in Table 10.2).

The chemistry of water encountered in the Mont Blanc tunnel is thus directly related to the lithological section. This was interpreted as indicating recharge along almost vertical joints, with little horizontal flow.

The tritium and stable isotopes results are given in Fig. 10.14, along with the surface profile and geological section. The tritium data indicate rapid recharge; all tritium values being post-bomb. By comparison with the Vienna line in Fig. 10.3, the values over 200 TU may be identified as post-1961 recharge. The recharge area in the central section is at an altitude of about 3200 m (top of Fig. 10.14) and the tunnel is at an altitude of about 1300 m. Thus, the recharged water descended 3200 − 1300 = 1900 m in 13 years or less, giving a recharge velocity of 150 m/year or more. The sharp variations in the tritium values indicate recharge flow in separated fractures, with varying flow velocities (reflected by different tritium concentrations, indicating different intake dates).

Comparison of the tritium data with the other parameters measured in the Mont Blanc tunnel is most instructive. The deuterium and ^{18}O lines follow the topography (note that the δD and $\delta^{18}O$ scales were reversed in Fig. 10.14 to facilitate comparison with the topographic profile). The isotopic values are less negative at each end of the tunnel, reflecting recharge from the lower parts of the respective landscape. The central part of the tunnel hosts water with more negative isotopic values, reflecting recharge from higher altitudes.

Tritium

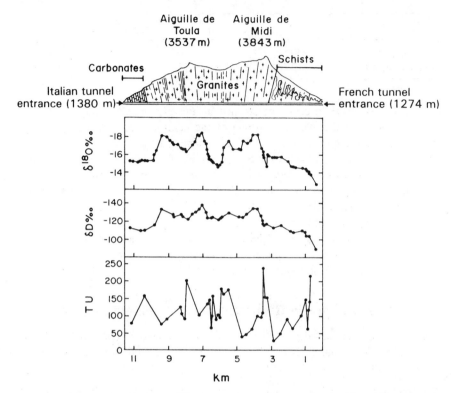

Fig. 10.14 Tritium, δD, and $\delta^{18}O$ in waters collected along the Mont Blanc tunnel during its construction in 1974. The δD and $\delta^{18}O$ axes have been plotted in a reverse mode to facilitate comparison with the topographic profile. The high tritium values indicate post-bomb recharge, and in the cases of over 200 TU one may even conclude post-1961 recharge (Fig. 10.3). (From Fontes et al., 1979.)

The average recharge altitude of the ends of the Mont Blanc tunnel and of the central part can be calculated using the isotopic altitude equations for the Maritime Alps (section 9.8):

$h = -320 \; \delta^{18}O - 2564$

$h = -40.2 \; \delta D - 2052$

The δD curve in Fig. 10.14 reveals average values of −110 ‰ for the edges of the tunnel and −130‰ for the central part. The respective values of $\delta^{18}O$ are −15.2‰ and −17.5‰. Applying these values to the altitude equations we get (in masl):

	hD	$h^{18}O$	Mean h
Tunnel ends	2300	2370	2330
Central section	3170	3040	3100

These results are in good agreement with the average altitude of the slopes of the ends, and the average altitude of the central part of the topographic profile above the tunnel (even better agreement could probably be obtained if local isotopic altitude equations were available). The stable isotopes thus indicate independently that recharged water descends in vertical fractures.

The combined application of dissolved ions, deuterium, and ^{18}O supplied redundancy in the indications of flow in separated vertical channels and the tritium provided the time scale or velocities.

The above study is an example of the modes by which recharge through fractures can be identified, apart from sponge-type recharge. Dripping water in caves and mines is a good potential target for local studies of this type.

10.8 The Tritium-^{3}He Groundwater Dating Method

The preceding sections dealt with a variety of hydrological applications of the man-made tritium signal. However, the tritium dating method has a number of shortcomings:

The method is based on the measurement of the concentration of tritium left in the sample, but the initial tritium concentration is taken from the records of tritium concentration in the local atmosphere; this concentration changes dramatically from one year to the next (Figs. 10.3, 10.4, 10.5, and 10.7). The uncertainty of the initial tritium concentration pertaining to a studied groundwater sample is so great that the method is only semiquantitative.

Mixing patterns of newly recharged water and older groundwater are complex and poorly known, making the assignment of a particular initial tritium concentration to a studied case difficult.

The tritium concentration in the atmosphere and in precipitation is almost back to the low natural concentrations, and the tritium concentrations in groundwater have also decreased significantly. This is good news from an environmental point of view, but not for water researchers.

During the last decade a different track has been followed for the tritium dating method, namely measurement of the decay product, ^{3}He (section 10.1), as a substitute for the initial tritium concentration. Tritium-derived ^{3}He

Tritium

(tritiogenic helium) is built up in the groundwater as the contained tritium disintegrates; hence, the ^{3}H:^{3}He ratio decreases with age. To facilitate the calculations both nuclides are expressed in TR (tritium ratio) units, that is, ^{3}H:H and ^{3}He:H ratios of 10^{-18}. In principle, the age of groundwater can be determined by measuring the tritium concentration left in the sample, the concentration of ^{3}He accumulated in the sample, and knowledge of the half life of the tritium integration, 12.3 years (Schlosser et al., 1988). This dating method is based on a number of assumptions:

The ^{3}He can be identified, based on the background of the atmospheric helium contained in every meteoric groundwater (Chapter 13).
The ^{3}He is retained in the sample, or the diffusive loss is known.
No ^{3}He was gained—for example, by diffusion from other groundwater bodies—or such gains can be estimated.

A major task in the tritium-^{3}He dating method is the identification of ^{3}He based on the background of the atmospheric helium, which has a ^{3}He:^{4}He ratio that varies from one place to another within a wide range of values.

The tritium-^{3}He groundwater dating method is still in the investigation stage, but the number of researchers that apply it is growing, and commercial laboratories will probably soon accept samples for routine measurements. It seems that this method will extend the tracing of the 1960s tritium signal, so it can be used by the next generation of hydrologists. Even when the tritium in groundwater is no longer measurable, the presence of ^{3}He will indicate post-1952 recharge.

10.9 Sample Collection for Tritium Measurements

Sample collection for tritium concentration testing is simple—samples of 1 l are adequate, exchange with the atmosphere has to be avoided by proper sealing, and it is preferred that samples be collected in glass bottles.

10.10 Summary Exercises

Exercise 10.1: In a well near Vienna, 30 TU were observed in 1965. What was the effective age of the water?

Exercise 10.2: Archive data reveal that a well in the United States contained 1 TU in 1975. What was the effective age of the water?

Exercise 10.3: An investigation was ordered to find out whether a certain well was polluted by an adjacent sewage pond. The well contained 0.3 TU in 1973. What is your expert opinion?

Exercise 10.4: What is the basic difference between the dating method based on the measured tritium peak and the developing tritium-^{3}He method?

11
RADIOCARBON AND ^{13}C

11.1 The Isotopes of Carbon

Carbon has three isotopes in nature: ^{12}C—common and stable; ^{13}C—rare and stable; and ^{14}C—very rare and radioactive. The heavy carbon isotope, ^{14}C, is unstable and decays radioactively into ^{14}N, emitting a beta (β^-) particle that can be measured in specialized laboratories. The half-life of ^{14}C is 5730 years. The above listed information can be summarized in the following way:

$$^{14}C \xrightarrow[T_{1/2}\ 5730\ \text{years}]{\beta^-} {}^{14}N$$

The radioactive decay curve of ^{14}C is given in Fig. 11.1. What fraction of an initial concentration of ^{14}C is left after 10,000 years? The answer, read from Fig. 11.1, is 27%.

11.2 Natural ^{14}C Production

Carbon-14 is formed in the upper parts of the atmosphere from secondary neutrons formed by cosmic ray interactions with the atmosphere. The neutrons interact with common nitrogen:

$$^{14}N + n \rightarrow {}^{14}C + {}^{1}H$$

The ^{14}C atoms are oxidized upon production and mixed with atmospheric CO_2. With the latter the ^{14}C is introduced into plants and into surface and groundwater. Most of the ^{14}C introduction into groundwater occurs through the soil

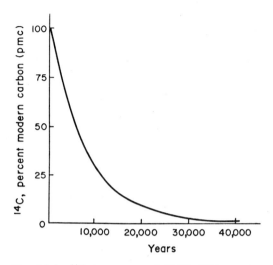

Fig. 11.1 ^{14}C decay curve (half-life-5730 years).

CO_2 (section 2.1), which in turn, is in constant exchange with the atmospheric CO_2 and, therefore, has similar concentrations of ^{14}C.

The concentration of ^{14}C is expressed in relation to the $^{14}C:^{12}C$ ratio in an international standard (oxalic acid). The ^{14}C concentration in the bulk carbon of the standard is defined as 100% modern carbon (pmc). Thus, a tree that died will contain only 50 pmc after 5730 years (one half-life). The ^{14}C concentration of a sample is measured in specialized counting laboratories.

What is the ^{14}C concentration of a tree trunk that fell 2000 years ago? The answer, read from Fig. 11.1, is 80 pmc. The analytical limit (including sample contamination) of the ^{14}C measurement is in many cases about 1 pmc. What is the time range of the ^{14}C dating method? The answer, again read from Fig. 11.1, is up to about to 40,000 years. This limit may be extended by more sensitive measuring and sampling techniques. The ^{14}C dating method has been carefully studied for archeological and paleoclimatic studies. Much effort has been invested in studying variations in the ^{14}C production over the last 100,000 years. Minor variations have been observed, but they are negligible in groundwater dating, compared to other complicating factors.

11.3 Man-Made ^{14}C Dilution and Addition

Since the industrial revolution of the early nineteenth century, large amounts of fossil fuels (oil, coal, gas) have been combusted, causing an increase of about 10% in the concentration of atmospheric CO_2. This added fossil CO_2 was devoid of ^{14}C and, correspondingly, lowered the $^{14}C:^{12}C$ ratio in the air by about 10%.

An anthropogenic addition of ^{14}C into the atmosphere occurred with the nuclear bomb testing from 1952 to 1963, along with the introduction of bomb tritium. As a result, the ^{14}C concentration also increased in plants (Fig. 11.2), in the soil CO_2, and in recharged groundwater. Values up to 200 pmc have been measured, but they decreased to about 120 pmc by 1987, and to about 110 pmc in 1995.

11.4 ^{14}C in Groundwater: An Introduction to Groundwater Dating

Rain and surface water dissolve small amounts of atmospheric CO_2. Significantly more CO_2 is added to water percolating through the soil layer, as soil air contains about 100 times more CO_2 as compared to free air. Soil CO_2 is produced by biological action such as root respiration and decay of plant material. This CO_2 was tagged by the atmospheric ^{14}C concentrations, that is, about 100 pmc in pre-nuclear bomb times (pre-1952), and up to 200 pmc in post-bomb years.

Once water reaches the saturated zone of an aquifer it is sealed from the atmosphere and its ^{14}C decays. Table 11.1 includes ^{14}C and tritium data from wells in the Kalahari desert. In two cases values above 100 pmc were seen, together with post-bomb tritium (samples 2 and 14). However, in other cases with post-bomb tritium, the corresponding ^{14}C values were found to be below 100 pmc (Fig. 11.3). This lowering is not the result of radioactive decay, since the ^{14}C half-life is too long for the decay to be noticeable in post-bomb samples. This lowering of ^{14}C was caused by interaction with carbonate rocks devoid of ^{14}C, a topic addressed in the next section.

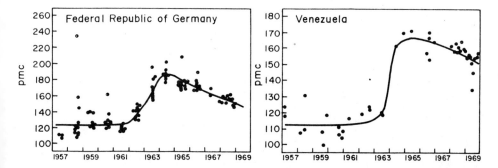

Fig. 11.2 ^{14}C in plants in Germany and in Venezuela (Tamers and Scharpenseel, 1970). Nuclear test's additions, above the natural 100 pmc, reached a peak value in 1964.

Table 11.1 Carbon-14 and tritium in phreatic groundwater in the northern Kalahari (Verhagen et al., 1974)

No.	Well	^{14}C (pmc)	Tritum (TU)
2	Makgaba 2	120.2 ± 2.6	4.3 ± 0.3
14	Rakops	106.1 ± 4.5	24.4 ± 1.7
54	Khodabis 1231	99.1 ± 2.2	5.1 ± 0.6
60	Toteng 1242	98.7 ± 1.8	16.9 ± 1.7
55	Z1183	97.7 ± 2.1	15.8 ± 1.3
63	Gweta	97.6 ± 1.5	1.2 ± 0.3
62	Bushman Pits	97.4 ± 2.2	8.0 ± 1.0
1	Makgaba 1	93.7 ± 1.7	5.0 ± 0.5
64	Zoroya 1606	91.9 ± 2.2	10.3 ± 1.1
8	Steinberg's BH	86.2 ± 1.6	2.8 ± 0.4
17	Tshepe	78.4 ± 1.3	1.6 ± 1.6
11	Makoba	69.1 ± 1.3	—
13	Mahata	69.1 ± 1.3	0.7 ± 0.3
58	Tsau	68.3 ± 1.3	0.9 ± 0.2
56	Kuke	66.8 ± 1.7	0.6 ± 0.6
59	Toteng 1320	59.5 ± 1.0	—
57	Sehitwa tribal borehole	54.3 ± 1.2	1.2 ± 0.3

Results of a hydrochemical study in basaltic aquifers of Hawaii (Hufen et al., 1974) are given in Table 11.2 and in a histogram in Fig. 11.4. This case study is of special importance because it deals with basaltic aquifers, in contrast to carbonate aquifers, to be discussed in the following section. The tritium

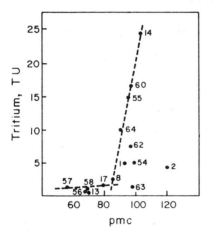

Fig. 11.3 Radiocarbon-tritium correlation, northern Kalahari groundwater (Table 11.1). Samples with values exceeding 85 pmc and 2 TU contain a post-nuclear bomb water component. Dashed lines are theoretical dividers between pre-bomb and post-bomb waters (Verhagen et al., 1974).

Radiocarbon and ^{13}C

Table 11.2 Carbon isotopes and tritium in recent groundwater in basaltic aquifers, Hawaii (from Hufen et al., 1974)

Pumping station	Draft (10^6 m³/a)	Cl (mg/l)	HCO_3^- (mg/l)	Tritium (TU)	Radiocarbon (pmc)	^{13}C (δ‰ PDB)
Aina Koa Wells	0.57	134	82	1.0	96.1	−19.2
Waialae Shaft	0.47	135	102	0.8	97.5	−17.2
Kaimuki Wells	7.43	76	76	0.6	96.1	−18.5
Wilder Wells	10.79	43	107	2.9	85.0	−15.5
Beretania Wells	12.76	52	73	0.3	91.4	−19.5
Kalihi Shaft	15.89	73	69	0.6	91.5	−18.3
Halawa Shaft	17.20	46	67	1.6	97.3	−19.2

values in the Hawaii study were 0.3–2.9 TU in 1974, indicating that most or all of the water had a pre-1952 age. The corresponding ^{14}C concentrations were 85–97 pmc. The pre-1952 ^{14}C in the atmosphere was about 100 pmc. Hence, part of the waters reported in Table 11.2 maintained all their initial ^{14}C, whereas others lost up to 15% of their initial value. This loss could occur for two reasons: (1) interaction with ^{14}C-devoid carbonate rocks (secondary in the basaltic terrain), and (2) aging of the water, reflected in radioactive decay of the ^{14}C. We will return to the Hawaii study, discussing additional information provided by the ^{13}C values.

A case study of groundwater in the Chad basin, conducted during 1967 and 1968 (Vogel, 1970), revealed a wide spectrum of ^{14}C concentrations from 0–150 pmc (Fig. 11.5). Waters with over 100 pmc were clearly post-1952, demonstrating the possible use of ^{14}C as an independent short-term age indicator, complementing the tritium method. The high ^{14}C values further indicated that little dilution by ^{14}C-devoid rocks occurred in water-rock interactions. Waters

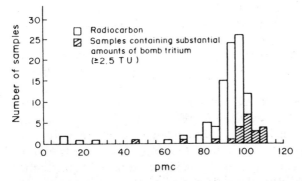

Fig. 11.4 Histogram of radiocarbon and tritium in groundwaters in basaltic aquifers, Oahu Island, Hawaii (Hufen et al., 1974). Pre-bomb (pre-1952) samples maintained nearly 100 pmc, indicating limited "dilution" by ^{14}C-devoid carbonates (see text).

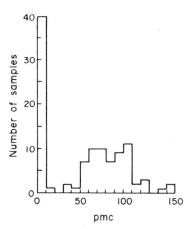

Fig. 11.5 ^{14}C values in groundwater samples, collected during 1967 and 1968 in the Chad basin, Africa. Post-nuclear bomb recharge is observed in several cases (greater than 100 pmc). Very old waters with low ^{14}C were also common (from Vogel, 1970).

with very low ^{14}C also prevailed in the Chad study area, indicating the presence of old groundwater, tens of thousands of years old (several ^{14}C half-lives).

11.5 Lowering of ^{14}C Content by Reactions with Rocks

Assume that water entered the saturated zone 4000 years ago with an initial concentration of 100 pmc, and emerges at present in a well. How much ^{14}C will it contain? The answer, read from Fig. 11.1, is 62 pmc.

What should be the ^{14}C concentration in a spring near Bonn, Germany, if the water was recharged in 1964 and emerged in 1982? From Fig. 11.2 it is seen that due to nuclear testing the atmospheric ^{14}C concentration rose to about 180 pmc in 1964. This concentration would not be effected by radioactive decay in the 18 years that passed between recharge and emergence (the half-life being 5730 years).

The last two hypothetical examples are correct only if the groundwater behaves as a closed system with regards to its $^{14}C{:}^{12}C$ ratio. Table 11.2 presents ^{14}C data of basalt aquifers in Hawaii, of mainly pre-bomb age, as reflected in the low tritium concentrations. The waters are seen to contain up to 100 mg HCO_3/l and have ^{14}C values exceeding 91 pmc in seven out of eight reported pre-bomb cases. It seems, thus, that little or no ^{14}C was lost due to reaction with rocks and little or no "dead" carbon (i.e., devoid of ^{14}C) was added by interactions with aquifer rocks, a fact reflected also in the low HCO_3 concentration.

In contrast, in the case of the Kalahari waters, reported in Table 11.1, notice the significant lowering of the ^{14}C content in recent, tritium containing waters. This is a common observation, mainly in carbonate terrains. Limestone and dolomite contain no ^{14}C, because they were formed much longer ago than the ^{14}C half-life of 5730 years. Recharged water containing CO_2 reacts with the carbonates to form dissolved bicarbonate:

$$H_2O + CO_2 + CaCO_3 \rightarrow Ca^{2+} + 2HCO_3^-$$

100 pmc   0 pmc   50 pmc

Thus, the dissolved CO_2 with 100 pmc reacts with the rock with 0 pmc to produce a 50 pmc bicarbonate.

Similar reactions occur with silicates:

$$H_2O + 2CO_2 + 2NaAlSi_3O_8 \rightarrow 2Na^+ + Al_2O_3 + 6SiO_2 + 2HCO_3^-$$

100 pmc   100 pmc

or

$$H_2O + 2CO_2 + 2KAlSi_3O_8 \rightarrow 2K + Al_2O_3 + 6SiO_2 + 2HCO_3^-$$

100 pmc   100 pmc

In such reactions, the initial ^{14}C concentration of 100 pmc (in pre-1952 samples) is closely maintained, as seen in the examples from Hawaii (Table 11.2).

A variety of reactions, of the kinds mentioned and other types, occur in the aerated and saturated zones, causing the initial ^{14}C concentration in pre-bomb groundwaters to vary between 100 pmc (in silicate rocks) and 50 pmc (in carbonate rocks). In reality, the water-rock interaction in carbonate rocks causes a change of only $60 \pm 5\%$ of the initial ^{14}C in the recharged water. Values of 60 ± 5 pmc seem to be common for recent, but pre-bomb, groundwater in calcareous aquifers (Kroitoru et al., 1987). The initial ^{14}C that typified old groundwater is best assessed by measuring the ^{14}C in recent local groundwater that is in contact with the same type of rocks. The obtained value can be applied to the age calculation of older water. In the absence of such local information, a drop to 65% of the initial ^{14}C can be attributed to water-rock interactions in carbonate aquifers, and a drop of 90% of the initial ^{14}C value can be attributed to silicate rocks (plutonic and volcanic rocks, sandstone, and quartzite).

Example: A local shallow water has been observed to contain 65 pmc and 5 TU. A sample from an adjacent deep well, in similar rocks, revealed 20 pmc and 0 TU. What is the deduced age of the water in the second well? The value of 65 pmc may be taken as the initial ^{14}C concentration of the groundwater when it first reached the deep aquifer. The observed 20 pmc represents 20 × 100/65 = 30.7% of the initial value, indicating an age of 10,000 years (Fig. 11.1)

The reader is referred to a number of basic articles dealing with chemical and isotopic aspects of groundwater dating by radiocarbon: Back and Hanshaw (1970), Fontes (1983), Geyh (1972, 1980), Pearson and Hanshaw (1970), Pearson and Swarzenski (1974).

11.6 ^{13}C Abundances and Their Relevance to ^{14}C Dating of Groundwater

So far we have discussed the occurrence of ^{14}C in hydrological systems. In a similar way one can follow ^{13}C. Its abundance in rocks, organic material, and groundwater is expressed in permil deviation of the ^{13}C:^{12}C ratio in the sample from that in a standard (PDB—a belamnite carbonate from the Pee Dee formation of South Carolina). Most marine carbonate rocks have δ^{13}C = -2 to 0‰, whereas frequent values for organic material and CO_2 in soil are -28‰ to -20‰. Most plants have values around -23 ± 3‰, but certain plants have more positive values, around -12 ± 2‰ (Tables 11.3-11.5 and Figs. 11.6 and 11.7).

The concentration of δ^{13}C in groundwater is determined by the input of recharged water and by reactions with rocks. For example, the reaction of water charged with CO_2 of δ^{13}C = -25‰ is

$$H_2O \;+\; CO_2 \;+\; CaCO_3 \;\rightarrow\; Ca^{2+} \;+\; 2HCO_3^-$$
$$\delta^{13}C = -25‰ \quad \delta^{13}C = 0‰ \qquad\qquad \delta^{13}C = -12.5‰$$

In contrast, in reactions with silicates, the original (organic) δ^{13}C values will be retained:

$$H_2O + 2CO_2 + 2NaAlSi_3O_8 \rightarrow 2Na^+ + Al_2O_3 + 6SiO_2 + 2HCO_3$$
$$\delta^{13}C = -25‰ \qquad\qquad\qquad\qquad\qquad\qquad \delta^{13}C = -25‰$$

Table 11.3 δ^{13}C values for selected samples from Saudi Arabia (Shampine et al., 1979)

Source	δ^{13}C (‰)
Tuwayq Mountain limestone	0.8
Upper Dhruma limestone	0.9
Upper Wasia Sandstone	-1.3
Carbonaceous shale, from 241 m	-25.7
Lignite, from 1450 m	-23.4
Thorn plant, living	-24.5

Table 11.4 Carbon isotopes of various samples from a forest near Heidelberg, Germany (from Vogel, 1970)

Analyses No.	Material	Collection date	$\delta^{13}C$(‰)	^{14}C (pmc)
H-415	Air CO_2	Apr 1958	−23.8	109
H−414	Soil CO_2	Apr 1958	−23.4	104
H-447	Growing leaves	Jun 1958	−29.5	109
H-655	Air CO_2	Apr 1959	−23.8	123
H-680	Air CO_2	Jun 1959	(−23)	129
H-701	Air CO_2	Jul 1959	−21.8	126
H-723	Soil CO_2	Mar 1959	(−23)	106
H-703	Growing leaves	Apr 1959	−28.1	112
H-702	Growing leaves	Jul 1959	−31.4	120
H-679	Humus 0-2 cm	Jun 1959	−25.7	94
H-749	Humus 2-5 cm	Jun 1959	−26.0	90

and

$$H_2O + CO_2 + 2KAlSi_3O_8 \rightarrow 2K^+ + Al_2O_3 + 3SiO_2 + 2HCO_3^-$$
$$\delta^{13}C = -25‰ \qquad\qquad\qquad\qquad \delta^{13}C = -25‰$$

At this point the $\delta^{13}C$ data of the Hawaii study (Table 11.2) warrants discussion. The tritium values indicated a pre-1952 age, the ^{14}C values of 85–97 pmc indicated no, or little, lowering of the initial ^{14}C value (100 pmc) by interaction with secondary calcite (section 11.4). The corresponding $\delta^{13}C$ values are −19 to −15‰. These values are closer to common plant material (−23 ± 3‰) and far from calcareous rocks (−2 to 0‰). Thus, the ^{13}C changed only slightly due to water-rock interactions in the basaltic aquifers.

11.7 Application of $\delta^{13}C$ to Correct Observed ^{14}C Values for Changes Caused by Interactions with Carbonate Rocks

It has been suggested by several researchers that $\delta^{13}C$ values can be applied to evaluate the extent to which ^{14}C in groundwater is altered by exchange with rocks. Three values have to be known:

Table 11.5 $\delta^{13}C$ in soil CO_2 and associated plants in northern Chile (from Fritz et al., 1979)

Station	$\delta^{13}C$ (‰), soil	$\delta^{13}C$ (‰), plant
1	−14.7	−14.0
2	−22.3	−28.1
3	−18.3	−20.0 to −23.6

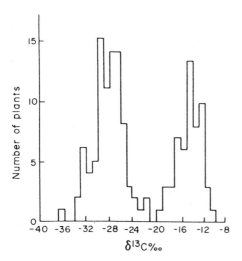

Fig. 11.6 ¹³C histogram in plants from Australia (following Rafter, 1974). Two major groups evolve, the total range being from -36‰ to -10‰.

The $\delta^{13}C$ value of the local soil material, representing the initial composition of groundwater, prior to the reaction with rocks.
The $\delta^{13}C$ value of the local aquifer rocks.
The $\delta^{13}C$ value in the studied groundwater.

For example, if the soil has $\delta^{13}C = -24‰$, the rock $= -1‰$, and the water $= -19‰$, then the initial ^{14}C in the water was "larger" (in absolute values) by a factor of

$$\frac{-1-(-24)}{-1-(-19)} = 1.3.$$

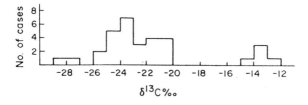

Fig. 11.7 ¹³C histogram in plants from Salar de Atacama, Chile (after Fritz et al., 1979). Two major populations emerge: $\delta^{13}C = -23 \pm 3‰$ and $\delta^{13}C = -14 \pm 2‰$, attributed to different systems in plant metabolism.

If the ^{14}C in the same water is observed to contain 30 pmc, then this value should be multiplied by 1.3 in order to correct for ^{14}C losses during reaction with aquifer carbonates: $30 \times 1.3 = 39$ pmc. This value is then applied to the decay curve (Fig. 11.1) and an age of 7500 years is obtained (instead of the 10,000 years indicated by the noncorrected ^{14}C value of 30 pmc).

This mode of correcting ^{14}C values for losses through interaction with rocks has serious drawbacks because the composition of ^{13}C in soil materials and aquifer rocks varies over a wide range. It is hard, and occasionally impossible, to collect relevant samples for measurement. In several studies the more common values of $-25‰$ for δ^{13}C of soil material, and $0‰$ for δ^{13}C of rock, are assumed. However, in certain cases such assumptions lead to severe overcorrecting.

For example, in well 2162 at Serowe, Botswana, the ^{14}C content was 95 pmc and δ^{13}C was $-11‰$. Hence, the correction factor would be

$$\frac{-0-(-25)}{-0-(-11)} = 2.3.$$

Applying this factor to the observed ^{14}C value in the water, a value of $95 \times 2.3 = 219$ pmc is obtained. This is clearly a heavy overcorrection (this is not a post-bomb case, as the accompanying tritium was only 0.7 TU).

In this example the overcorrection is obvious because it leads to a value that is higher than possible. But if, for example, the observed ^{14}C value is 12 pmc, then such a correction would yield a value of 28 pmc, which is in the "permitted" range, and hence the overcorrection would not be noticed.

The solution in the cited case from Botswana may be that the local plant community was not of the type that has a δ^{13}C value of $-25‰$, but a plant type that has a δ^{13}C value of $-13‰$ (Fig. 11.6). The problem of how to select the right paleo-input values is discussed in Chapter 15, which deals with paleohydrology, paleoclimate, and groundwater dating.

As will be discussed later, the main importance of the carbon isotopes in hydrological studies lies in:

Providing relative ages.
Indicating aquifer flow velocities.
Checking the continuity of proposed regional aquifers.
Studying mixed systems.
Establishing modes of flow.

In these applications the δ^{13}C and ^{14}C values serve as most useful tracers.

11.8 Direction of Down-Gradient Flow and Groundwater Age Studied by ^{14}C: Case Studies

Regional aquifers consist, in certain cases, of a recharged outcrop of a conductive rock layer, that dips below other rocks and is thus partially confined (section 2.8). Occasionally, wells tap only the confined section of such an aquifer, but in other cases wells also exist in the recharge area. In such systems carbon isotope studies are relevant for:

Checking the suggested continuity of water flow in the aquifer.
Checking additional recharge routes.
Calculating flow velocities.

11.8.1 Artesian Aquifer, South Coast of South Africa

An often quoted case study (Vogel, 1970) is that of an artesian aquifer in an area near the south coast of South Africa (Fig. 11.8). The decrease in ^{14}C downslope from the aquifer has been taken by the researcher to indicate continuity. One can even calculate the velocity of groundwater flow in the aquifer by selecting two points on the lines of Fig. 11.8, for example, 2 km 4000 years, and 18 km, 28,000 years. The average flow velocity in the aquifer is

$$\frac{18 - 2}{28,000 - 4000} \times 1000 = 0.66 \text{ m/year}.$$

This case study is a good example for demonstrating the general idea of "expected" behavior and the application of ^{14}C in aquifer studies. However, more data are needed in such cases:

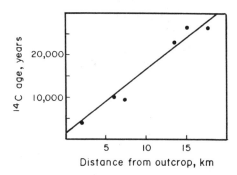

Fig. 11.8 ^{14}C ages (assuming initial concentration of 85 pmc) of water samples from artesian wells near the south coast of South Africa, as a function of the distance from the outcrop of water-bearing strata (Vogel, 1970).

Points of sampling (in Fig. 11.8 a gap of 7 km exists between the central wells).
Piezometric levels.
Temperatures.
Chemical compositions.
Isotopic compositions (e.g., δD, $\delta^{18}O$, $\delta^{13}C$, and tritium).

The continuity of the aquifer, and active flow in it, are independently checked by each of these parameters. This is essential because ^{14}C may decrease with depth (and distance) also in cases of discontinuous flow. As a matter of fact, the obtained average flow velocity of 0.66 m/year is extremely slow, and it might indicate lack of flow in parts of the system.

11.8.2 Paris Basin

A classical hydrochemical study of the Paris basin, 400 km across, has been presented by Evin and Vuillaume (1970). Their data have been compiled in Table 11.6 and the location of studied wells is plotted in Fig. 11.9. Important information, included in Table 11.6, relates to the type of aquifer (phreatic or free versus confined), the nature of the water source, (spring, well, artesian flow), and the stratigraphic unit at which the respective water was tapped. A hydraulic map of piezometric level contours has been constructed (Fig. 11.10) for water in the Albian aquifer. The distinction of aquifers has to be remembered in the processing of hydrological and hydrochemical data. Too often piezometric level contour maps are prepared of all the wells in a study area, ignoring the stratigraphic position of the sampled water systems. In cases where water is abstracted from aquifers in different stratigraphic units, water level contour maps have to be produced for each aquifer separately (not mixing apples with bananas). The ages applied to draw Fig. 11.11 were calculated by the researchers assuming an initial ^{14}C concentration of 80 pmc. The contours were drawn in ^{14}C half-life intervals. A general aging of water from the recharge areas in the peripheries down-dip in the confined Albian aquifer is seen in Fig. 11.11, supporting the original hypothesis that was based on the piezometric data (Fig. 11.10).

The Paris basin study warrants a closer look. In Fig. 11.11 it is seen that the spacing between studied wells was several kilometers near Paris, tens of kilometers in the periphery, and almost 100 km between the southern wells and the central ones. The conclusion that may be drawn is that the aging of water toward the central part of the Albian aquifer should be regarded with extreme caution: many more wells have to be studied and local deviations may be discovered.

At this point it might be asked whether a $\delta^{13}C$ correction should be applied to the ^{14}C values in order to calculate groundwater ages in the Albian aquifer of the Paris basin. To check this point the data for the Albian aquifer (Table

Table 11.6 Hydrological and isotopic data of groundwaters in the Paris basin (extracted from Evin and Vuillaume, 1970)

No.	Region	Place	Aquifer	Source	Stratigraphic unit tapped	Date	HCO_3^- meq/l	Tritium TU	$\delta^{13}C‰$	^{14}C pmc
1	Southeast	Parly-Chenons	free	well	Upp. Albian	10/66	0.3	77		94.7
2		Parly-Bernier	free	spring	Upp. Albian	10/66	3.1	27		86.7
3		Poilly	free	spring	Mid. Albian	10/66	5.1	9		91.7
4		Dracy	confined	drill	Upp. Albian	7/69	2.4	10		66.7
5		Chichery	confined	drill	Upp. & Mid. Albian	7/69	2.4	l.d.*	−13.0	58.0
		Chichery	confined	drill		4/66		l.d.	−17.7	53.2
6		Migennes	confined	drill	Upp. Albian	10/67	4.0	l.d		46.6
		Migennes	confined	drill	Upp. Albian	7/69	3.8	21		67.3
7		Neuilly	confined	artesian	Upp. Albian	3/69		l.d.	−15.0	15.9
8		Fleury	confined	artesian	Lower Albian	11/67	2.8	l.d		14.9
9		Champvallon	confined	artesian	Upp. Albian	3/69		l.d.	−12.8	13.6
10		Grand-Chaumont	free	well	Cenomanian Chalk	7/69	4.7	l.d.		95.6
11		Froville	free	well	Turonian	7/69	6			72.5
12		Appoigny	confined	artesian	Barremien	3/69	3.6	l.d.	−18.3	0.6
		Appoigny	confined	artesian	Barremien	10/67		l.d.		11.6
13		Grande Paroisse	confined	artesian	Albian	7/69	2.8	l.d.		3.2
14		Montbouy	confined	artesian	Upp. Albian	10/67	2.4	l.d.		28.9
15	East	Nuisement	free	drill	Albian	6/69	5.3	77		84.3
16		Humbecourt	free	drill	Albian & Aptian	6/69	4.48	208		76.9
17		Chaudefontaine	confined	drill	Albian & Aptian	6/69	6.8	10		74.8
18		Voillecomte	free	well	Albian & Aptian	6/69	2.2	l.d.		66.3
19		Montier-en-Der	confined	drill	Albian & Aptian	6/69	5.6	37		52.8
20		Dompremy	confined	artesian	Albian	6/69	5.2	l.d.	−9.0	17.9
21		Ste Ménéhoule	confined	artesian	Albian & Jurassic	6/69	l.d.			1.4
22	South	Barlieu	confined	drill	Albian & Cenomanien	10/67	4.7	l.d.		82.1

23		Blancafort	confined	drill	Albian & Cenomanien	10/67	3.9	l.d.		42.2
24		Chapelle Angilon	confined	drill	Cenomanien	11/67	4.1	l.d.		34.2
25	West	Bernecourt	confined	drill	Rauracian	5/69	2.8		−13.2	75.5
26		Thiberville	confined	drill	Albian & Kimmeridgian	5/69	5.2		−13.0	65.3
27		Brou	confined	drill	Albian	5/69	3.6		−9.0	52.0
28		Châteadun	confined	artesian	Albian	6/69	4.9		−5.7	40.9
29	Northwest	Gauciel	confined	drill	Upp. Albian	7/68	4.1		−9.8	37.5
30		Marais-Verniet	confined	well	Upp. Albian	7/68	4.4	l.d.	−14.2	36.9
31		Vernon	confined	artesian	Albian	7/68	3.2		−8.2	21.7
32		Les Logs	confined	drill	Albian & Cenomanian	8/68	8/68	4.9	−9.6	15.9
33		St. Pierre-en-Port	confined	artesian	Upp. Albian	8/68	3.3		−11.4	16.0
34		Le Trait	confined	artesian	Upp. Albian	7/68	3.4		−10.9	12.4
35		Mantes	confined	artesian	Upp. Albian	3/66	3.0		−10.9	12.4
36		Pont del'Arche	confined	artesian	Up. Albian	11/67	4.1	l.d.		7.2
37		Honfleur	free	spring	Cenomanian	7/68	5.8	12	−13.3	78.6
38		Incarville	confined	artesian	Jurassic to Cenomanian	5/69	7.5		+2.9	1.4
39		Epinay	confined	artesian	Albian	3/68	3.4		−13.2	9.7
40		Villeneuve	confined	drill	Albian	6/67	2.7	l.d.		9.1
41	Central	Achères	confined	drill	Albian	3/68	2.9		−16.3	8.9
42		Ivry	confined	drill	Albian	1/68	3.2	8.5	−13.3	
43		Orsay	confined	drill	Albian	1/68	3.5		−10.6	7.8
44		Le Pecq	confined	drill	Upp. Albian	3/68	2.6		−14.3	7.3
		Pantin	confined	drill	Albian & Barremian	12/66	2.6		−15.3	6.4
47		Issy	confined	drill	Albian	7/69	2.5	l.d.	−15.0	3.7
48		Noisy le Grand	confined	artesian	Albian	3/68	3.1		−14.1	3.5
49		Noisy le Grand	confined	artesian	Albian	8/69	3.1			3.7
		Paris ORTF	confined	drill	Albian	2/67	2.6		−12.2	3.2
		Aulnay sous Bois	confined	drill	Albian	3/68	2.9		−16.4	2.7
		Aulnay sous Bois	confined	drill	Albian	3/68	2.9	l.d.	−95	1.0
		Aulnay sous Bois	confined	drill	Albian	10/69	2.9			2.6
50		ViryChatillion	confined	artesian	Albian	8/69	3.1			2.1

*l.d.: limit of detection

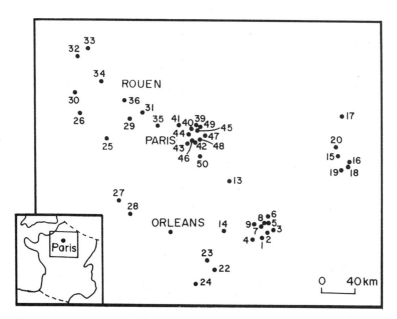

Fig. 11.9 Location map of wells studied in the Paris basin; well numbers as in Table 11.6. (From Evin and Vuillaume, 1970.)

11.6) have been drawn in a $\delta^{13}C$-^{14}C diagram (Fig. 11.12a). A random scatter is seen, indicating that ^{14}C dilution by interaction with carbonate rocks is not a continuous process along the down-gradient flow in the aquifer. Hence, in the Paris basin no grounds exist for $\delta^{13}C$ corrections.

The tritium-^{14}C values of the Albian aquifer of the Paris basin are drawn in Fig. 11.12b. Carbon-14 values in the free water table samples are in the range of 66–96 pmc, with an average of 80 pmc. This range is rather high, and an initial post-bomb ^{14}C of more than 130 pmc must have dominated most recharge waters in the free water-table aquifer, as may also be deduced from the accompanying post-bomb tritium values. It seems, thus, that immediate reactions with soil and rock carbonate lowered the ^{14}C concentration to a value of about 60% of the recharge concentration. A lower limit to the initial value can be defined by the highest ^{14}C values observed in the confined (and tritium devoid) aquifer sample. This value is 58 pmc (sample 5). Thus, the initial ^{14}C value in the studied aquifer was around 60 pmc for pre-bomb recharge and 80 pmc for post-bomb recharge. The researchers of the Paris basin selected the 80 pmc value (Fig. 11.11).

Four samples (no. 4, the July 1969 sample no. 6, no. 17, and no. 19; Table 11.6) have post-bomb tritium, yet they are from the confined part of the aquifer. Does this necessarily mean rapid flow into the confined aquifer, or is there

Radiocarbon and ^{13}C

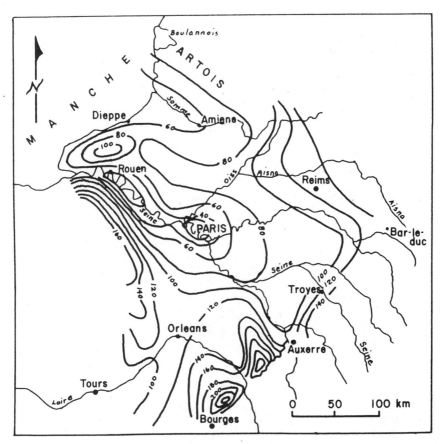

Fig. 11.10 Piezometric levels (in masl) of the Albian aquifer (following Evin and Vuillaume, 1970). Recharge seems to occur in the peripheries, giving rise to down-gradient flow toward Paris in the confined Albian aquifer. This hydrological hypothesis has been tested by ^{14}C dating (Fig. 11.11).

another explanation? These cases may represent *mixed pumping* of the confined Albian aquifer with shallow free-table water. The Migennes (no. 6) well revealed the following sets of data:

October 1967: 0 TU and 46.6 pmc
July 1969: 21 TU and 67.3 pmc

Thus, in the 1969 pumping, evidently post-bomb shallow water was added.

These examples demonstrate the importance of tritium measurements in every ^{14}C study. In the Paris basin a relatively large number of tritium measurements

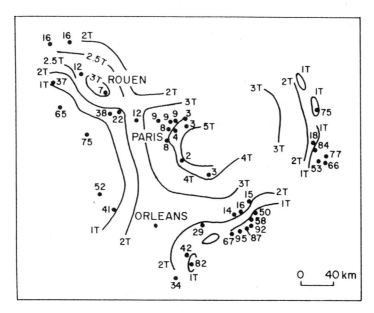

Fig. 11.11 ^{14}C values (in pmc) and iso-^{14}C lines in half-life intervals (assuming an initial ^{14}C content of 80 pmc). 1T: 40 pmc; 2T: 20 pmc; 3T: 10 pmc; 4T: 5 pmc, and 5T: 2.5 pmc. The observed increasing age toward the center of the basin may agree with the hypothesis of continuous flow as postulated by the piezometric map (Fig. 11.10). (From data by Evin and Vuillaume, 1970.)

were done and, except for the four samples mentioned, no tritium was detected in the samples assigned to the confined aquifer.

11.8.3 Watrak Shedi Basin, Western India

Borole et al. (1979) presented another example of a regional aquifer with ^{14}C ages increasing down-gradient. Figure 11.13 reveals a water table map of the 500 km² subbasin of Watrak Shedi in western India. The upper half of the basin was hypothesized to act as an intake area, feeding a confined aquifer existing in the downflow lower half of the basin. Carbon-14 results are given in Fig. 11.14 in terms of "apparent radiocarbon age" (i.e., an age calculated assuming an initial value of 100 pmc). Thus, waters with an initial ^{14}C value of 80 pmc will have an apparent age of 2000 years (Fig. 11.14). The upper, northeast part is dominated by apparent ages of about 2000 years, representing the phreatic part of the aquifer. In the lower, southwest part ages become progressively older, interpreted as reflecting the confinement of the aquifer. What is the down-gradient flow velocity in the confined part of the aquifer shown in

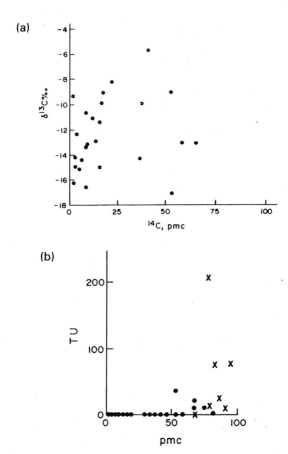

Fig. 11.12 ^{13}C and tritium as a function of ^{14}C in the Albian aquifer of the Paris basin (Table 11.6). ×: free water table; dots: confined. Initial ^{14}C value is 58–80 pmc (see text).

Fig. 11.14? It seems that the confinement begins with the lower edge of the 2000-year apparent age zone. The distance between the lowest 2000-year contour and the 6000-year contour is about 24 km. Hence, the average velocity is

$$\frac{24 \times 1000}{6000 - 2000} \times 6 \text{ m/year.}$$

The apparent age contours in Fig. 11.14 were based on data from shallow (30–80 m) wells. A few values for deeper wells are given as well (circles). The

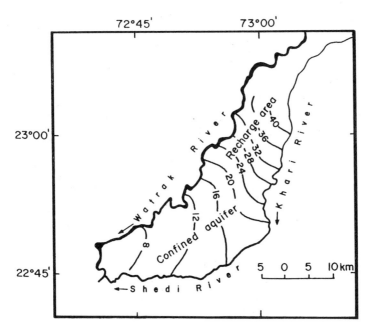

Fig. 11.13 Water-level contours of the Watrak Shedi subbasin, western India (April 1978). A gradient of 0.5–0.6 m/km is deduced. Down-gradient flow in this 500 km² basin was hypothesized (Borole et al., 1979).

age values in these deeper wells are significantly higher than in the neighboring shallow wells. This was interpreted by the researchers as indicating little or no downward mixing of the subaquifers. A good agreement is, thus, observed between the hydrological model, based on the water table data (Fig. 11.13), and the observed ^{14}C data (Fig. 11.14). The down-gradient flow in the confined aquifer was further tested by electrical conductivity data (Fig. 11.15). The values indicate fresh water (1000 μmho/cm) in the phreatic aquifer, and gradual salinization in the confined section.

11.9 Flow Discontinuities Between Adjacent Phreatic and Confined Aquifers Indicated by ^{14}C and Other Parameters

Phreatic aquifers are often regarded as recharge zones feeding adjacent confined systems. A continuous through-flow is commonly envisaged, controlled by (and deduced from) water level gradients and transmissivities. However, in certain

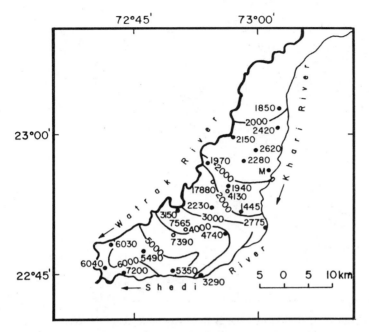

Fig. 11.14 Apparent radiocarbon ages (see text) in shallow (dots) and deeper (circles) wells and age contours for the Watrak Shedi subbasin (Borole et al., 1979). The 2000-year values correspond to the recharge area (equivalent to an initial ^{14}C of 80 pmc). The higher apparent ages indicate confinement of the aquifer in the lower half of the subbasin. A flow velocity of 6 m/year is calculated in the confined aquifer (see text).

cases a discontinuity is observed between the phreatic and confined parts of a system, reflected in abrupt changes in the chemical and isotopic compositions and, especially, in tritium, ^{14}C, ^{4}He, and other age indicators. Six case studies that reveal phreatic-confined discontinuities are briefly discussed below. They were described in detail in Mazor and Kroitoru (1987).

11.9.1 The Judean Mountains, Central Israel

This is an anticlinal structure of marine carbonates of the Judean group (Cenomanian-Turonian), covered on the flanks by chalk and marl of the Mt. Scopus group (Senonian) (Figs. 11.16 and 11.17). Average annual rainfall on the mountain crest is 550 mm/year, on the Hashphela foothills to the west the value is 450 mm/year, and in the Judean Desert, on the eastern slope, the value drops to 200 mm/year. Limestone, dolomite, and marl beds of the Judean group are exposed on the mountain crest, acting as the main recharge area and host-

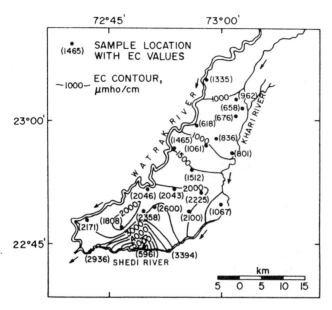

Fig. 11.15 Contours of electrical conductivity of groundwaters of the Watrack Shedi subbasin (Borole et al., 1979). The phreatic section reveals freshwater values around 1000 μmho/cm, whereas gradual salinization is observed down-gradient in the confined section.

ing a phreatic groundwater system penetrated by many wells. The Judean group rocks are covered on the anticlinal flanks by chalk and marl of the Mt. Scopus group. The latter formation hosts limited phreatic water systems of its own, but confines the Judean group water system. Study (Mazor and Kroitoru, 1987) revealed post-bomb tritium and ^{14}C values in the phreatic Judean group system, and no tritium and low ^{14}C in the confined Judean group system. The transitions are abrupt, both on the western and eastern flanks, as seen in Fig. 11.17.

Constancy in the δD and $\delta^{18}O$ data (Fig. 11.17) indicates that the phreatic and confined waters both originate from the same mountain crest recharge zone. Thus, the phreatic and confined zones seem to communicate hydraulically, but the drainage rate of the confined section seems to be restricted, as indicated by the high ^{14}C ages.

The validity of ^{14}C as an age indicator in the described study area has been established by the observed constancy of $\delta^{13}C$ values ($-12 \pm 1‰$; Fig. 11.17) and of HCO_3 and other components. The confined water has calculated ages of several thousand years (Kroitoru et al., 1987).

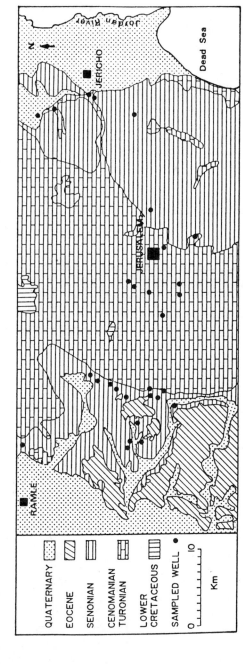

Fig. 11.16 Study area of the Judean Mountains, central Israel (Mazor and Kroitoru, 1987). Limestone and dolomite (Cenomanian-Turonian) outcrops serve as recharge areas into a phreatic aquifer, confined on the flanks by younger chalk (Senonian).

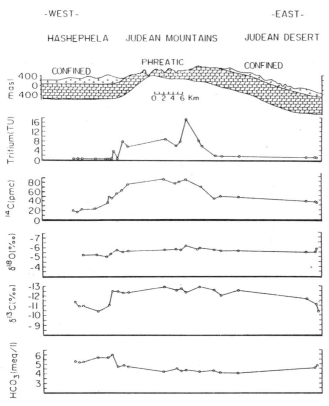

Fig. 11.17 East-west transects through the Judean Mountains of Judean group rocks, covered by Mt. Scopus group rocks. Tritium and ^{14}C values in well samples reveal recent (post-bomb) waters in the phreatic recharge zone and water several thousand years old in the confined zones. The transition is abrupt, indicating that the confined zones are poorly drained. Constancy in $\delta^{18}O$ values (–5.5‰) indicates common recharge at the mountain crest rainy area. Constancy of the $\delta^{13}C$ values (–12 ± 1‰) indicates absence of secondary isotopic exchange reactions, making ^{14}C a reliable tool for dating in this system (Mazor and Kroitoru, 1987).

11.9.2 Bunter Sandstone Aquifer, Eastern England

The Bunter sandstone aquifer has been studied over an area of 2000 km² in eastern England (Fig. 11.18). A large body of data has been published by Andrews and Lee (1979) and Bath et al. (1979). The regional geology, shown in Fig. 11.19, reveals a sequence of formations from the Permian in the west to the Jurassic in the east, the strata dipping eastward. A more detailed geology of the study area, with sampled well locations, is given in Fig. 11.20, and

Radiocarbon and ^{13}C

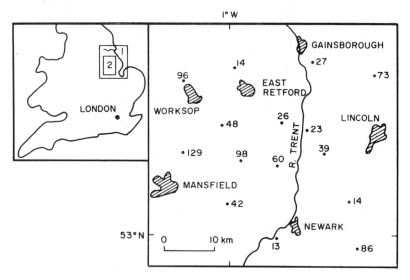

Fig. 11.18 Study area of the Bunter sandstone in eastern England: 1—area shown in Fig. 11.19; 2—area shown in Fig. 11.20.

a cross section is given in Fig. 11.21. The researchers suggested that water is recharged at the Bunter sandstone outcrops and moves eastward, down-gradient, into a confined section.

The reported data have been plotted in composition diagrams in order to determine whether it is really one continuous aquifer. The Cl-$\delta^{18}O$ plot (Fig. 11.22) reveals three data clusters, indicating that three specific water groups—A, B, and C—occur. The Cl-^{14}C plot (Fig. 11.23) also reveals three groups,

	LEGEND	
8	Kimmeridge Clay	
7	Corallian	
6	Oxford Clay	Jurassic
5	Oolite Series	
4	Lower Lias	
3	Keuper Marl	Triassic
2	Bunter Sandstone	
1	Magnesium Limestone	Permian

Fig. 11.19 Generalized geological map of a section of east England.

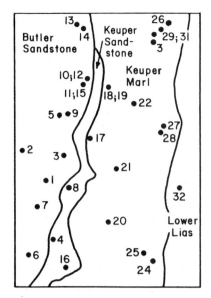

Fig. 11.20 Geology and location of sampled wells in the study area of Bunter sandstone aquifer. (Following Bath et al., 1979.)

but the data of group A are spread along a line, interpreted as a mixing line of a recent saline component (possibly contaminated with anthropogenic chlorine) and a several thousand year old, chlorine-poor, end member. The three geochemical groups have a simple geographical distribution pattern of north-south strips, as seen in Fig. 11.24. Group A coincides with the belt of phreatic Bunter sandstone outcrops, whereas groups B and C are in the confined, eastward-dipping section. The ^{14}C, Cl, and He data maps of Fig. 11.24 reveal two discontinuity lines—I and II. These lines are parallel with the strike of the rock

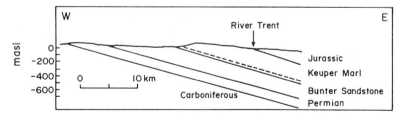

Fig. 11.21 Suggested geological cross section of the Bunter sandstone aquifer (Bath et al., 1979). The researchers suggested that water recharged in the western outcrops flows eastward down-gradient into the confined section.

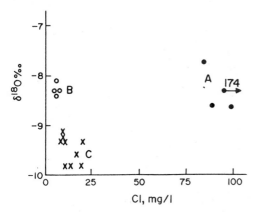

Fig. 11.22 Cl-δ^{18}O plot of the Bunter sandstone wells. Three distinct geochemical groups emerge: A, B, and C. (Data from Bath et al., 1979.)

strata and formation contacts (Figs. 11.20 and 11.21). Thus, there is little or no active through-flow from the phreatic zone of group A to the confined zone of group B. Furthermore, a discontinuity also occurs in the confined zone, and the group B section does not communicate freely with the group C section.

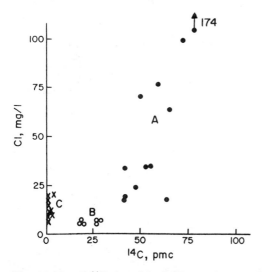

Fig. 11.23 Cl-^{14}C plot of the Bunter sandstone wells. Again, three groups emerge. The group A points plot along a line, interpreted as a mixing line of recent (post-bomb, also tritium containing) saline (polluted?) water with several thousand year old fresh water. (Data from Bath et al., 1979.)

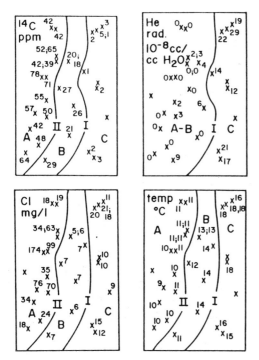

Fig. 11.24 ^{14}C, He, Cl, and temperature data marked on the well location map (Fig. 11.20). The three distinct geochemical groups—A, B, and C—already defined in Figs. 11.22 and 11.23, have a clear geographical pattern: group A coincides with the phreatic wells in the Bunter sandstone outcrop area, whereas groups B and C are in the confined section. The data mark two discontinuity lines, I and II, discussed in the text. (Data from Bath et al., 1979.)

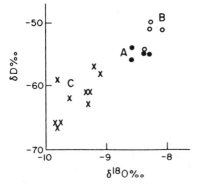

Fig. 11.25 δD-$\delta^{18}O$ plot of the Bunter sandstone wells. The obtained pattern is discussed in the text. (Data from Bath et al., 1979.)

Radiocarbon and ^{13}C

The Bunter sandstone δD and δ^{18}O data provide further support for the lack of communication between the phreatic and confined aquifer sections. The data plot in Fig. 11.25 along a line, but the data reveal an order of B→A→C. In other words, one is dealing with three distinct water groups, placed on the meteoric line.

The three water groups of the Bunter sandstone have distinct ages, as revealed by two independent age indicators, ^{14}C and ^{4}He (Fig. 11.26). The positive ^{4}He-^{14}C ages correlation gives the age grouping a high degree of confidence. From Fig. 11.26 it seems that water group C has for a long time been separated from group B, but the discontinuity between A and B is also maintained in Fig. 11.26.

Bearing in mind the concept of stagnant groundwater prevailing beneath the level of the terminal base of drainage (section 2.13), let us have a second look at the geological cross section of the Bunter sandstone aquifer (Fig. 11.21). The rock strata dip eastward in the direction of the sea, attaining depths of hundreds of meters below sea level. Thus, stagnation is to be expected, and is supported by the discussed chemical and isotopic data. A glance in Fig. 11.18 reveals that the topographic relief of the suggested recharge area is very shallow, only a few tens of meters above sea level. Thus, no extra-high hydraulic pressures can be developed, that could push groundwater through theoretical U-tubes in the Bunter sandstone. It seems plausible that the wells of zone A (Figs. 11.22–11.26) tap a through-flow system (phreatic aquifer), whereas the wells of zones B and C tap stagnant systems (fully confined aquifers).

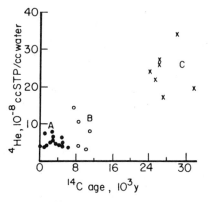

Fig. 11.26 ^{4}He concentration-^{14}C ages of waters in the Bunter sandstone wells. ^{14}C ages were calculated from the data of Bath et al. (1979), applying 70 pmc as an initial ^{14}C concentration. The positive correlation validates the relative dating by the two independent methods. Waters of groups A, B, and C have been recharged at different times but do not represent continuous down-gradient flow (see text).

11.9.3 Lincolnshire Limestone Jurassic Aquifer, Eastern England

The Lincolnshire limestone Jurassic aquifer in eastern England provides another insight into a phreatic-confined system. Isotopic data were reported (Dowing et al., 1977) from a number of wells in this system, located in the intake area and in part of the confined westward-inclined section. A distinct discontinuity (dashed line) is observed in ^{14}C (Fig. 11.27), supported by four other parameters: tritium, δ^{13}C, δD, and δ^{18}O. Thus, little or no hydraulic communication takes place between the phreatic and confined sections of the described Lincolnshire limestone aquifer.

11.9.4 Gwandu Formation, Sokoto Basin, Northern Nigeria

This system has been studied by Geyh and Wirth (1980). Figure 11.28 depicts the outcrop areas and the suggested groundwater flow direction. The ^{14}C data reveal a sharp discontinuity of waters, with 32–70 pmc near the outcrops (A) and 2–7 pmc in the downflow section of the confined aquifer (B)—another example of lack of hydraulic communication between adjacent phreatic and confined systems.

11.9.5 The Florida Limestone Aquifer

This system provides another example of a phreatic-confined interrelationship. Data are given in Table 11.7. A clear phreatic-confined discontinuity is established by ^{14}C, δ^{13}C, and tritium.

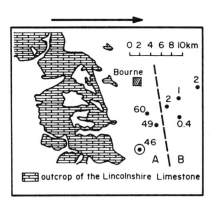

Fig. 11.27 Set of ^{14}C data (pmc) marked on well locations in the Lincolnshire limestone aquifer. The arrow depicts the orginally suggested direction of groundwater flow. Dashed line marks discontinuity between phreatic (A) and confined (B) aquifers. (Data from Dowing et al., 1977.)

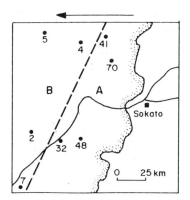

Fig. 11.28 ^{14}C values (pmc) in wells of the Sokoto basin, northern Nigeria. The arrow denotes the orginally suggested groundwater flow direction. The phreatic aquifer (A) coincides with the rock outcrops (stippled) and has high ^{14}C values. An abrupt drop in ^{14}C values is observed in the adjacent confined section of the system (B). (Data from Geyh and Wirth, 1980.)

11.9.6 The Blumau Aquifer, Southeastern Austria

This complex has been intensively studied (Andrews et al., 1984). Drastic discontinuities were reported for groundwater properties passing from the phreatic part of the aquifer to the confined part. The transition is observed between two wells, about 400 m apart. For example, ^{14}C drops from 114 pmc to 48 pmc, tritium drops from 49 TU to 0.8 TU, δ^{18}O drops from -8.6‰ to -9.3‰, and bicarbonate rises from 100 mg/l to 320 mg/l.

11.9.7 Conclusions Drawn from the Case Studies

The examples of phreatic-confined discontinuities shed light on a number of basic hydrological features:

Table 11.7 Isotope data for wells in the Florida limestone aquifer (Pearson and Hanshaw, 1970)

Well	^{14}C (pmc)	δ^{13}C (‰)	Tritium (TU)	Remarks
Weeki Vachee	62.4	-13.7	103 ± 10	Recharge zone
Lecanto #5	50.0	-11.5	36 ± 4	Recharge zone
Lecanto #6	51.1	-11.8	15 ± 1	Recharge zone
Frost proof	6.7	-9.8	0 ± 1	Confined zone
Holopane	5.2	-9.6	1 ± 1	Confined zone
Arcadia	3.0	-8.3	0 ± 1	Confined zone

Flow velocities of water in an aquifer, calculated by gradients and transmissivities, provide the maximum possible values. These are subject to limitations imposed by stagnation conditions. In extreme cases, confined systems may be rich in fossil karstic conduits, but with no through-flow, due to complete confinement and/or burial beneath the level of the terminal base of drainage.

Restricted drainage may, in certain cases, explain discrepancies between hydrologically calculated ages, using gradients and transmissivities, and much higher ages obtained from isotopic age indicators such as ^{14}C, ^{4}He, or ^{36}Cl. In fact, such discrepancies might be used as indicators of poor drainage of partially confined systems.

Phreatic systems that do not drain into adjacent confined sections must have their own outputs to maintain the observed young groundwater ages. These outputs have to be identified to gain a full understanding of the regional hydrology.

Of special interest is the degree of abruptness of the discontinuities. These may be enhanced by clogging of rock pores in the transition zone. In other cases, a geological barrier may be present, coinciding with a suggested phreatic-confined boundary.

11.10 Mixing of Groundwaters Revealed by Joint Interpretation of Tritium and ^{14}C Data

Mixing of groundwater was discussed in sections 6.6 and 6.7. There, mixing patterns were recognized by processing data from numerous samples (Figs. 6.6, 6.9, 6.17, and 6.18). Mixing may also be recognized from data obtained from a single sample—in cases of "forbidden" combinations of tritium and ^{14}C data.

Example: 1.2 TU and 48.3 pmc were found in well 1572, Hanahai, Botswana. What is the age of this water? The values represent a "forbidden" combination: 1.2 TU indicates post-bomb (an age of a few decades), whereas 48.3 pmc indicates water a few thousand years old. In other words, the tritium and ^{14}C ages are discordant (disagreeing), and neither is correct. The combination could be formed only by intermixing of old water (zero tritium and <48.3 pmc) with recent water (post-bomb tritium and post-bomb ^{14}C). This example is of great importance. Concordant ages (agreeing ages) confirm each other and discordant ages indicate mixing of water of different ages.

In the identification of intermixing water types and the calculation of mixing percentages, ^{14}C is useful because it is often the only parameter that occurs in a linear negative correlation with other parameters needed to extrapolate end member properties (section 6.7). Deep groundwaters commonly have high parametric values: high dissolved ion concentrations and elevated temperatures (as compared to the intermixing shallow and cold end member). However, ^{14}C is commonly low in the deep end member (which is old) and is high in the shallow end member (which is young), thus providing the needed negative correlations. Tritium can be used in the same way (section 10.6), but its use is limited to mixtures of old water with post-bomb water. If the young end member is pre-1952, it contains no measurable tritium, but high concentrations of ^{14}C, making the latter useful in studying mixed waters.

An example of the use of ^{14}C in studying a mixed groundwater system is provided by the Hammat Gader spring complex in the Jordan Rift Valley, Southeast of the Tiberias Sea (Table 11.8). Four springs emerge at Hammat Gader along a stretch of a few hundred meters. Various parameters have been plotted on compositional diagrams as a function of Cl concentration (Fig. 11.29). Linear positive correlations are exhibited by Li, K, Na, Mg, Ca, Sr, SO_4, and Br. These positive linear correlations indicate mixing of a fresh water with a more saline water. Positive correlations with chlorine were observed for three additional parameters: temperature, dissolved helium, and radium (Fig. 11.30). A negative correlation was provided in the Hammat Gader study by ^{14}C alone (Fig. 11.31). The vertical axis in Fig. 11.31 deserves some discussion: the ^{14}C data (reported in pmc) provide the value of the $^{14}C:^{12}C$ ratio in the sample, compared to a standard (section 11.2). In the hydrochemical jargon the pmc values are often called ^{14}C concentrations. However, to obtain values that are really concentration-related, the pmc value has to be multiplied by the corresponding HCO_3 concentration, as shown in Fig. 11.31. The best-fit line is seen in Fig. 11.31 to extrapolate to 69°C, providing the maximum possible temperature of the warm end member. The real temperature of the warm end member lies in the range of the highest temperature measured (52°C) and the extrapolated value (69°C). The extrapolated temperature of the cold end mem-

Table 11.8 Dissolved ions (mg/l) in springs A–D, Hammat Gader, and extrapolated fresh (fem) and saline (sem) end members (Mazor et al., 1973)

Spring	Li	Sr	K	Na	Mg	Ca	Br	Cl	SO_4	HCO_3
sem	0.23	17	33	375	48	196	8.8	800	262	272
A	0.143	10.4	21.8	245	39.8	152	5.4	506	177	330
B	0.079	6.0	13.1	140	44.5	105	3.7	300	121	344
C	0.057	4.1	10.5	120	37.5	124	—	218	100	361
D	0.010	1.3	3.9	55	32.7	89	2	87	58	381
fem	0.0	0.0	0.9	18	34	78	0.0	10	37	385

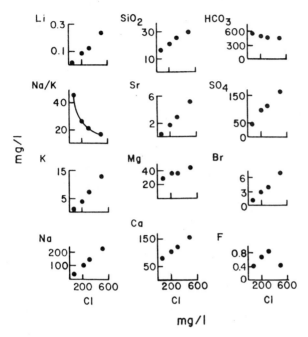

Fig. 11.29 Composition diagrams of four springs issuing a few hundred meters apart at Hammat Gader, Jordan Rift Valley, Israel (Mazor et al., 1980). The mixing of two water types is evident.

ber was obtained from Fig. 11.30, assuming a chlorine concentration as low as that of local rainwater, 10 mg/l. An extrapolated temperature of 25°C was obtained for the cold end member. The real value lies between this extrapolated value of 25°C and the lowest measured value, 29°C. The extrapolated chlorine values could then be applied to the positive correlation lines of Fig. 11.29 to calculate the concentration of the dissolved ions in the two end members.

11.11 Piston Flow Versus Karstic Flow Revealed by ^{14}C Data

In piston flow, groundwater is "layered" in the saturated zone of the intake area, younger water layers over older water layers. Such systems provide downward decreasing ^{14}C values in depth profiles. In contrast, in a karstic system of water descending in conduits crossing porous rocks, occasionally "disor-

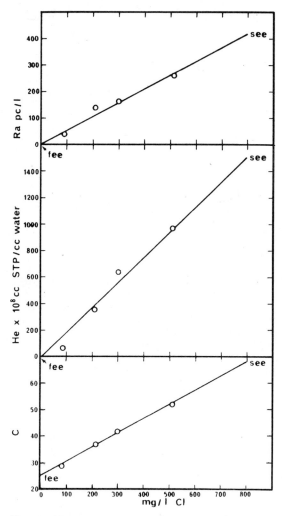

Fig. 11.30 Temperature, helium, and radium as a function of chlorine at Hammat Gader, springs A–D, and extrapolated end members (see text) (Mazor et al., 1973).

dered" depth profiles occur, with water of relatively high ^{14}C values encountered beneath water with lower ^{14}C values.

Case studies near Nyirseg, in eastern Hungary, provide examples for the discussed types of depth profiles (Table 11.9). The first two wells, of Fehergyarmat and Nyirlugos, reveal "normal" depth profiles, with ^{14}C decreasing

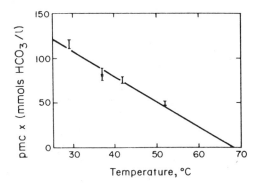

Fig. 11.31 ^{14}C versus temperature in the spring complex of Hammat Gader. The negative correlation could be used to extrapolate to 0 pmc, obtaining a value of 69°C. The real hot end member lies between the highest value measured (52°C) and the extrapolated value (69°C) (Mazor et al., 1973).

Table 11.9 Carbon-14 depth profiles in wells near Nyirseg, eastern Hungary (from Deák, 1979)

Location	Perforation (m under surface)	$\delta^{13}C$ (‰) PDB	Tritium (TU)	$^{14}C \pm \sigma$ (pmc)
Fehérgyarmat	÷ 10	− 15.3		78.7 ± 6.3
	41 − 56	− 16.1	<6	12.8 ± 1.2
	81 − 96	− 17.4		12.1 ± 0.9
Nyirlugos	<10	− 12.9	150	57.9 ± 1.0
	31 − 34	− 13.8	<6	27.6 ± 1.1
	110 − 175	− 15.1		24.9 ± 1.1
Nyirbátor	<10	− 17.2	119	91.2 ± 4.2
	46 − 73	− 3.1	<6	54.7 ± 3.1
	161 − 193	− 13.7		14.0 ± 0.8
	229 − 268	− 13.2		16.3 ± 0.6
	297 − 313	− 13.0		22.6 ± 0.5
Nyiradony	<10	− 13.5	240	72.3 ± 3.1
	55 − 61	− 15.3	<6	23.6 ± 2.3
	127 − 133	− 13.1		16.4 ± 0.9
	172 − 219	− 13.2		1.8 ± 1.8
Debrecen	70 − 93	− 14.5		3.4 ± 1.3
	138 − 181	− 15.6		8.0 ± 0.5
	160 − 190	− 12.7		12.6 ± 1.2
Derecske	154 − 159	− 14.9		4.5 ± 0.9
	233 − 253	− 13.5		7.9 ± 1.4
Berettyóujfalu	<10	− 15.2	189·	85.1 ± 2.4
	131 − 137	− 22.6		1.2 ± 0.5
	296 − 355	− 20.7		15.6 ± 2.2

downward, indicating water layering, which in turn indicates a piston flow of recharged water. The remaining five wells reveal disordered depth profiles, for example, Berettyoujfalu, which has in a sequence of depth profile samples the following ^{14}C values: 85.1, 1.2, and 15.6 pmc. These five wells encountered karstic conduits or zones of preferred conductivity with waters of varying ^{14}C values.

For a review of other methods that have been developed to date groundwaters, the reader is referred to an article by Davis and Bentley (1982).

11.12 Sample Collection for ^{14}C and ^{13}C Analyses and Contact with Relevant Laboratories

Carbon-14 is measured in specialized radioactive counting laboratories and in accelerator-mass spectrometer laboratories. In the first case, relatively large water samples of about 100 l have to be treated, and in the latter case samples of about 1 l are sufficient. There is a difference in the price, and each researcher has to determine the track he will follow. The larger samples can be reduced by precipitation of the dissolved carbon-containing compounds (mainly HCO_3) in the field, a process that needs some expertise. In any case, care has to be taken to avoid exchange with atmospheric CO_2. Carbon-13 is measured mass spectrometrically, either on the sample collected for the radiocarbon or on separately collected samples. You should get your sampling instructions from the laboratory that will conduct the measurements.

11.13 Summary Exercises

Exercise 11.1: What is the age of water with 25 TU and 102 pmc?

Exercise 11.2: What can be said of water in which 0 TU and 70 pmc have been detected?

Exercise 11.3: When was water with 0 TU and 0 pmc recharged?

Exercise 11.4: Is there something special in groundwater that contains 6 TU and 6 pmc?

Exercise 11.5: Based on the data provided in Table 11.6 for groundwaters in the Paris basin, give your expert opinion on the hydrology of well 1, Parly-Chenons.

Exercise 11.6: Well 6, Migennes, is reported in Table 11.6 as being confined. Check this conclusion in light of the given data.

Exercise 11.7: Assess the hydrology of well 19, Montier-en-Der, in light of the data given in Table 11.6.

Exercise 11.8: Assess the age of the water in well 46, Issy, applying the data given in Table 11.6.

Exercise 11.9: Compare Fig. 11.3 with Fig. 11.12. What is the common thread?

12
CHLORINE-36

12.1 The Radioactive Isotope of Chlorine and Its Production

Chlorine has three isotopes in nature: ^{35}Cl—common and stable; ^{36}Cl—very rare and radioactive, with a 301,000-year half-life; and ^{36}Cl—less common and stable. Chlorine-36 concentrations are commonly expressed in units of 10^7 atoms/l of water. The measurement is performed in specialized laboratories with dedicated accelerators. Measurements can be done on 1-l water samples.

A few background remarks on the formation of ^{36}Cl are needed, the reader being referred to a historical discussion by Bentley et al. (1986). Chlorine-36 is constantly formed by cosmic rays interacting with the atmosphere, and it is incorporated, along with stable atmospheric chlorine (mainly sea-derived and airborne) into the hydrological cycle. In addition to the atmospheric source, ^{36}Cl is also formed by in situ reactions in rocks, its production rate being proportional to the uranium, thorium, and chlorine concentrations prevailing in the rocks. Chlorine-36 has also been introduced into the atmosphere by nuclear bomb testing. The major portion of ^{36}Cl is atmospheric, the in situ production amounts to a few percent of the atmospheric contribution, and nuclear bomb testing added a few tenths of percents.

12.2 A Potential Tool for Groundwater Dating

Every radioactive isotope detected in groundwater serves as a potential tool for dating. The half-life of ^{36}Cl—3×10^5 years—is especially promising because it fills in a most important time span; that is, beyond the limits of groundwater dating by radio carbon.

Fig. 12.1 depicts a ^{36}Cl decay curve by plotting P_{36} versus time. P_{36} is the percent of atmospheric ^{36}Cl left in the sample, which is the parameter measured in groundwater investigations. The time is indicated in half-lives and translated into years before present by multiplying by the ^{36}Cl half-life. In a straightforward mode, one may read from the decay curve the groundwater age that corresponds to a given P_{36}. But there is a long list of complications that have to be overcome, as discussed in the following sections. The literature on ^{36}Cl groundwater dating evolved quickly in the last decade, and a variety of data presentation and interpretation approaches have been applied. In this chapter a simple approach is presented (Mazor, 1992) in order to provide a clear overall picture, equipping the field hydrologist with a most important semiquantitative dating tool.

12.3 Processes Controlling the ^{36}Cl Concentration in Groundwater

12.3.1 Geographical Variations in the Production of Atmospheric ^{36}Cl

The flux of cosmic rays varies with sun spot activity and other parameters, and the flux reaching the earth's surface varies with latitude and altitude. Furthermore, reactions with ^{35}Cl produce some ^{36}Cl, which causes a dependence in the distance from the ocean. Altogether, the production of ^{36}Cl varies spatially by

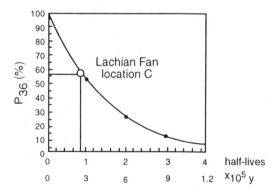

Fig. 12.1 A radioactive decay curve of ^{36}Cl. P_{36} is the percentage of atmospheric ^{36}Cl left in the sample. The time axis is expressed in half-lives and in 10^5-year units. The marked point of 56% P_{36}, taken from the Lachlan case study, reveals water with an age of about 290,000 years. The mode of defining the P_{36} is explained in later sections.

a factor of up to five. Hence, the ^{36}Cl production rate is site specific and has to be determined for each study site, as outlined later.

12.3.2 Evapotranspiration Affecting the Cl and ^{36}Cl Concentration in Recharged Groundwater

Evaporation and plant transpiration operate on water prior to its arrival in recharged groundwater systems. Evapotranspiration returns part of the precipitated water into the atmosphere, while the dissolved salts, including chlorine, stay behind and become enriched in the water fraction that reaches the groundwater systems. Fig. 12.2 shows chlorine and bromine concentrations observed in various wells in a stretch of several hundred square kilometers in the flat Murray-Malee region of Australia. Several conclusions can be drawn from the observed linear positive correlation, and from the fact that the marine value (Cl:Br = 293) fits the correlation line: (1) both chlorine and bromine are atmospheric (i.e., sea derived and airborne); (2) the formation of sea spray and its atmospheric transportation do not involve elemental fractionation; (3) chlorine and bromine reveal a conservative hydrochemical behavior, as is well established worldwide (Mazor and George, 1992; Mazor et al., 1992); (4) the concentration of these ions is controlled by a process that acts on both in the same mode, this process being evapotranspiration; (5) the range of several orders of magnitude of observed ion concentration reveals frequent differences in the surface conditions, resulting in different concentrations of chlorine in groundwaters. Similar chlorine-bromine correlations have been observed in

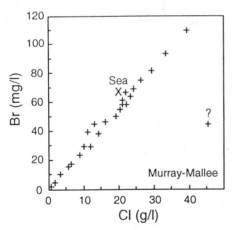

Fig. 12.2 Bromide as a function of chlorine in unconfined wells of the Murray-Malee region, Australia (following Mazor, 1992). The observed positive linear correlation, and the fact that the sea value (x) falls on the same line, lead to several conclusions that are discussed in the text in the context of groundwater dating by ^{36}Cl.

groundwaters all over the world, indicating the widespread nature of the discussed processes.

Evapotranspiration effects ^{36}Cl and stable Cl in the same way. Thus, the initial concentration of ^{36}Cl in groundwater must vary locally, in the mode observed for Cl in Fig. 12.2. This leads, in turn, to the conclusion that the initial ^{36}Cl concentration in newly recharged groundwater is site specific and has to be determined for each study site, as is outlined later.

12.3.3 Halite Dissolution

To overcome the large effect of evapotranspiration on the initial ^{36}Cl concentration in recharged water, several researchers applied the ^{36}Cl:Cl ratio in their dating process. The production rates by the different processes, and the measurement in the sample, were expressed in ^{36}Cl:Cl ratios (Bentley, et al., 1986). This procedure ran into difficulties because of possible dissolution of halite. Halite (NaCl) occurs occasionally in sequences of old rocks and contains no atmospheric ^{36}Cl. Thus, dissolution of old halite produces water with lower ^{36}Cl:Cl ratios, leading to the computation of apparent, too old, water ages. This obstacle can be overcome by plotting the chlorine data versus bromine (e.g., Fig. 12.2)—the Cl/Br ratio in seawater and airborne sea spray is 290, whereas in halite it is over 3000 (section 6.14), hence samples that plot on a line that contains the marine Cl/Br value are freed from the possible involvement of old halite dissolution, and their data can be applied to age calculations.

12.3.4 In Situ Formation of ^{36}Cl

Two opposing processes control the concentration of ^{36}Cl as a function of the age of the water: the initial atmospheric ^{36}Cl decays, whereas in situ produced ^{36}Cl accumulates. The in situ production of ^{36}Cl varies with rock composition and is thus another site-specific parameter. In old groundwaters, with ages of 10^6 years or more, the in situ produced ^{36}Cl reaches a steady state at which the rate of production equals the rate of radioactive decay. The steady-state in situ production of ^{36}Cl places an upper limit of about 10^6 years on the range of water dating via ^{36}Cl. In other words, in cases of low ^{36}Cl concentration one cannot establish whether any of the original atmospheric ^{36}Cl is left, or all the observed ^{36}Cl stems from in situ production. Whenever a small ^{36}Cl concentration is measured in a water sample, the age can still be calculated via the decay graph (Fig. 12.1), but this will be a minimum age and the real age may be significantly older.

12.3.5 Nuclear Bomb ^{36}Cl

Along with tritium and ^{14}C, ^{36}Cl has been released into the atmosphere by nuclear bomb testing, increasing the concentration in newly recharged water by up to a few tenths of percents. This anthropogenic ^{36}Cl does not interfere with groundwater dating, since the ^{36}Cl method can be applied only to waters that are over 10^4 years old, that is, pre-bomb. Water that is free from measurable tritium and/or contains ^{14}C in concentrations that are lower than the initial ^{14}C concentration—for example, e.g. >50 pmc (see Chapter 11)—do not contain bomb ^{36}Cl.

12.4 Groundwater Dating by ^{36}Cl

12.4.1 Establishing the Initial $^{36}Cl:Cl$ Ratio in Local Groundwater

As in the case of ^3H and ^{14}C, the application of ^{36}Cl for groundwater dating is based on the knowledge of the input values at age zero. In the cases of ^3H and ^{14}C, research efforts were directed to the establishment of the concentrations in precipitation. In the case of ^{36}Cl, the concentration varies appreciably at different altitudes and latitudes, and is significantly increased by evapotranspiration. Hence, the input values have to be established for the water as it entered the aquifer. This is best done by selecting several local springs or wells that comply with the following requirements:

Waters that are clean of bomb ^{36}Cl contributions. The best criterion for this requirement is that no ^3H or other man-made contamination is found in significant concentrations (<1 TU).

Groundwater samples that are recent in terms of the half-life of ^{36}Cl (i.e., ^{36}Cl decay being negligible). Samples that comply with this criterion are best selected with the aid of ^{14}C measurements. They should be lower than pre-bomb concentrations in groundwater (>60 pmc), but still significantly above the limit of the ^{14}C dating method (>20 pmc).

There should be good representation of local recent groundwaters, for example, the full spectrum of salinity, chemical composition groups, and temperatures.

The selected recent water samples should be analyzed for the listed parameters, and the data should be plotted on composition diagrams as functions of the chlorine concentration. The chlorine-bromine diagram will serve to select water samples that have marine $^{36}Cl:Cl$ ratios (ensuring no ancient halite was dissolved). The line observed on the ^{36}Cl-Cl diagram serves as the local intake line (demonstrated later in Fig. 12.7 for the Lachlan Fan case study).

12.4.2 Groundwater Age Read from Isochrones on a ^{36}C-Cl Diagram

Figure 12.3 is a schematic version of a ^{36}Cl-Cl diagram of data obtained in field studies. The local intake line is traced to the zero values through the data points

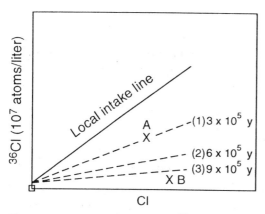

Fig. 12.3 Isochrones plotted on a ^{36}Cl-Cl composition diagram. Calculation of the age for samples A and B is discussed in the text.

of the selected local recent waters. The data of the recent waters may be spread along an oblique line in the case of regions where evapotranspiration varies from one surface cell to the other, or the data may cluster, revealing uniform and possibly low evapotranspiration effects. At the second stage, isochrone lines can be plotted by half-lives. The age of a water sample can be read directly. For example, sample A in Fig. 12.3 has an (semiquantitative) age of 3×10^5 years, and sample B, which lies in the range of the in situ ^{36}Cl production, has a minimum age of 10^6 years.

12.4.3 Calculation of P_{36} and Its Application to Estimate Groundwater Age

The percentage of ^{36}Cl left in a sample (P_{36}) equals

$$P_{36}(\%) = (^{36}Cl_{measured} / \,^{36}Cl_{input}) \times 100$$

A local intake line (from the case study of Lachlan, Australia) is drawn in Fig. 12.4, along with point X, representing a sample that was found to contain 530×10^7 atoms ^{36}Cl/l, accompanied by 3180 mg Cl/l. The corresponding point Y on the local intake line reveals the initial ^{36}Cl input value of 950×10^7 atoms/l. Hence, $P_{36} = (530/950) \times 100 = 55.8\%$. Applying this value to the decay curve of Fig. 12.1, point O is obtained, and an age of 2.9×10^5 years is read on the lower scale of the horizontal axis.

Chlorine-36

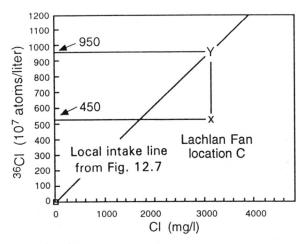

Fig. 12.4 Mode of calculating P_{36}, the percentage of ^{36}Cl left in the sample, demonstrated for a case discussed in the text (site C, the Lachlan Fan, Australia).

12.5 A Number of Case Studies

Following are a number of case studies examined in the light of the previous discussion. The reader is encouraged to read the original articles to become familiar with the mode of data processing and interpretation presented by the various researchers.

12.5.1 The Lachlan Fan, Australia

Bird et al. (1989) and Calf et al. (1988) reported a set of ^{36}Cl and Cl data from the Lachlan Fan, southern Australia (Figs. 12.5 and 12.6). The distribution of the ^{36}Cl data is seen in Fig. 12.6 to be uneven; for example, a well with 12×10^7 atoms/l is located near a well with 114×10^7 atoms/l in location A, or a well with 530×10^7 atoms/l is seen near a well with 1175×10^7 atoms/l in location C. The chlorine values reveal a similar distribution. Observations of this nature indicate that the groundwater system in the study area is composed of a number of groundwater cells, or compartments, that have little lateral hydraulic interconnection—an expected phenomenon in flat terrains. The data from location A are plotted in Fig. 12.7 and the perfect correlation line of these fresh waters is regarded as representing the local intake line. The spread of the points along the intake line reflects degrees of water removal by evapotranspiration prior to recharge into the groundwater system. The complete

Fig. 12.5 Key map of case studies from Australia discussed in this chapter: 1—Lachlan Fan; 2—Murray-Malee Basin; 3—Great Artesian Basin.

set of data from the Lachlan Fan region is plotted in Fig. 12.8, along with the intake line, extrapolated from Fig. 12.7. The data of the more saline samples fall below the intake line, indicating a rather old age. The age calculation for one well in location C has shown these waters to be around 290,000 years (around 1 half-life; Figs. 12.4 and 12.1).

12.5.2 The Murray-Malee Region, South Australia

Calf et al. (1988) and Davie et al. (1989) presented data from the Malee region in the Murray Basin, southern Australia (Figs. 12.5 and 12.9). The data reveal again the existence of groundwater compartments with distinct compositions: (1) a ^{36}Cl value of 9×10^7 atoms/l at the southern edge of the study area is followed to the north by a value of 21×10^7 atoms/l; (2) a value of 28×10^7 atoms/l is seen near a value of 145×10^7 atoms/l at the northern end of Fig. 12.9.

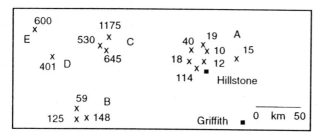

Fig. 12.6 Map of a study in the Lachlan Fan, providing sampling points and ^{36}C concentrations in units of 10^7 atoms/l. The letters designate sampling sites marked in following figures and are discussed in the text. [Figs. 12.6–12.10 are based on data published by Bird et al. (1989) and Calf et al. (1988), and processed in Mazor (1992)].

Chlorine-36

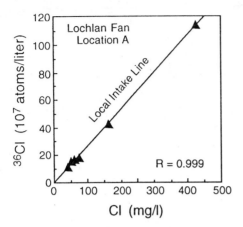

Fig. 12.7 The ^{36}Cl-Cl diagram for data from location A of the Lachlan case study. The waters at this site are seen to be fresh, as reflected by the low chlorine concentrations. The observed perfect correlation line is suggested to represent the local intake line.

A local intake line is drawn through the data points plotted on a ^{36}Cl-Cl diagram (Fig. 12.10), applying the Murray group unconfined water samples. One Renmark group sample and two of the discharge area samples of the Murray group also lie on the local intake line in Fig. 12.10. These wells thus

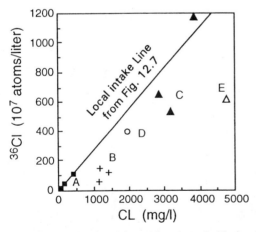

Fig. 12.8 Chlorine-36 as a function of chlorine for the Lachlan case study. The intake line is based on the freshwater samples from location A (Fig. 12.7). All the data points, except one, lie on or beneath the local intake line, indicating ages on the order of 1 half-life, that is, about 300,000 years (see text). The single point above the intake line may indicate contributions of bomb ^{36}Cl.

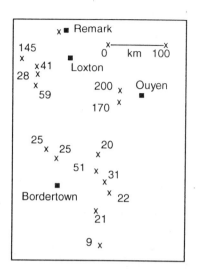

Fig. 12.9 Sampling points and ^{36}Cl concentrations (10^7 atoms/l) for the Murray-Malee study.

seem to belong to the presently active unconfined system. Other samples plot beneath the local intake line, indicating groundwaters with high ages.

12.5.3 The Great Artesian Basin

Bentley et al. (1986a) presented a pioneering study of ^{36}Cl in the Great Artesian Basin (GAB) in eastern Australia (Fig. 12.5). The ^{36}Cl data are plotted in

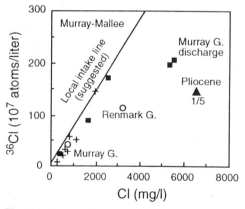

Fig. 12.10 Chlorine-36 as a function of chlorine for the Murray-Malee study. Symbols designate the reported formation from which the water was sampled: Murray group (+), Murray group at discharge areas (squares), Renmark group (circles), Pliocene aquifer (triangle). The local intake line was based on unconfined water samples. The data fall on this line or beneath it, indicating old ages.

Fig. 12.11, along with the Cl, Ca, and SO_4 data. Two distinct groups of values are seen in the ^{36}Cl map: a group of wells in the northeast corner of the study area, at the mountainous Great Dividing Range (GDR), contains (17–42) × 10^7 atoms/l, whereas the confined artesian wells of the GAB contain only (1–5) × 10^7 atoms/l. The first group of samples was collected from the active recharge area, whereas the second group of samples was collected from artesian wells drilled up to 1800 m into the confined Jurassic (J) aquifer. The dashed line DD marks the abrupt discontinuity inferred from the ^{36}Cl data (Mazor, 1992).

A similar pattern is seen in the distribution of the Ca and SO_4 concentrations (Fig. 12.11): the values of the confined J aquifer of the GAB are an order of magnitude lower than the values found in the unconfined aquifer of the GDR. The distinction is less clear for the chlorine distribution, the higher values in the GDR wells possibly reflecting anthropogenic chlorine pollution of the unconfined system. The abrupt decrease of ^{36}Cl, Ca, and SO_4, and in some of the locations also of Cl, indicates that the unconfined groundwater system of the GDR does not communicate with the buried J aquifer. The GDR system may, however, be hydraulically interconnected with the shallow groundwater system of the eastern part of the GAB.

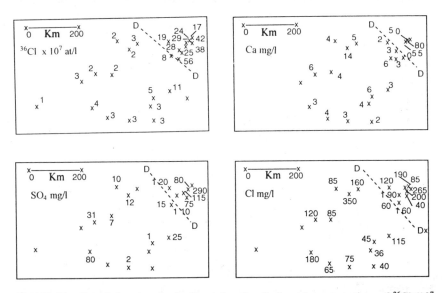

Fig. 12.11 Sample locations in the Great Artesian Basin and concentrations of ^{36}Cl (10^7 atoms/l; data from Bentley et al., 1986a), and Cl, Ca, and SO_4 (mg/l; data from Water Resources Commission of Queensland). Line DD marks a discontinuity between the ^{36}Cl, Ca, and SO_4 results from unconfined wells in the Great Dividing Range recharge area (located at the northeast corner) and the results from the buried confined J aquifer of the GAB.

The ^{36}Cl-Cl diagram of the GAB (Fig. 12.12) is a fine summary of ^{36}Cl dating. The data of the GDR unconfined groundwaters and those of the confined J aquifer of the GAB plot in two distinct locations. The GDR data spread along a line that indicates variability in evapotranspiration water losses prior to recharge at distinct surface cells. A suggested intake line has been drawn through the unconfined groundwaters and the origin of the axes. Its location could not be selected precisely because there were no tritium data available to identify wells with bomb contributions (one well at the top of the diagram clearly has a bomb component). The confined GAB groundwaters reveal the same large range of chlorine concentrations, but all are low in ^{36}Cl. This is interpreted as indicating that all the confined waters are very old, and the bulk of the observed ^{36}Cl is in situ produced. Thus, the confined waters are of an age that is at the limit of the ^{36}Cl dating method, that is, $\geq 10^6$ years.

This high age is of direct bearing to the question of the nature of the J aquifer—whether the artesian waters are part of a basin-wide through-flow system, as portrayed in Fig. 2.21, or whether they are composed of a multitude of hydraulically isolated compartments that are pressurized by compaction (section 2.14). The age of the confined waters, studied in the GAB, is at the limit of the dating method, which is 1 million years. This leaves open the possibility that the real age is much higher and other dating methods are needed to

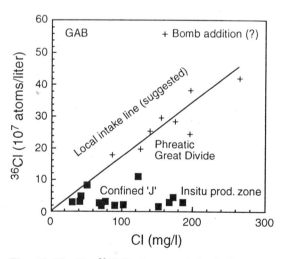

Fig. 12.12 The ^{36}Cl-Cl diagram of the GAB study. The local intake line and the in situ production zone are discussed in the text, along with the conclusion that the artesian groundwaters of the Jurassic rock system are $\geq 10^6$ years old (following Mazor, 1992).

Chlorine-36

Fig. 12.13 Location of the Milk River study area in Canada.

check this point. High groundwater ages support the isolated compartment model and rule out the through-flow model.

12.5.4 The Milk River Aquifer, Alberta, Canada

Bentley et al. (1986b) and Phillips et al. (1986) published ^{36}Cl and Cl data for the Milk River Aquifer in southern Alberta (Fig. 12.13). The ^{36}Cl data are plotted in Fig. 12.14. A group of wells in the southwest corner of Fig. 12.14 has ^{36}Cl values of $(23-50) \times 10^7$ atoms/l, two wells in the northeast corner have $(13-15) \times 10^7$ atoms/l, and the central part contains water with only $(2-8) \times 10^7$ atoms/l. Thus, different water groups seem to be tapped by the wells, and no lateral flow across the entire study area is indicated by the ^{36}Cl data.

The Milk River data are plotted in Fig. 12.15 on a ^{36}Cl-Cl diagram. A number of samples with low Cl concentrations reveal rather high ^{36}Cl concen-

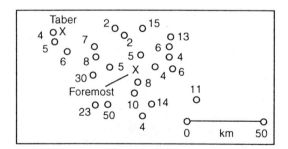

Fig. 12.14 Sample locations and ^{36}Cl concentrations (10^7 atoms/l) for the Milk River study (data from Phillips et al., 1986). Geographical groups of concentration values indicate that several distinct water types are involved (see text).

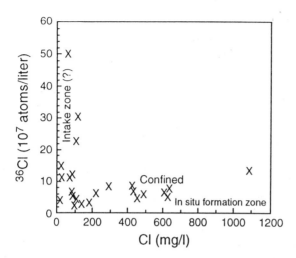

Fig. 12.15 The data of the Milk River study plotted on a ^{36}Cl-Cl diagram (following Mazor, 1992).

trations. They seem to belong to recently recharged unconfined waters, but they do not trace a local intake line, since the high ^{36}Cl values may reflect bomb contributions. The rest of the data occupy a band of significantly low ^{36}Cl concentrations (6 × 10^7 atoms/l). By analogy with the GAB study (Fig. 12.12), it seems that here too one deals with old confined groundwaters with an age on the order of ≥ 0.5 × 10^6 years. These groundwaters were exposed to varying degrees of evapotranspiration prior to their being recharged.

12.6 Summary

Chlorine-36 is an isotope that can be measured by specialized laboratories and its sample collection in the field is simple. It is a radioactive isotope with a half-life of 3 × 10^5 years, making it useful for groundwater age determinations in the range of 10^5–10^6 years.

No conclusions can be drawn on the basis of ^{36}Cl determinations alone—auxiliary data are needed, such as general chemistry, including Cl, Br, as well as tritium and ^{14}C.

Samples have to be demonstrated to represent a single water type, and not a last-minute mixture of two or more waters (sections 6.6 and 6.7). The local intake line, that is, the local ^{36}Cl:Cl ratio, has to be determined on local samples that are relatively young, but pre-bomb.

The age determination has to be regarded as semiquantitative, but its value cannot be overestimated in striving to understand groundwater systems scientifically and in relation the pollution aspects. Groundwater older than 10^5–10^6 years is most probably stagnant, and in any case, it is not renewed, but it is isolated from the surface, and thus immune to anthropogenic pollution.

12.7 Summary Exercises

Exercise 12.1: What is the age of a groundwater sample that contains 20% of the initial atmospheric ^{36}Cl?

Exercise 12.2: In the center of Fig. 12.10 is a sample marked "Renmark G." What is its ^{36}Cl concentration? What was its initial ^{36}Cl concentration? What is the respective P_{36}? What is the age?

Exercise 12.3: A water sample is found to have a P_{36} of 25%. How many half-lives have passed since its recharge? What is the age of this water?

Exercise 12.4: The points on Fig. 12.7 lie on the intake line. Thus, all these water samples have a recent age. What could be the reason why they do not cluster around a single point?

Exercise 12.5: Chlorine-36 measurements are a highly recommended parameter of groundwater studies, especially if confined systems are possibly included. Which other parameters have to be measured in order to establish ^{36}Cl-based water ages?

Exercise 12.6: Which sampling strategy will you recommend if a study area includes 70 wells but the budget is enough for only 25 full laboratory analyses?

13
NOBLE GASES

13.1 Rare, Inert, or Noble?

Helium, neon, argon, krypton, and xenon occupy a special place in the periodic table—their outer shell of electrons is complete (2, 8, or 18) and therefore they do not interact with other atoms (section 5.4). This property gives these elements the name *inert* or *noble*. They are gases, and they are present in small quantities in air (Table 13.1); hence, they are also called *rare gases*. It has been pointed out that one of them, argon, is present in air in a concentration of nearly 1%, and therefore the term "rare" was challenged. In the last decade a few xenon and krypton compounds have been prepared under specific laboratory conditions and the term "inert" has been challenged as well. In nature, however, their chemical nobility is well preserved, making them ideal tracers in hydrological systems. They have a number of advantages, including:

They are not involved in any chemical or biological activity.
They enter groundwater from two distinct major sources: by equilibration with air during infiltration, or from deep-seated origins, which include flushing of radiogenic products from aquifer rocks, and from mantle-derived gases. The origin of these different sources is identifiable by the isotopic composition.
The initial concentrations of atmospheric noble gases in recharge water can be predicted from the ambient annual temperature and altitude of the suggested recharge area.
The fact that there are five noble gases provides valuable redundancy.
Let us begin the discussion with the atmospheric input.

Noble Gases

Table 13.1 Noble gases (ppm volume) in dry air at 1 atm (sea level)

He	Ne	Ar	Kr	Xe
5.24	18.18	9340	1.14	0.086

13.2 Atmospheric Noble Gases and Their Dissolution in Water

The atmosphere is a well-mixed reservoir with known concentrations of noble gases (Table 13.1). These atmospheric noble gases have characteristic isotopic abundances that are given in Table 13.2. The solubility of the noble gases is given in Fig 13.1, expressed in cc STP noble gas/cc water. STP stands for standard temperature (0°C) and pressure (760 mm Hg = 1 atmosphere). What is the solubility of argon in distilled water, at sea level, at 15°C? The answer, from Fig. 13.1, is 3.5×10^{-4} cc STP Ar/cc water.

The lack of chemical interactions in the noble gases makes them ideal gases, and as such their solubility in water is directly proportional to their partial pressure in air. This, in turn, depends on the barometric pressure, which is linearly correlated to the altitude. The higher the altitude, the less noble gases dissolve in water. This relation is expressed in Fig. 13.2. The correction factor, read on the right vertical axis in Fig. 13.2, is multiplied by a solubility value observed at a given altitude in order to get the equivalent solubility (at the same ambient temperature) at sea level. Or, the solubility at a given temperature at sea level (Fig. 13.1) can be divided by the correction factor in order to get the solubility, at the same temperature, at a selected altitude. The need for these procedures will be soon explained, but let us first get familiar with them: What is the solubility of krypton at 18°C on a mountain that is 2500 masl? The answer is worked out in steps:

Table 13.2 Isotopic abundances of atmospheric noble gases (chart of the nuclides, 1977)

Element	He		Ne			Ar		
Isotope	3	4	20	21	22	37	38	40
(%)	0.0014	100	90.51	0.277	9.22	0.337	0.063	99.60

Element	Kr					Xe					
Isotope	80	82	83	84	86	128	129	130	131	132	134	136
(%)	2.25	11.6	11.5	57.0	17.3	1.91	26.4	4.1	21.2	26.9	10.4	8.9

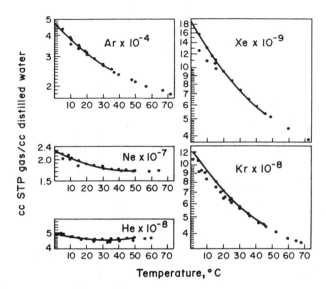

Fig. 13.1 Solubility of atmospheric noble gases in fresh water at sea level (1 atm) as a function of the ambient temperature. (From Mazor, 1979.)

The solubility of krypton at 18°C at sea level is (from Fig. 13.1) 7.5×10^{-8} cc STP/cc water.
The correction factor for 2500 masl is (from Fig. 13.2) 1.35.
The solubility of krypton at 18°C and 2500 masl is

$$\frac{7.5 \times 10^{-8}}{1.35} = 5.5 \times 10^{-8} \text{ cc STP/cc water.}$$

The solubility of the noble gases depends on a third parameter: the concentration of dissolved ions in the water. The data in Fig. 13.1 are for salt-free water. Seawater, for example, dissolves 30% less. This effect has only seldom to be regarded in groundwater tracing, since recharge is in most cases relatively fresh.

13.3 Groundwater as a Closed System for Atmospheric Noble Gases

Recharge water equilibrates with air during its infiltration and it dissolves atmospheric He, Ne, Ar, Kr, and Xe in amounts defined by the local altitude and the ambient temperature. Once water enters the saturated zone of groundwater, no further additions of atmospheric gases are possible since the rock pores

Noble Gases

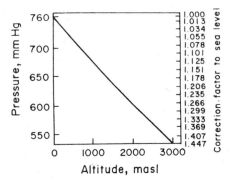

Fig. 13.2 Atmospheric pressure variations as a function of altitude. (From the U.S. Standard Atmosphere, 1962.) Correction factors on the right axis serve to convert values at sea level (Fig. 13.1) to solubility values at desired altitude (dividing by the factor), or to normalize data at a given altitude to the corresponding value at 0 masl (multiplying by the factor). The last conversion is needed to read intake (recharge) temperatures from Fig. 13.1, which is for sea level (i.e., 760 mm Hg) (Mazor, 1975).

contain only water. Release of dissolved gases in the saturated zone is not possible, either, because of the hydrostatic pressure of the system. Because of these arguments it has been postulated that the nobles gases are kept in closed system conditions in groundwater, and therefore, one can measure the noble gas concentrations in the water of a spring or well and calculate from them the ambient temperature at the recharge area (Mazor, 1972).

A coastal spring was found to contain 12×10^{-9} cc STP Xe/cc water. What was the ambient temperature at the point of recharge (assuming the system remained closed)? From Fig. 13.1 it is seen that 12×10^{-9} cc STP Xe/cc water are dissolved in water that equilibrates with air at 12°C. Hence, if recharge was at sea level, the ambient temperature would have been 12°C.

A spring emerges at an altitude of 200 masl. The highest point in the potential recharge area is 400 masl, and the average altitude of the recharge area is $(400 + 200)/2 = 300$ m. In a water sample of this spring 10×10^{-9} ccSTP Xe/cc water were found. What is the recharge temperature? Let's do the calculation in steps:

The altitude correction factor for 300 m is read from Fig. 13.2 to be 1.05.
Multiplying the observed xenon concentration by this factor, the equivalent concentration at sea level is obtained: $10 \times 10^{-9} \times 1.05 = 9.5 \times 10^{-9}$ cc STP/cc water.
Using Fig. 13.1, read the equivalent temperature: 16°C.

In summary, it might be concluded that:

The initial noble gas concentrations in recharged groundwater can be calculated from the ambient temperature and recharge altitude.

The concentration of dissolved noble gases in groundwater (a spring or a well) can be used to calculate the ambient intake temperature, if the recharge altitude is known or assumed.

The scope of noble gas applications in groundwater studies is much greater than described thus far, but first we have to check the basic assumption that groundwater provides closed system conditions for dissolved atmospheric noble gases.

13.4 Studies on Atmospheric Noble Gas Retention in Groundwater Systems: Cold Groundwater

Groundwater samples from various locations around the world have been analyzed for their noble gas isotopic composition and elemental concentrations. One such study was conducted by Herzberg and Mazor (1979) on four springs of simple hydrological structures (Fig. 13.3 and Table 13.3). The isotopic composition of the Ne, Ar, Kr, and Xe was found to equal that of air, revealing the atmospheric origin. The noble gas concentrations are given in Table 13.4. These data were multiplied by the respective altitude correction factors (Fig.

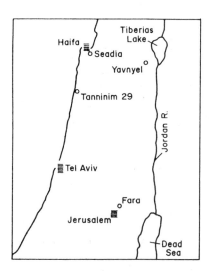

Fig. 13.3 Location of springs (circles) of simple hydrological systems in Israel selected for noble gas retention studies (Table 13.3).

Table 13.3 General data on springs selected for noble gas retention studies (Fig. 13.3) (from Herzberg and Mazor, 1979)

Name of spring	Sample no.	Date	Temperature (°C)	Altitude (m) Recharge area			Description
				maximum	average	minimum	
Saadia	OH-10	2.10.74	22.0	520	270	20	issues on the fault of Mount Carmel, on the contact with the Yzreal Plain
	OH-13	3.3.75	22.0				
Yavniel	H-6	14.9.74	20.5	300	225	150	a small spring in a basaltic region, draining two adjacent hills
Taninim 29	OH-16	19.3.75	23.5	900	450	10	one spring, out of several, issuing in the Kabara plain; seems to be recharged from the Shomron Mountains
Fara	OH-5	22.7.74	20.5	800	550	290	issues on the eastern slopes of the Judean Mountains

Table 13.4 Noble gas concentrations[a] (cc STP/cc water) in samples from springs in Israel (Table 13.3) (Herzberg and Mazor, 1979)

No.	Source	Date	Spring temp. (°C)	Ambient temp. (°C)	Ne ($\times 10^8$)	Ar ($\times 10^4$)	Kr ($\times 10^8$)	Xe ($\times 10^9$)
OH-10	Saadia	2.10.74	22.0	17	20.7 ± 1.0(4)	3.00 ± 0.10 (4)	6.49 ± 0.24 (4)	8.67 ± 0.60 (4)
OH-13	Saadia	4.3.75	22.0	12	20.4 ± 0.4(4)	2.99 ± 0.08 (4)	6.40 ± 0.25 (4)	8.56 ± 0.79 (4)
OH-6	Yavniel	14.9.74	20.5	21	21.3 ± 0.9(3)	3.01 ± 0.06 (3)	6.33 ± 0.33 (3)	8.65 ± 0.33 (3)
OH-16	Taninim 29	19.3.75	23.5	18	21.3 ± 0.3(3)	2.98 ± 0.04 (3)	6.31 ± 0.09 (3)	8.47 ± 0.31 (3)
OH-5	Fara	22.7.74	20.5	23	18.8 ± 0.4(3)	2.72 ± 0.11 (3)	7.59 ± 0.39 (3)	8.47 ± 0.31 (3)

[a] Errors cited are mean errors for several samples collected at the same time (their number is given in parentheses). STP denotes 760 mm Hg and 0°C.

13.2) to obtain the noble gas concentrations normalized to 1 atm (sea level), as seen in Table 13.5. These values were then applied to calculate intake temperatures, which are given in Table 13.6, along with temperatures measured in the field.

Six points of temperature measurements seem to be especially informative along the hydrological cycle. These points are (Fig. 13.4):

T_1—the average ambient air temperature during the rainy season at the recharge location; obtained from meteorological surveys.

T_2—the average annual air temperature at the recharge location, obtained from meteorological studies.

T_3—the temperature at the base of the aerated zone, just above the water table. This temperature is deducible from the noble gas concentration.

T_4—the highest temperature reached during the groundwater cycle, presumably obtained at the deepest point of circulation.

T_5—the temperature observed at the spring or well head.

T_6—the ambient air temperature at the site and time of sample collection.

T_4 is the only value we cannot measure, but if flow rates are high enough, T_5 will be equal to or slightly lower than T_4. T_5, the spring or well temperature, may thus be regarded as a minimum value of T_4.

The data in Table 13.6 reveal the following observations:

Agreement between the intake temperatures calculated via Ar, Kr, and Xe is $\pm 1°C$. This in itself indicates the systems are closed, since losses or gains would most probably be connected to fractionation that would introduce disagreements between the various calculated intake temperatures.

In the first three cases (Sa'adia, Yavniel, and Taninim 29) the noble gas deduced temperatures (T_3) are equal, or 2°C higher, than the average annual temperature. In contrast, they differ by about 10°C from the average temperature of the cold rainy season. Hence, it seems that the temperature documented by the noble gases during equilibration with air at the base of the aerated zone equals the local average annual temperature. This makes sense because infiltrating recharge water often descends at rates of a few meters per year or less. Hence, infiltrating water is delayed in the aerated zone, adopting the average temperature prevailing underground. This topic will be further discussed in the next section.

In the first three samples of Table 13.6 the noble gas deduced temperature, T_3, differs from the ambient air temperature at the time of sample collection, T_6, revealing no equilibration at the point of emergence (in agreement with the previous paragraph). The last spring (Fara), however, shows $T_3 = T_6$. This water issues below a slope of tallus and equilibration of noble gases could take place prior to sample collection.

Table 13.5 The noble gas concentrations of Table 13.4 normalized to 1 atm (sea level) (cc STP/cc water)

No.	Source	Av. recharge altitude (m)	Altitude correction factor	Ne ($\times 10^8$)	Ar ($\times 10^4$)	Kr ($\times 10^8$)	Xe ($\times 10^9$)
OH-10	Saadia	270	1.030	21.3	3.09	6.68	8.93
OH-13	Saadia			21.0	3.08	6.59	8.82
OH-6	Yavniel	225	1.024	21.8	3.08	6.48	8.86
OH-16	Taninim 24	450	1.049	22.3	3.13	6.62	8.89
OH-5	Fara	550	1.068	20.1	2.90	6.23	8.11

Table 13.6 Intake (recharge) temperatures deduced from noble gases[a] and measured in the field[b] (°C)

No.	Spring	Date	Recharge area		Noble gases, T_3				Sampling site	
			T_1^c average rainy season	T_2^c average	Ar	Kr	Xe	average[d]	T_5 spring	T_6 ambient air
OH-10	Saadia	2.10.74	10	19	21	21	21	21 ± 1	22.0	17
OH-13	Saadia	4.3.74	10	19	21	22	21	21 ± 1	22.0	12
OH-6	Yavniel	14.9.74	14	21	21	23	21	21 ± 2	20.5	21
OH-16	Taninim 19	19.3.75	14	19	21	22	22	21 ± 1	23.5	23
OH-5	Fara	22.7.74	10	16	23	24	23	23 ± 1	20.5	23

[a] Applying the data of Table 13.5 and the solubility curves (Fig. 13.1).
[b] T_1–T_6 are defined in the text and in Fig. 13.4.
[c] From various meteorological sources.
[d] The cited errors were calculated from the errors of the individual runs as given in Table 13.4.

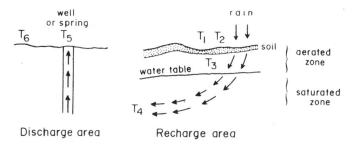

Fig. 13.4 Definition of temperatures relevant to hydrological studies: T_1 = average temperature in the rainy season at recharge area; T_2 = average annual temperature at recharge area; T_3 = temperature at the base of the aerated zone, above the water table (deduced from the Ar, Kr, and Xe concentrations); T_4 = maximum temperature reached at the deepest point of the water path; T_5 = observed spring or well temperature at the time of sampling; T_6 = ambient air temperature at the time of sampling. (From Herzberg and Mazor, 1979.)

Good agreement, obtained for various cold groundwaters, between noble gas deduced intake temperature and average annual temperature at the recharge area prove the validity of the original assumption, that groundwater is a closed system for the dissolved atmospheric noble gases.

13.5 Further Checks on Atmospheric Noble Gas Retention: Warm Groundwater

The closed system conditions seen in cold groundwater systems may be further checked in warm springs, that is, in waters that percolate to 1–2 km depth and get heated by local (normal) heat gradients. The first study of this type was done on warm springs and wells (up to 60°C) of the Jordan Rift Valley (Mazor, 1972). The isotopic compositions of the encountered dissolved Ne, Ar, Kr, and Xe were found to be atmospheric, that is, very close (in the range of the analytical accuracy) to the values given in Table 13.2.

The atmospheric noble gases are presented in Fig. 13.5 in a fingerprint diagram. An overall similarity of the line patterns to those of air-saturated water at 15°C–25°C is clear. Thus, the relative abundances of the four noble gases Ne, Ar, Kr, and Xe indicate they originated from air dissolved in the recharge water. The fingerprint lines in Fig. 13.5 are parallel to each other, but vary in their height on the concentration axis. What might be the explanation for this? The answer is differences in the recharge areas and, hence, in the initial noble gas concentrations due to differences in altitude (temperature and pressure). In one case, the Tiberias Hot Springs, the initial salinity played a significant role.

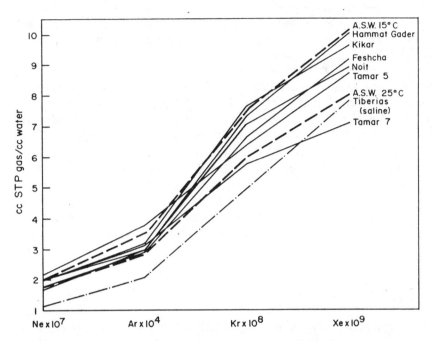

Fig. 13.5 Noble gases dissolved in thermal waters in the Jordan Rift Valley. (Data from Mazor, 1972.) The nonsaline waters lie between the values of air-saturated water (ASW) at 15°C and 25°C, demonstrating the meteoric (atmospheric) origin and the closed-system conditions that prevailed in the ground. The saline Tiberias hot water is suggested to have originated by seawater entrapment, a hypothesis supported by the lower noble gas content (close to sea water saturation at about 15°C) (Mazor, 1972).

Let's have a closer look at the Tiberias Hot Springs case. It is lowest in Fig. 13.5, revealing the lowest noble gas concentrations. However, it has a salt concentration almost equal to seawater and, based on geochemical and hydrological arguments, it has been suggested that this water originated from infiltration of seawater when the Mediterranean Sea invaded the Rift Valley for a short period (Mazor and Mero, 1969a, 1969b). If so, the initial noble gas concentration was as in seawater—30% less than in fresh water under similar conditions. Hence, to normalize the Tiberias Hot Springs line in Fig. 13.5 and compare it with freshwater samples, the values have to be increased by 30%. In doing so, the Tiberias Hot Springs line falls right in the middle of the other Rift Valley warm waters, close to A.S.W. 15°C, agreeing with closed system conditions. The noble gases thus provide independent support of the origin of the Tiberias Hot Springs from seawater.

As noble gas concentrations are obtained, the hydrochemist is always confronted with the need to establish whether the data reflect the indigenous noble gas concentrations, or whether reequilibration with the atmosphere occurred prior to sampling. The problem is acute in springs: the samples are collected in the spring as deep as possible below the water surface, but did the water reequilibrate while flowing below gravel, for example? or did the water circulate in the spring's outlet? An answer may be obtained by comparing the observed noble gas concentration with the concentration calculated for equilibration with air at the emergence temperature of the spring.

Example: A spring emerges at 34°C and is found to contain 3.8×10^{-4} cc STP Ar/cc water. Is it oversaturated, that is, does it contain more gas than expected from air equilibration at the emergence temperature? The air-water solubility at 34°C is seen in Fig. 13.1 to be 2.5×10^{-4} cc STP Ar/cc water. Hence, the observed value of 3.8×10^{-4} cc STP Ar/cc water is oversaturated by

$$\frac{3.8 \times 10^{-4}}{2.5 \times 10^{-4}} = 100 = 152\%$$

of the equilibration value at the temperature of emergence (the observed value of Ar concentration indicates a recharge temperature of 10°C–Is this right?).

The observed krypton and xenon concentrations in each water source in the Rift Valley study were converted in this way to percent air saturation in Fig. 13.6. It is seen that the samples contain more than 100% air saturation—they are oversaturated, indicating reequilibration subsequent to heating did not occur. This diagram shows the importance of duplicate or triplicate sampling. The one low value of sample 3 is clearly wrong due to gas loss during sampling, the higher sample 3 values being more representative (and not the average).

Warm springs are a rigorous check of closed-system conditions for the atmospheric noble gases, as their waters circulate deep and their ages are relatively high. Figures 13.7 and 13.8 show results for another case, in Swaziland.

13.6 Identification of the Nature of the Unsaturated Recharge Zone: A Porous Medium or a Karstic System

In Table 13.7, data for two karstic springs from northern Israel and one from eastern Switzerland are given. The table includes:

1. Ne, Ar, Kr, and Xe concentrations.
2. The noble gas concentration in air-saturated water at the respective recharge temperatures and altitudes.
3. Division of (1) by (2), multiplied by 100, provides the percent concentration relative to the expected air-saturated water at the recharge location.

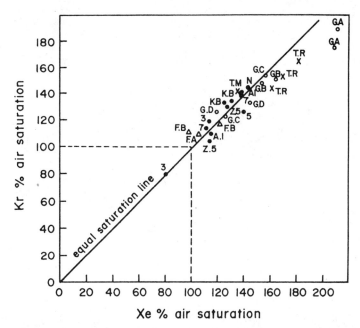

Fig. 13.6 Percentage of air saturation of krypton and xenon for the Jordan Rift Valley waters. The values were obtained by dividing the measured amounts by those expected for water equilibrated with air at the temperature at which each sample was collected. All samples, except one, were found to be air supersaturated, indicating that the rare gases were retained under closed system conditions. Differences in duplicate samples are attributed to gas losses to the atmosphere prior to sampling (Mazor, 1975).

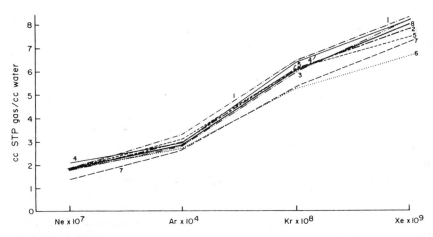

Fig. 13.7 Noble gas pattern of atmospheric noble gases in warm springs in Swaziland. The isotopic compositions were found to be atmospheric, and the abundance patterns, similar to ASW 25°C, confirm this. The variations in concentrations are caused by different recharge altitudes (Mazor et al., 1974).

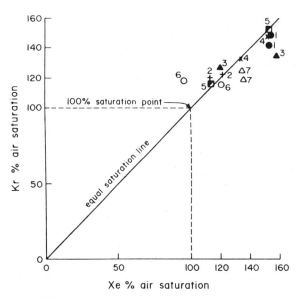

Fig. 13.8 Percent air saturation of krypton and xenon for warm springs in Swaziland (same spring numbers as in Fig. 13.7). All samples were oversaturated at the temperature of emergence (35°C–52°C), indicating the systems were closed (Mazor et al., 1974).

Table 13.7 Stages in the calculation of percentage noble gas retention in karstic water samples

Data	Karstic spring	Ne	Ar	Kr	Xe
Measured values	Dan, Israel	28.8	36700	7.85	1.02
10^{-8} cc STP/cc water[a]	Banias, Israel	24.1	33800	6.97	0.97
	Source de la Doux, Switzerland	28.6	39600	8.81	1.13
ASW, 14°C, 1700 m[b]	recharge of Dan and Banias springs	16.7	28000	6.70	0.96
ASW, 10°C, 1000 m	recharge of Source de la Doux	18.3	32000	7.00	1.09
% retention, relative	Dan	172	131	116	106
to AWS at recharge point[c]	Banias	144	121	104	101
	Source de la Doux	156	124	116	104

[a]Data from Herzberg and Mazor (1979).
[b]The values, in 10^{-8} cc STP/cc water, were calculated from the relevant solubility values, divided by the corresponding altitude correction factor.
[c]$(a/b) \times 100$.

Noble Gases

The values (Fig. 13.9) are significantly different from what we have seen so far:

Neon concentrations are 144–172%. Thus, excess atmospheric neon (recognized as atmospheric by its isotopic composition) is present.

The other noble gases are also present in concentrations exceeding 100%, but the excess decreases towards xenon. In other words, the retention percentage pattern is

Ne > Ar > Kr > Xe.

The explanation is that free air has been siphoned into the karstic water, in addition to the initial dissolved air. The solubility of the noble gases is smallest for helium and increases towards xenon. This should not be confused with Fig. 13.1, which was obtained by taking the solubility of each noble gas and multiplying it by its abundance in air (Table 13.1). (One can obtain the noble gas solubility for a selected temperature by dividing the respective values in Fig. 13.1 by the abundance values of table 13.1). Free air is relatively richer in the less soluble gases, therefore its addition to already saturated water in the karstic springs causes (1) the excesses over 100% saturation and (2) the excess pattern of Ne > Ar > Kr > Xe.

The excess atmospheric air was observed to be negatively correlated to spring discharges, reflecting that larger parts of the system contain air in the low flow season, and when the water flows through narrowing conduits the air is sucked in. In the high flow season most of a karstic system is filled with water and less air can be siphoned (Herzberg and Mazor, 1979).

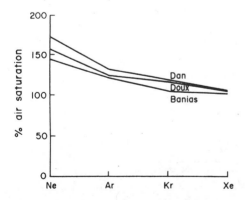

Fig. 13.9 Percent air saturation for karstic springs (Table 13.7). Values are above 100% and the lines reveal a "reversed" pattern of Ne > Ar > Kr > Xe.

Excess air in groundwater has been reported by Heaton and Vogel (1981) for cases of special soil conditions observed in South Africa. Hence, periodic noble gas measurements are desired: a negative correlation with discharge will establish the karstic nature of a studied spring or well.

13.7 Depth of Circulation and Location of the Recharge Zone

Infiltrating water is, in many cases, delayed in the aerated zone long enough to equilibrate to the ambient temperature. In shallow depths—less than 10-15 m—this equilibrated temperature will be close to the local average annual temperature. Daily temperature fluctuations are damped in less than 0.5 m (Fig. 13.10) and seasonal fluctuations are damped in a few meters.

If the temperature of a spring, or well, is higher than the noble gas deduced temperature ($T_5 > T_3$ in Fig. 13.4), then the additional temperature reflects heating by the local geothermal gradient. If this gradient is known, then dividing

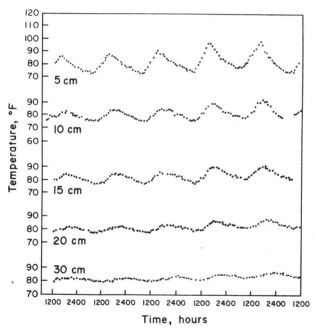

Fig. 13.10 Daily temperatures in soil profiles, measured in 1970 at the Santa Rita Experimental Range, Tucson, Arizona. The daily fluctuations are seen to be damped at 30 cm (after Foster and Fogel, 1973).

the temperature gain by the gradient will provide the depth to which the water is circulated in the saturated zone.

Example: The water of the Taninim spring is seen in Table 13.6 to emerge at 23.5°C, whereas the noble gas deduced temperature at the base of the aerated zone is 21 ± 1°C. How can this difference be explained? The additional 2.5°C resulted from equilibration of the aquifer temperature. The local heat gradient is not known, so let us use the common value of 3°C/100 m. Accordingly, the water circulated to a depth of 80 m or more in the saturated zone. (It might be more because partial cooling during ascent could occur).

Locating recharge areas. A well in a coastal plain provides fresh water at 26°C. The question came up of whether it is recharged by local rain or snowmelt on the 20-km distant mountains. To resolve this problem the noble gases were measured and the following concentrations were found: 2.1×10^{-7} cc STP Ne/cc water and 3.0×10^{-4} cc STP Ar/cc water. The local heat gradient has been established to be 2.5°C/100 m. Where is the recharge location? The neon curve in Fig. 13.1 is seen to vary little with temperature, and therefore it is of little use to the calculation of the intake temperature. The argon curve is, in contrast, sensitive to temperature changes. An intake (base of aerated zone) temperature of 19°C is obtained. Hence, 26°C – 19°C = 7°C is the temperature gain obtained by the depth of circulation. Applying the local heat gradient, the depth of circulation is $7 \times 100/2.5 = 280$ m. The depth of circulation indicates remote recharge at a relatively high altitude. If so, a respective altitude correction should be applied to the raw noble gas data, resulting in an intake temperature lower than 19°C, further supporting mountain recharge.

This example demonstrates the potential use of the atmospheric noble gases in identifying a recharge location. Combined with stable isotope measurements of hydrogen and oxygen in water, that is, δD and $\delta^{18}O$, a data matching, is obtained. One can calculate the recharge altitude from the δD and $\delta^{18}O$ values (section 9.8) and then, applying the respective altitude correction, one can calculate the intake temperature, thus checking the validity of the suggested recharge location. Furthermore, the difference between observed emergence temperature and noble gas deduced intake temperature will give the minimum depth of circulation and, thus, tell whether the water moves underground in a shallow path or a deep path.

13.8 Paleotemperatures

13.8.1 Principles and Importance

Old groundwaters contain dissolved noble gases in concentrations that were defined by the ambient temperature that prevailed at the time of recharge. Coupling of groundwater dating methods, for example ^{14}C, with intake temperatures calculated from observed Ar, Kr, and Xe concentrations (section 13.3) may provide relevant paleotemperatures (Mazor, 1972). This noble gas based method is the only method that can be used to directly reconstruct paleotemperatures. Other techniques of paleoclimatic reconstruction are more general, combining humidity and temperature in relative terms. Other methods are based on the width of tree rings, the types of pollen in dated paleosoils, the composition of stable hydrogen and oxygen isotopes in dated groundwaters, or variations in the distribution of ancient human settlements.

The application of noble gas based paleotemperature determination is demonstrated in the following sections by case studies, and thereafter some aspects of paleoclimatic feedback are discussed as a requirement of a better understanding of the reconstructed paleotemperatures. Finally, the relevance of paleotemperatures to the topic of global temperature changes is addressed.

13.8.2 Case Studies

An example of a noble gas paleotemperature study was conducted by Andrews and Lee (1979) in the Bunter sandstone aquifer in eastern England. The noble gases were studied as part of an intensive hydrochemical study, discussed in section 11.9. Table 13.8 contains the noble gas concentrations in three sections of the studied groundwater system: group A—recharge zone, unconfined system; group B—an adjacent confined system; group C—another confined system (section 11.9). The observed noble gas concentrations were used to calculate intake temperatures. The paleotemperatures are plotted versus emergence temperatures (indicating depth) in Fig. 13.11. The results: $10 \pm 1°C$ for the shallow group A; $9 \pm 1°C$ for the medium depth group B; and $4 \pm 1°C$ for the deepest group C.

Other examples of noble gas paleotemperatures have been reported by Rudolph et al. (1983). They calculated temperatures as low as $1°C-2°C$ 22,000 years ago (dated by ^{14}C) in Germany, accompanied by relatively negative δD and $\delta^{18}O$ values and a measurable excess of radiogenic helium.

13.8.3 Paleotemperatures During the Last Glacial Maximum: Real or Apparent Differences Between Terrestrial and Oceanic Evidences

The last glacial maximum occurred about 18,000 years ago, well within the dating range of ^{14}C. The event was quite dramatic, and field evidence reveals that the snow lines of the higher mountains were up to 1000 meters lower than their present positions. In connection with speculations connected with possible man-induced global climatic changes in the future, it is of much interest to establish how many degrees lower the temperature was during the last glacial maximum. Fauna abundances and ^{18}O studies of dated oceanic sediment cores

Table 13.8 Noble gases (cc STP/cc water) of groundwaters in the Triassic sandstone aquifer in eastern England (from Andrews and Lee, 1979), deduced intake temperatures (read from solubility curves, Fig. 13.1), and radiogenic He[a]

Group	No	Well	$He \times 10^{-8}$	$Ne \times 10^{-7}$	$Ar \times 10^{-4}$	$Kr \times 10^{-8}$	Rad. He* 10^{-8}	Deduced intake temp. (°C) Ne	Ar	Kr	Average
A	5	Elkesley, No.6	4.0	2.09	4.02	8.65	−0.8	10	9	12	10 ± 1
A	9	Elkesley, No.5	4.3	2.09	4.02	8.68	−0.5	10	9	12	10 ± 1
A	12	Elkesley, Clark's No.1	7.4	2.09	3.87	8.69	2.6	10	10	12	11 ± 1
A	6	Far Baulker, No.3	4.3	2.09	4.05	8.78	−0.5	10	9	12	10 ± 1
A	1	Amen Corner, No.1	4.2	2.09	3.83	9.02	−0.6	10	11	11	11 ± 1
A	3	Boughton, No.2	6.2	2.01	4.07	8.71	1.5	15	8	12	12 ± 2
A	10	Retford, Clark's No.2	6.5	2.07	3.94	9.14	1.7	11	9	11	10 ± 1
A	8	Ompton, No.2	7.8	2.12	4.16	9.12	2.9	9	7	11	9 ± 2
A	4	Fransfield	5.1	2.09	4.09	8.70	0.3	10	8	12	10 ± 2
A	7	Rufford, No.3	4.9	2.02	3.89	8.74	−0.2	(15)	10	11	11 ± 1
A	11	Retford, Orsdall No.1	3.5	2.09	4.04	8.64	−1.4	8	7	12	9 ± 2
A	13	Everton, No.1	4.3	2.07	3.76	8.84	−0.5	11	11	11	11 ± 1
A	14	Everton, No.3	4.2	2.05	4.28	9.13	−0.6	10	11	11	11 ± 1
A	15	Retford, Whisker Hill	4.6	2.08	3.89	8.80	−0.2	10	10	12	11 ± 1
B	16	Halam, No.1	14.1	2.23	4.15	8.82 9.2		6	7	12	8 ± 2
B	17	Markham Clinton, No.1	4.3	2.09	3.99	8.74	−0.5	10	9	12	10 ± 1
B	21	Egmanton, B.P.	10.8	2.11	4.07	9.13	5.9	9	8	11	9 ± 1
B	20	Caunton	3.6	1.84	3.65	7.93	−1.1	(25)	12	15	9 ±
B	19	Retford, Grove No.2	8.4	2.14	4.19	9.13	3.6	5	7	11	8 ± 2
C	26	Gainsborough, B.P.	23.7	2.20	4.58	9.85	18.7	5	3	8	5 ± 2
C	24	Newark, British Gypsum	21.9	2.29	4.86	10.63	16.9	3	2	5	3 ± 1
C	25	Newark, Castle Brewery	25.9	2.27	4.78	10.50	20.9	4	2	5	4 ± 1
C	30	Gainsborough, Humble Carr	27.2	2.18	4.49	10.10	22.2	5	4	7	5 ± 1
C	28	Newton, No.2	17.3	2.27	4.72	10.54	12.3	4	3	5	4 ± 1
C	31	Gainsborough, No.3	34.3	2.20	4.52	10.28	29.3	5	4	6	5 ± 1
C	22	Rampton	19.3	2.27	4.68	10.48	14.3	4	3	5	4 ± 1
C	32	South Scarle	17.6	2.20	4.51	9.96	12.6	5	4	8	6 ± 2

[a] Observed He minus atmospheric He, read from the solubility curve (Fig. 13.1), applying the average deduced intake temperture (last column of present table).

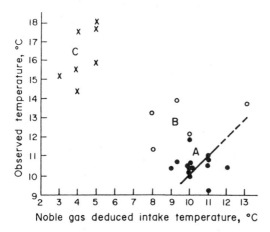

Fig. 13.11 Noble gas deduced intake temperatures versus observed emergence temperatures (Table 13.8) in wells of the Bunter sandstone aquifer in eastern England. The dashed line is an equal temperature line, that is, noble gas temperature equals observed temperature. Group A data are around this line; group B reveals slightly colder paleotemperatures (8°C–10°C); and group C reveals significantly colder values (3°C–5°C).

revealed that the low-latitude ocean surface water was colder by less than 2°C (Brocker, 1986).

In contrast, noble gas paleotemperatures, defined in groundwaters from the Carizzo aquifer (Texas), revealed that during the last glacial maximum the temperature was 5.2 ± 0.7°C lower than the local present temperature (Stute et al., 1992). A comprehensive study of noble gases dissolved in groundwater conducted in the San Juan basin, New Mexico, revealed that the temperature was 5.5 ± 0.7°C lower than the local present temperature (Stute et al., 1995). A similar study in tropical Brazil revealed a drop of 5.4 ± 0.6°C during the last glacial maximum (Stute et al., 1995).

The meaning of the presented paleotemperatures during the last glacial maximum warrant some discussion. For the global change issue are of interest multi years average temperatures, that prevailed in different places all over the earth. However, there seems to exist a paleohydrological effect that is of importance: even a slight drop in the average temperature seems to cause significant lowering of the snow line, which increases the significance of snowmelt in the local recharge. Snowmelt recharge is, in turn, coupled with increasing importance of karstic recharge, on account of porous medium recharge. Thus, the recharge water entered the saturated zone at relatively low temperatures, that are somewhat between the snowmelt temperature and the multiyear average ambient temperature. In other words, the cooling of the last glacial maximum yielded an increased temperature drop in the studied paleowaters—the latter reflecting a seasonal paleotemperature—and the average annual temperature was somewhat higher.

Brocker (1986) and Stute et al. (1995) pointed out that there seems to be a disagreement between the paleotemperature deduced for the last glacial, based on oceanographic sediments (less than 2°C), and the significantly larger temperature drop (5.2°C) deduced from noble gases in dated groundwaters. The discussed accentuation of the noble gas paleotemperature record, as a result of increasing importance of recharge by snowmelt, may reconcile the oceanic and continental paleotemperature records related to the last glacial maximum: the marine record may reflect the multiyear average ambient temperature, whereas the studied paleowaters may reflect temperatures closer to the snow melt recharge. This hypothesis can be checked by measuring noble gases in groundwaters for locations that were remote of any paleo-snow line and had porous medium recharge.

13.8.4 The Relevance of Paleotemperatures to the Question of Possible Man-Made Global Climatic Changes

Recently, much research has been devoted to the possible climatic consequences of man-induced changes in atmospheric compounds and landscape features. Examples are increases in the concentration of CO_2 and other greenhouse gases and dust or the thickness of the ozone layer, as well as large-scale deforestation and desertification. Opinions vary as to the expected climatic changes, and a knowledge of the natural fluctuations is of great relevance: will the man-induced changes be in the range of the natural fluctuations or higher? This, in turn, will tell researchers to what degree man-made global temperature changes can be read from temperature records. For this reason, as continuous paleotemperature records as possible are needed, of as long a duration as possible (mainly limited by the available dating methods).

13.9 Sample Collection for Noble Gas Measurements

Noble gases are measured with special mass spectrometers. The size of the water samples needed is 5–10 cc. The major difficulty in noble gas work is the absolute need to avoid air contamination: a bubble of air present in the collected sample may contain more noble gases than the entire sample. Two techniques are in use to sample for noble gases: (1) filling water in glass tubes with special glass stopcocks, and (2) filling copper tubes, closed at their ends with special clamps. The sampling method should be talked over with the laboratory staff, who should provide the collection vessels and detailed sampling procedures. Since sample collecting has to be done with care, it is recommended that laboratory personnel join the field team to do the sample collections, if possible.

The number of laboratories that measure noble gases in water samples is slowly, but steadily, increasing. The know-how is actually available in most laboratories that do potassium-argon dating of rocks. Because of increased demand from field hydrochemists, more laboratories are willing to enter the domain of noble gas hydrology. Eventually, noble gas measurements will become as routine as δD, $\delta^{18}O$, tritium, or ^{14}C measurements.

13.10 Summary Exercises

Exercise 13.1: Fill in the missing data of air solubility, at sea level (in cc STP/cc water):

Temperature	He	Ne	Ar	Kr	Xe
15°C			3.5×10^{-4}		
22°C					
51°C					

Exercise 13.2: What is the solubility of the five noble gases at 5°C and 1500 masl?

Exercise 13.3: Calculate the intake (recharge) temperature for (a) a well with 3.2×10^{-4} cc STP Ar/cc, located at 250 masl, with the highest point in the recharge area at 650 masl; (b) a spring with 10×10^{-8} cc STP Kr/cc water, emerging at 1000 masl, the highest point of recharge being 1300 masl; and (c) an artesian well with 12×10^{-9} cc STP Xe/cc water, with an average recharge intake altitude of 850 m.

Exercise 13.4: A coastal spring emerges at 32°C and contains 3.7×10^{-4} ccSTP Ar/cc water. What is the minimum depth of circulation of the water in the saturated zone?

Exercise 13.5: A spring issues in a plain, the temperature being 24°C, and its water was found to contain 2.1×10^{-7} cc STP Ne/cc water, and 3.8×10^{-4} cc STP Ar/cc water. What is the intake temperature, and what is the depth of water circulation? Is the recharge location nearby in the plain or distant in the mountains?

14
HELIUM-4 AS A TOOL FOR GROUNDWATER DATING

14.1 Sources of Radiogenic ^{4}He Dissolved in Groundwater

The radioactive decay of uranium and thorium results in the formation of a series of isotopes that are radiogenic by themselves and keep disintegrating into stable lead isotopes. These radioactive disintegrations are accompanied by the emissions of ^{4}He atoms. Three such radioactive series exist:

^{238}U $\rightarrow$ ^{206}Pb + 8 ^{4}He

^{235}U $\rightarrow$ ^{207}Pb + 7 ^{4}He

^{232}Th $\rightarrow$ ^{208}Pb + 6 ^{4}He

The production rate of the radiogenic helium in each of these decay series is determined by the respective half-lives, and the overall helium production per unit rock is determined by the concentration of uranium and thorium in the rock—common rocks containing uranium and thorium in parts per million (ppm) concentrations.

The radiogenic helium is partially released from the crystal lattices of rocks and is dissolved in water that is in contact with the rocks. The production of helium is constant in any given set of rocks, and the amount dissolved in stored groundwater is proportional to the duration of water storage. This point is demonstrated by the comparison of groundwater ages deduced from ^{14}C and helium concentrations observed in three groups of groundwater of the Bunter sandstone in eastern England (Fig. 11.26). A linear correlation is observed, supporting the ^{14}C-deduced ages and demonstrating the applicability of radiogenic helium as an age indicator. The term *age indicator* is used in regards to radiogenic helium and not *dating method*. This is to emphasize the semi-quantitative, yet hydrologically valuable, nature of the helium dating method.

The ^{14}C dating method is based on a known concentration of ^{14}C in the initial water and depletion of ^{14}C with age. Thus, the older the water gets, the more difficult it is to measure the ^{14}C that is left in it. The ^{4}He method is just the opposite: the initial concentration of radiogenic helium is zero in newly recharged groundwater, but it accumulates with age, and the older the water is, the easier it is to measure the concentration of the radiogenic ^{4}He. Thus, the two dating methods are complementary. A practical limit of ^{14}C dating of groundwater is about 25,000 years, whereas for radiogenic ^{4}He it is 10^4–10^8 years.

14.2 Parameters Determining Helium Concentration in Groundwater

Originally, researchers (Andrews and Lee, 1979; Marine, 1979) assumed that radiogenic helium formed in rocks is dissolved in the associated water. Accordingly, the amount of radiogenic helium accumulating in groundwater is proportional to the time in underground storage, the emanation efficiency of helium from the rocks (nearly 1 in aged rocks), the concentration of uranium and thorium in the hosting rocks, and the rock:water ratio. The amount of helium accumulating in groundwater (expressed in cc STP/cc water) is thus determined by the following equation (Bosch, 1990):

$$He = 10^{-7} t \ \xi He \ (1.21U + 0.287Th) \ d \ (\text{rock:water})$$

Where U and Th are the concentration of uranium and thorium in the host rocks, here expressed in g/g (the respective coefficients incorporate the half-lives and number of helium atoms formed along each disintigration chain (Zartman et al., 1961). The rock:water ratio in common rocks ranges between 4 and 20, and is calculable from the effective porosity. ξHe is the emanation efficiency of helium from the rocks (nearly 1); t is the storage time, or age, of the water, in years; and d is the rock density.

Example: The yearly accumulation of radiogenic helium in a typical groundwater can be calculated applying $\xi He = 1$, $U = 3$ ppm, and $Th = 12$ ppm (as in the average crust), $d = 2.5$ g/cc, and a rock:water ratio of 4. The helium accumulation for 1 year is

$$He = 10^{-7} \text{ (l) (l) } (1.21 \ (3 \times 10^{-6}) + 0.287 \ (2 \times 10^{-6}) \ (2.5) \quad (4)$$

$$= 7.10^{-12} \text{ cc STP/cc water}$$

The equation of the accumulation of radiogenic helium in groundwater is based on a number of assumptions:

All the emanated helium is dissolved in the associated water.
No oversaturation occurs, and therefore no helium is lost.
No helium is added by diffusion from rocks outside the aquifer.
Mantle-derived helium is negligible, or may be subtracted, applying ^{3}He:^{4}He ratios.
No mixing of groundwaters of different ages takes place.
No helium (and other gases) was lost or gained by phase seperation underground.
No helium (and other gases) was lost during ascent to the sample collection point.

These assumptions are discussed in the following sections. Examples of helium concentrations in groundwaters of the world are given in Table 14.1.

14.3 Case Studies and Observed Trends in the Distribution of Helium Concentrations in Groundwater Systems

Helium-rich groundwaters are as a rule encountered only in confined systems situated deep below sea level, an observation that in itself indicates that one is dealing with stagnant systems (section 2.13). Examples, included in Table 14.1 (from Mazor and Bosch, 1992), are:

The Bunter sandstone aquifer, England, reaching a depth of up to 400 m (Andrews et al., 1984).
The Molasse Basin, Austria, in which Oligocene, Cretaceous, and Jurassic aquifers reach depths from sea level to 2000 m (Andrews et al., 1985, 1987).
The Paris Basin, in which Jurassic aquifers are encountered at depths of 1600–1850 m (Marty et al., 1988).
The Great Hungarian Plain, in which Plio-Miocene aquifers reach a depth of 1000 m (Deak et al., 1987).
The Savannah River Plant site, United States, in which a confined Cretaceous aquifer reaches a depth of 300 m (Marine, 1979).
Uitenhage, South Africa, where a quartzite aquifer attains a depth of 70–470 m below the surface and up to 222 m below sea level (Heaton, 1984).

Table 14.1 ^{4}He concentrations in groundwaters around the world (10^{-8} cc STP/cc water)

Sample	He	Remarks	References
Retford, U.K.	3.6	Triassic, sandstone	Andrews and Lee (1979)
Gainsborough, U.K.	29.3	Triassic, sandstone	Andrews and Lee (1979)
Blumau 11, Austria	61.7	Tertiary, sedimentary	Andrews et al. (1984)
Blumau 32, Austria	1152		Andrews et al. (1984)
Uitenhage 14, South Africa	2	Ord., quarzite	Heaton (1984), Talma et al. (1984)
Uitenhage 11, South Africa	425	Ord., quarzite	Heaton (1984), Talma et al. (1984)
Uitenhage 22, South Africa	3970	Ord., quarzite	Heaton (1984), Talma et al. (1984)
Stampriet 45, South Africa	115	Sandstone	Heaton (1984), Talma et al. (1984)
Stampriet 6, South Africa	6630	Sandstone	Heaton (1984), Talma et al. (1984)
Stampriet 50, South Africa	9330	Sandstone	Heaton (1984), Talma et al. (1984)
Stripa, Sweden	39,000	345 m, granite	Andrews et al. (1982)
Stripa, Sweden	182,000	820 m, granite	Andrews et al. (1982)
Lincolnshire, U.K.	1400	Jurassic, limestone	Andrews & Lee (1980)
Molasse, Austria:			
Ried 3	14	Miocene, various sedimentary	Andrews et al. (1985)
Aurolzmunster	330	Miocene, various sedimentary	Andrews et al. (1985)
Schallerbach 2	1900	Oligocene, various sedimentary	Andrews et al. (1985)
Geinberg	5900	Jurassic, limestone	Andrews et al. (1985)

Location	Value	Description	Reference
Bottstein, Switzerland	21,000	621 m, granite	Balderer (1985)
Great Hungarian Plain:			
Heves	22	370 m, Pliocene, sedimentary	Deak et al. (1987)
Heves	164	720 m Pliocene, sedimentary	Deak et al. (1987)
Kaba	1000	Quaternary, sedimentary	Deak et al. (1987)
Savannah River Project, U.S.	20,000	Cretaceous, sedimentary	Marine (1979)
Paris Basin	90,000	Jurassic, limestone	Marty et al. (1988)
Great Artesian Basin:			
The Gap	13,000	Jurassic, sandstone	Marty et al. (1988)
Tiberias, Israel	2700	Rift Valley fault	Mazor (1972)
Zin 5, Israel	7	Rirt Valley fill	Mazor (1972)
Hammat Gader, Israel	980	Rift Valley system	Mazor et al. (1973)
Die Bad, S.A.	21	Fault, Cape System sedimentary	Mazor et al. (1983)
Lilani, S.A.	3500	Archaean metamorphic	Mazor et al. (1983)
Natal Spa, S.A.	14,100	Archaean metamorphic	Mazor et al. (1983)

The Great Artesian Basin, Australia, in which a Jurassic aquifer is up to 2000 m deep (Habermehl, 1980).

14.4 Mantle Helium

The concentration of helium in air-saturated water is at most 5×10^{-8} cc STP/cc water, as seen in Fig. 13.1. Any excess above this concentration is of nonatmospheric origin. In terrains located on stable parts of the crustal plates, radiogenic helium is the sole nonatmospheric source. However, along plate margins and in volcanic systems, mantle helium is added to groundwater as well. These mantle helium contributions are noticeable by their relatively high ^{3}He:^{4}He ratios, and rarely reach 50% of the observed helium. In addition, mantle helium is mainly added by release from mantle-produced rocks. This release is in itself time correlated, and interferes only slightly, if at all, with the application of helium as an age indicator.

14.5 The Helium-Based Groundwater Age Equation

During migration and subsurface storage groundwater comes into contact with crustal rocks that continuously release helium from the decay of uranium and thorium. The basic assumption is made that the water acts as a sink for the helium evolved from the local rocks. The age of groundwater, t, is calculable from the equation

$$t = \mathrm{He}/[10^{-7} \, \xi \, \mathrm{H} \, (1.21\mathrm{U} + 0.287\mathrm{Th}) \, d \, (\mathrm{rock:water})]$$

where t is in years, He is in cc STP/cc water (and neglecting compressibility of the water, the He concentration may also be expressed in cc STP/g water), ξHe is dimensionless, d is in g/cc, U and Th are in g/g and the rock:water ratio is in cc/cc.

Example: In the Milo Holdings 3 well, 1421 m deep, in the Jurassic rock sequence of the Great Artesian Basic, east Australia, the helium concentration of 1000×10^{-8} cc/cc water was found (Mazor and Bosch, 1990). The following local parametric values were applied to calculate the water age: ξH = 1; U = 1.7 ppm; Th = 6.1 ppm; d = 2.6; effective porosity of 0.2 or a rock:water ratio of 4. Thus, the age of this groundwater was found to be

$$t = 1000 \times 10^{-8}/[10^{-7}(1)(1.21 \, (1.7 \times 10^{-6}) + 0.287 \, (6.1 \times 10^{-6})) \, (2.6) \, (4)]$$

$$t = 2.5 \times 10^6 \text{ years}$$

The obtained age for the deep water in the artesian Milo Holdings 3 well is semiquantitative because of uncertainties in the parameters applied in the calculation. Yet, the order of magnitude—millions of years—is of outmost importence: the water resource is not renewed, but it is entirely shielded from the surface and, hence, is immune to anthropogenic pollution.

The low helium concentrations of rocks of known ages indicate emanation efficiencies of 0.8-1 (Tolstkhin and Drubstkoy, 1975; Podosek et al, 1980; and Zaikowski et al, 1987). In fact, helium diffuses through silicates even at room temperature. The concentrations of uranium and thorium vary by a factor of 10 in common rocks, but knowledge of the lithologies of the main aquifer rocks can narrow the uncertainty to a factor of three or less. The rock:water ratio is more difficult to assess, but reasonable values are 4-20. Local information from drilling cores and logs, and hydrological considerations may reduce the estimate of this parameter to a factor of less than three. Thus, uncertainties in the parameters of the age equation introduce errors of at most one order of magnitude to the helium-based groundwater age calculation.

Helium can be reliably measured in waters with an age of 10^4 years, but the age of groundwaters varies widely and may reach 10^8 years. Thus, the measureable helium concentrations in groundwater may vary over four orders of magnitude due to differing ages; therefore, helium concentration data are the most variable parameter of the age equation.

In summary, the various parameters of the helium-based age equation can be resonably well estimated for each case study, and the water age can be calculated semiquantitatively. Thus the available methods cover the whole range of groundwater occurrences, with some overlaps: decades (tritium dating), 10^2-10^3 years (^{14}C dating), 10^5-10^6 years (^{36}Cl method), 10^4-10^8 (^4He dating), and 10^5-10^8 (^{40}Ar dating, still in development).

14.6 Comparison of Helium-Based Water Ages and Hydraulically Calculated Ages

Helium ages in confined groundwater systems are frequently greater than the corresponding hydraulic ages (section 2.10). Some examples include:

- Andrews et al. (1984) calculated flow velocities of 0.7-4 m/year for the confined Bunter sandstone groundwater system in eastern England. Hence, along the 20-km down-gradient flow in the study area the water should attain a hydraulic age of up to 80,000 years, whereas the corresponding helium age is 300,000 years.
- At Blumau, Austria, Andrews et al. (1984, 1985) calculated a hydraulic age of 700-7000 years but a helium age of 400,000 years for well 32 in the Molasse Basin; and the Birnbach well, in the deep Jurassic aquifer, yielded a helium age of ~10 million years.
- Heaton (1984), discussing the confined groundwater systems of Uitenhage and Stampriet, South Africa, concluded that the helium ages are 70 times greater than the hydraulic ages.

Torgersen and Clarke (1985) found for the confined Jurassic aquifer in the Great Artesian Basin, Australia, that the observed helium concentrations were 70 times higher than expected for their calculation of the hydraulic ages.

Discrepancies of this type led the above mentioned investigators to the conclusion that helium-based water ages are exaggerated. It was, thus, suggested by these researchers that the assumption that all the formed helium is dissolved in the rock-associated groundwater is wrong, and that the measured waters harvested helium that diffused from a thick sequence of the crust. This crustal helium diffusion model is ruled out by observations discussed in the following two sections (14.7 and 14.8). The helium-based groundwater dates seem to be valid, as can be checked by means discussed in sections (14.11 and 14.12). In contrast, it has been suggested that the hydraulic calculations of groundwater age have their shortcomings, often producing too young groundwater ages (Mazor and Nativ, 1992, 1994). The hydraulic travel time or age calculation applies Darcy's law, which assumes hydraulic interconnection between wells from which data are applied. Failure to identify hydraulic barriers that separate different groundwater systems will result in apparent too young ages to be calculated.

14.7 Does Helium Diffuse Through the Upper Crust?

Water has a high helium dissolution capacity, and groundwaters in nature contain helium in a concentration range spanning over five orders of magnitude, as seen in Table 14.1. In all these cases the partial helium pressure in the aquifers was orders of magnitude below the hydrostatic pressure, not to mention the lithostatic pressure, ensuring no gas losses took place.

The water in each aquifer dissolves all the available helium and "shields" the overlying aquifers from helium fluxes of the type suggested by the crustal helium diffusion model.

14.8 Helium Concentration Increases with Depth: Groundwater as a Sink of Helium

In confined groundwater systems deeper water strata often contain higher helium concentrations. This pattern is observed with depth increase along the dip of confined aquifers, or in samples obtained from a sequence of aquifers of different depths sampled at one location. In some examples, presented in Figure 14.1, the depth of the water is known from direct measurements, and in others it is reflected in the temperature, which reflects the depth.

The increase in helium concentration with depth is taken as an indication that the deeper waters are older, that is, they have been stored for longer times as

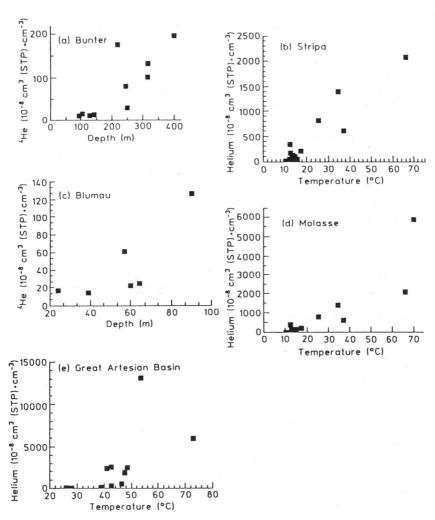

Fig. 14.1 Helium concentrations as a function of aquifer depth or temperature (following Mazor and Bosch, 1992a); Bunter sandstone aquifer, eastern England (data from Andrews et al., 1984); Stripa granite, Sweden (data from Andrews et al., 1982); Blumau, Austria (data from Andrews et al., 1984); Molasse Basin, Austria (data from Andrews et al., 1981); Great Artesian Basin, Australia (data from Torgersen and Clarke, 1985).

compared to shallower waters. In this context it should be remembered that the cited groundwater systems are confined below sea level and are devoid of drainage or have restricted drainage (along major faults or due to abstraction). The water in such deep basins may have been stored there since the burial of the confining rocks, or since subsidence of these rocks to below the depth of the former base of drainage. In certain cases one may deal with genuine connate water, stored in the rock voids since their formation. The increase in helium concentration with depth is in many cases distinct, as seen in Table 14.1 for the examples of Blumau, Uitenhage, Stampriet, the Molasse Basin, the Great Hungarian Plain, and the Great Artesian Basin.

The increase of helium concentration with depth in these examples cannot be explained by helium diffusion from deep parts of the crust. This would require unrealistic high diffusion gradients. As the examples of Table 14.1 reveal, helium increases by factors of 2–20 over depth increases of only a few hundred meters. The existence of neighboring groundwater units, or compartments, with different helium concentrations indicates a high degree of helium fixation and a negligible amount of diffusion. Thus, helium dissolution in water seems to be the major mechanism of this fixation, as will be discussed later.

14.9 Effect of Mixing of Groundwaters of Different Ages on the Observed Helium Concentration

Mixing of groundwaters before emergence at the surface occurs naturally or as a result of abstraction operations, and is rather common (Mazor, 1985, 1986; Mazor et al., 1973). The helium concentration in a groundwater mixture has no meaning in terms of age. Hence, a check for possible mixing of waters should be performed before helium concentration is applied for groundwater dating. This can be done with chemical, physical, and isotopic data obtained from a number of hydraulically relevant water sources or observed in periodically collected samples of springs or wells (Mazor et al., 1985). Mixing can also be established by discordant groundwater ages (Mazor, 1986). Examples of discordant ages, obtained by applying tritium, ^{39}Ar, ^{14}C, and He data, are given in Table 14.2 for wells in the East Midlands Triassic aquifer in England. Thus, it is highly recommended that water measured for helium also be measured for other age indicators. Samples with distinct tritium concentrations but low ^{14}C concentrations represent mixtures. Similarly, ^{14}C ages that are lower than the corresponding helium ages indicate that one deals with a mixture. The age pattern of mixed waters is intrinsically tritium age $<$ ^{39}Ar age $<$ ^{14}C age $<$ ^{4}He age.

Observation of discordant ages provides information that is crucial to the understanding of a studied groundwater system. Information on the age of the intermixing water end members may be obtained from their reconstructed parameters (Mazor et al., 1986). In any case, the age of the young end member is less than the lowest age obtained, and the age of the oldest end member is higher than the highest age calculated by applying the raw data.

Table 14.2 Discordant groundwater ages (years) in the East Midlands Triassic aquifer, U.K.

Well	Tritium	^{39}Ar	^{14}C	Hydraulic	He
Far Baulker 2	~15		0–2000		
Ompton 3	~40	25	1500–5000		
Halam 3		220–285	0–4500	3000–8000	
Markham 3		240–320	3500–7000	5000–14,000	
Grove 1			7000–11,000	5000–14,000	
Newton 1			23,000–28,000	6000–16,000	
Newark 1			>29,000	20,000–56,000	×10 hydraulic age
				21,000–60,000	×10 hydraulic age

Source: Data from Andrews et al., 1984.

14.10 Helium Losses or Gains Due to Phase Separation Underground or During Ascent to the Sampling Point: Modes of Identification and Correction

Water in confined basins is characterized by high temperature, pressure, and dissolved gas concentrations. With the formation of a vapor or steam phase, helium is preferentially transferred into the gas phase, along with atmospheric Ne, Ar, Kr, and Xe. At the same time, the residual water is depleted in the noble gases. An example from the Great Artesian Basin is given in some detail: The samples were collected by the same field team that provided samples for the work of Torgersen and Clarke (1985), our samples being collected in helium-tight glass tubes with spring-loaded high-vacuum stopcocks at their ends. The noble gas concentrations are seen to vary over a wide range (Fig. 14.2). Percent retention of atmospheric Ne, Ar, Kr, and Xe was calculated compared to 15°C air-saturated recharge water (Fig. 14.3). Retentions as low as 16% are seen, indicating gas was lost. In contrast, cases with over 100% retention are observed, indicating gas was gained.

These significant changes in gas concentrations are not surprising, as the confined waters of the Great Artesian Basin are rich in CO_2 and methane, and attain temperatures of over 90°C. Hence, samples collected at the (free-flowing) well head are likely to lose gases during ascent, and gas seperation may also occur underground as a result of pressure drops related to the intensive abstraction. The separated gases, forming gas pockets or compartments, are tapped by some wells, producing water enriched by noble gases.

The original helium concentration in the confined water can be reconstructed by dividing the observed concentration by the the fraction of gas retained, as calculated from the accompanying atmospheric gases. These corrected helium concentrations, plotted as a function of the distance from the recharge area (Fig. 14.4), reveal a reasonable correlation, which provides confirmation of the reliability of the corrected helium values.

The observation of older groundwaters in the deeper parts of the Great Artesian Basin should not be mistaken as indicating inward flow. The older ages are in good agreement with the evolution of a subsidence basin—the inner parts are buried and disconnected from the groundwater through-flow regime for longer time periods.

14.11 Internal Checks for Helium Ages

A number of internal checks is available in specific case studies:

Good correlation of He concentration with ^{14}C ages, as obtained for Blumau (Fig. 14.5), the Bunter sandstone in eastern England, or at Uitenhage. This

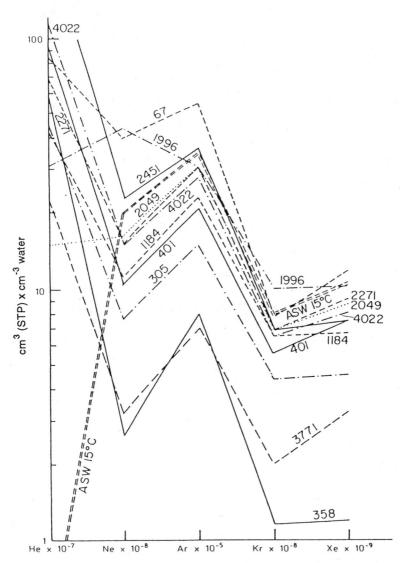

Fig. 14.2 Noble gas concentrations in water samples from the confined part of the Great Artesian Basin, with well numbers (following Mazor and Bosch, 1992b). The concentrations of atmospheric Ne, Ar, Kr, and Xe and of nonatmospheric He vary over a wide range.

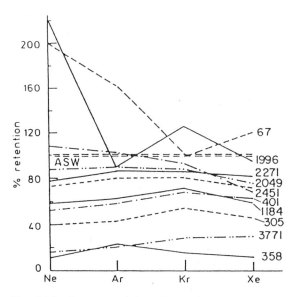

Fig. 14.3 Percent retention of the atmospheric noble gases of the Great Artesian Basin samples included in Fig. 14.2 (compared to air-saturated water at 15°C at sea level). Significant gas losses are seen.

check is limited by the half-life of ^{14}C to rather recent shallow aquifers. In addition, this correlation can be taken as an approval of the deduced age only if absence of mixing is demonstrated.

Good correlation with depth of confinement, as seen for the Great Artesian Basin (Fig. 14.4), and as indicated by Andrews et al. (1984) for the Bunter sandstone aquifer.

Good correlation with local depth, as seen in Blumau, the Bunter sandstone aquifer of eastern England, the Stripa granite in Sweden (Fig. 14.1), and in other examples included in Table 14.1, or good correlation with temperature, as seen at the Molasse Basin in Austria and the Great Artesian Basin in Australia (Fig. 14.1).

Positive correlation of radiogenic helium and radiogenic ^{40}Ar concentrations provides another check, observed in the geothermal fields of Larderellow, Italy (Mazor, 1979b), Yellowstone and Lassen Volcanic (Mazor and Wasserburg 1965), or Cerro Prieto, Mexico (Mazor and Truesdell, 1984).

Agreement with the geological setting. The obtained groundwater age should be between the age of the aquifer rock and the time of disconnection from the local base of drainage.

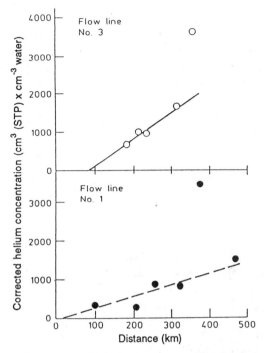

Fig. 14.4 Helium concentration (corrected for gas losses) as a function of distance from the recharge area along two transects through the Great Artesian Basin (following Mazor and Bosch, 1992a). The good correlation provides a positive check on the validty of the corrected helium data, as the deeper inland groundwater must have been trapped earlier.

14.12 The Issue of Very Old Groundwaters

As was stated earlier, researchers are confronted with the observation that hydraulic ages, calculated in a straightforward way for cases of confined systems, are significantly shorter than the corresponding helium ages. The occurrence of old and even very old groundwaters, with ages up to 10^8 years, seems plausible in light of the local geological and hydrological histories of most deep basins. Much interest is to be found, in this context, in the results reported by Zaikowski et al. (1987) for deep Permian and Pennsylvanian brine aquifers at the Palo Duro Basin, Texas. The researchers found 90,000–590,000 × 10^{-8} cc STP He/cc brine, and computed water ages in the range of 100–300 million years, comparable with the age of the host rocks. Marty et al. (1988) observed in 1600–1850 m deep Jurassic limestone aquifers of the Paris Basin up to 90,000 × 10^{-8} cc STP He/cc water, and calculated ages of 10–140 million

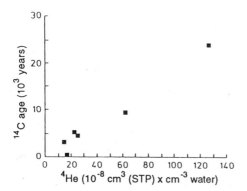

Fig. 14.5 ^{4}He concentration–^{14}C age correlation in the Blumau wells (following Mazor and Bosch, 1992a; data from Andrews et al., 1984).

years, with a grouping around 70 million years. This may well be the age of the water entrapment due to tectonic events that disconnected the system from its former drainage system.

14.13 Conclusions

The accumulation of radiogenic helium in confined groundwaters is mainly determined by the duration of water entrapment. Applying the data available on uranium and thorium concentrations in the host rocks and rock:water ratios, reliable though semiquantitative dating is possible in the range of 10^4–10^8 years.

Interference of helium diffusion from rocks outside the aquifers seems to be negligible because (1) diffusion, if dominant, would tend to cause waters in different deep confined systems to have similar helium concentrations, but nature reveals in these systems values that range over five orders of magnitude; and (2) the diffusion coefficient of argon is orders of magnitudes lower than that of helium. If diffusing helium is substantially added to confined water, but practically no argon diffuses into the aquifer, then no correlation is expected between radiogenic ^{4}He and ^{40}Ar in water samples collected at different parts of a groundwater system. The data on the joint occurrence of helium and radiogenic argon in confined water systems is scarce, but if such a positive correlation turns out to be common, it will further demonstrate the limited interference of helium diffusion in dating groundwater with helium concentrations.

Contributions of mantle helium (mainly via magmas and water-dominated fluids) may be detected and corrected by ^{3}He:^{4}He ratios.

In order to date groundwaters, they have to be studied for possible mixing of waters of different ages. Helium concentrations should be applied for the calculation of meaningful ages only for nonmixed water, or for reconstructed end members.

Observed helium concentrations should be applied for water dating only in conjunction with the ANG, to correct for gas losses or gains.

Internal checks are necessary for reliable water dating—for example, correlations of helium concentration with distance from the recharge area, water depth, temperature, or concentration of accompanying radiogenic ^{49}Ar.

14.14 Summary Exercises

Exercise 14.1: Study Fig. 11.26. Take the average helium concentration in group C as 28×10^{-8} cc He/cc water, and assume that the hosting rocks contain 2 ppm U, 2 ppm Th, the density of the rocks is 2.5, the rock:water ratio is 10, and the helium emanates from the rocks to the groundwater by a coefficient of 0.8. Applying the age equation, calculate the age of the groundwater. Does it seem reasonable, in light of Fig. 11.26?

Exercise 14.2: Sort the isotopic groundwater dating methods into decaying and cumulative methods. What are the advantages of each group?

Exercise 14.3: How can the radiogenic helium be sorted out of the total helium measured in a sample?

Exercise 14.4: On what arguments is it claimed that most of the produced radiogenic helium is released from the rocks and dissolved in the associated groundwater (i.e., the helium emanation coefficient is close to 1)?

Exercise 14.5: Which mechanisms may cause losses of helium from sampled groundwater? How can we correct for these losses so that groundwater dating is feasible?

Exercise 14.6: How can the patterns seen in Fig. 14.1 be understood in terms of the parameters controlling radiogenic helium contribution to groundwater?

Exercise 14.7: Figure 14.2 incorporates a large body of noble gas data obtained from a huge subsidence basin. What can be learned from the reported noble gas fingerprints?

Exercise 14.8: How was the data plotted in Fig. 14.2 transferred into the data plotted in Fig. 14.3? What patterns are observable in the Fig. 14.3?

15
INTERRELATIONS BETWEEN GROUNDWATER DATING, PALEOCLIMATE, AND PALEOHYDROLOGY

15.1 Scope of the Problem

The glacial phases were events of extreme climatic conditions, manifested by a significant increase in the global snow cover. By how many degrees was the local ambient temperature lower? What were the temperatures in the intermediate warmer phases? How abrupt were the climatic changes? Were there any phases of temperatures higher than the current ones?

Answers on these questions will help us understand global climate dynamics. This topic is of much interest in connection with the discussion of possible climatic changes induced by anthropogenic activity such as release of CO_2 and other greenhouse gases, release of dust and aerosols, changes in the ozone layer, deforestation, and desertification. Knowledge of natural temperature fluctuations is essential in order to properly interpret fluctuations in the temperature record available for the last century. Are they in the range of the natural noise, or are they outstanding and alarming?

The concentration of the atmospheric noble gases dissolved in groundwaters, coupled with the dating of the respective groundwaters, provides a most valuable method to reconstruct paleotemperatures. The work done so far in this field by various researchers (section 13.8) demonstrates the applicability of the method, and thus opens a most promising field of research. The future of this field encompasses several tasks:

Development of a technique to securely collect the water samples at specified depths in boreholes in order to prevent gas losses during ascent.
Routine collection of the auxiliary data needed for proper interpretation of the noble gas temperatures and isotopic dating.

Development of strategies to select the best sampling sites, for example, preference of stagnant (naturally capsulated) groundwater systems.

Global coverage of the sampling locations.

Application of paleo-input functions in the final calculations of both the temperatures and water ages.

15.2 The Need to Reconstruct Paleo-Input Values

Practically all computations discussed in this book so far are based on data measured in the field, combined with certain coefficients and certain assumptions pertinent to the nature of the studied system: for example, the soil $\delta^{13}C$ value, applied to correct the measured ^{14}C value, that is applied in the groundwater age calculation (sections 11.6 and 11.7). This soil $\delta^{13}C$ value is dependent on the type of growing plants, and these are climatically controlled. Thus, applying a local present day $\delta^{13}C$ value to water that contains only 10 pmc is improper. During the age that is indicated by the low ^{14}C concentration, a significantly different climate may have prevailed, and therefore a different plant community predominated, along with a different soil $\delta^{13}C$ value.

A second example: The $^{36}Cl:Cl$ ratio in local precipitation is dependent on, among other parameters, the distance from the sea. Thus, applying the present local $^{36}Cl:Cl$ ratio to calculate the age of groundwater may be wrong if the shoreline was located at a significantly different distance.

A third example: At certain geological and hydrological setups, located in low lands, one can conclude that the local unconfined groundwater system is recharged through a porous aerated zone, resulting in slow inflow. In such cases the concentration of the atmospheric noble gases reflects the average local ambient temperature (section 13.6). However, a deeper groundwater in the same site may be old and was recharged during glacial times when the snow lines were substantially lower and closer. At that time, recharge may have been mainly from snowmelt, and the concentration of dissolved atmospheric noble gases may reflect a temperature that is close to the snowmelt temperature, rather than the average ambient temperature (section 13.8).

A fourth example: The recharge altitude of an unconfined water system is, under suitable conditions, reconstructed from the composition of the stable isotopes (the altitude effect; sections 9.8–9.10). The relevant formula is locally calibrated by measurements of springs of known recharge altitudes. However, ancient groundwater may have been recharged during a different climate, expressed by a different ambient temperature, and a different recharge altitude, as a result of a different snow line altitude. Thus, the current coefficient of the altitude effect cannot be applied to old groundwater, and the respective paleocoefficient has to be estimated.

15.3 Stagnant Systems as Fossil Through-Flow Systems

Stagnant groundwater systems are fossil through-flow systems that subsided into the zone of zero hydraulic potential (section 2.13, Fig. 2.14). The deeper stagnant systems became hydraulically sealed and isolated by overlying sediments that accumulated as subsidence continued (section 2.14, Fig. 2.16). Stagnant groundwaters are, in a way, samples that were naturally collected in the past and kept underground in closed system conditions. It is up to us to locate them and incorporate them in the global paleoclimate survey networks.

The resolution of groundwater dating (e.g., by ^{14}C) in through-flow systems is reduced by dispersion and other flow-induced disturbances, such as deviation from piston flow. In this respect stagnant systems are advantageous because they carry the age signal of the time of burial with no further dispersion effects.

15.4 Screening of Data for Samples of "Last Minute" Mixing of Groundwaters of Different Ages

Mixing of different groundwaters is always a possibility. Data from a sample that is mixed from different water sources is meaningless for both the dating of the water and the paleoclimate reconstruction. For example, a water sample from Baden, Switzerland, revealed 6 TU and only 7 pmc (Mazor et al., 1986). This set of data makes no sense, because it means the water has a low age (post-bomb) and a high age (several half-lives of ^{14}C) at the same time. The Baden sample is an example of a mixture of very young water (post-1954) and very old water (>30,000 years). Any two age indicators that have different half-lives can serve to identify mixed water samples.

A second example is a well at the Lachlan Fan, Australia. It tapped confined water in Pliocene rocks, and revealed 36 pmc (indicating an age of about 1 half-life; about 6000 years), along with a $^{36}Cl:Cl$ ratio of 78×10^{-15}, which is significantly lower than the ratio reported for the active recharge area ($^{36}Cl:Cl$ ratio of 170×10^{-15}), indicating a much older age of about 1 ^{36}Cl half-life, or about 300,000 years. Clearly, the accompanying δD and $\delta^{18}O$ values of such samples are a mixture too, not reflecting the composition of the young water or that of the old water, and the same holds true for the noble gas concentrations, ion concentrations, or concentrations of other age indicators. Data from mixed groundwater samples are misleading if applied for paleoclimate reconstruction or dating. Hence, the importance of screening data prior to their interpretation in terms of paleoclimate and age.

The mentioned examples of contradicting tritium and ^{14}C concentrations, or ^{14}C and ^{36}Cl concentrations, provide a clear indication that one deals with a

Interrelations in Groundwater Dating

mixture of waters. However, in many cases of mixed water samples, the age indicators are not that contradicting (e.g., mixing of two pre-bomb waters with 0 TU, or two waters older than 30,000 years with 0 pmc), and other methods for screening mixed water samples are needed, such as variations in ion concentration, observed in samples periodically collected from the same spring or well, or mixing lines obtained in ionic scatter diagrams of water samples collected in many sources at the same area (sections 6.6 and 6.7).

In some cases the composition of the intermixing water end members can be retrieved and applied for dating and reconstruction of paleoclimate. Application of data of samples of single (nonmixed) groundwaters is always preferable, and in order to identify them the concentration of the various age indicators must always be identified, and chemical analyses of repeatedly collected samples are needed.

15.5 Paleohydrological Changes

15.5.1 Mode of Flow of Recharged Water in the Aerated Zone

Recharge through a porous medium is relatively slow, and the water flowing down to the saturated zone undergoes two important processes:

Partial evaporation and transpiration from the uppermost region, the former process influencing the stable isotope composition.
Constant temperature reequilibration, resulting in water entrance into the saturated zone at the temperature prevailing at the base of the aerated zone.

The temperature prevailing at the base of the aerated zone equals the local average ambient temperature if the aerated zone is at a depth on at least 15 m. At greater depths heating by the geothermal gradient is added. In contrast, recharge intake through fractures and karstic conduits is fast, and:

No evaporation takes place from the upper aerated zone, preserving the isotopic composition of the recharged precipitation water.
The water enters the saturated zone with a "memory" of the temperature prevailing during the recharge season.

Thus, the nature of the recharge intake of a studied paleowater has to be defined by comparison with the present nature of the relevant recharge zone. The effect in terms of noble gas paleotemperatures can be demonstrated by an example from northern Israel: The average temperature in the rainy winter season is about 10°C lower than the average annual temperature. This effect has been observed in noble gas temperatures of recent (tritium containing) groundwaters that reflected the average annual temperatures in springs fed by porous aerated zones, in contrast to about 6°C lower noble gas temperatures obtained for karstic springs fed by winter rains and snowmelt water (Herzberg and Mazor, 1979).

15.5.2 Sea Level Changes

Water levels in the oceans were lower by over 150 m during the ice ages, and similar water level fluctuations occurred in lakes and inland seas due to climatic changes and tectonic events. The sea level changes influenced the near-shore regions in the following ways:

The depth of the water table increased with the lowering of the sea level, and decreased with the rise in the sea level.
The depth of the aerated zone varied with the changes in the sea level.
The intensity of the respective hydraulic gradient varied.
The depth of karstic activity changed, forming paths of preferred flow.
The rate of chlorine fallout, determining the paleo-input value of the relevant initial $^{36}Cl:Cl$ ratio (needed for ^{36}Cl-based groundwater dating) could vary from the value calculated (Bentley et al., 1986), or locally measured, in recent groundwaters (Mazor, 1992).

15.5.3 The Special Case of the Drying up Event of the Mediterranean Sea

Since the 1960s more and more reports have been made describing the occurrence of deep Miocenian paleochannels detected in Mediterranean countries— in Africa, the Middle East, and Europe. The great depth of the paleochannels made it clear that the Mediterranean was, at that time, thousands of meters lower. Deep sea research drilling conducted in the Mediterranean Sea encountered gypsum among the bottom sediments, which is also exposed at the surface in Sicily. Gypsum is precipitated out of seawater whenever it is evaporated to less than one-third of its initial volume. Thus, the finding of Messinian (Miocene) gypsum at the bottom of the Mediterranean Sea indicated that the sea was closed and highly evaporated, lowering its water level by 2000 m or more, as compared to its present level. Rock salt was found as well, and this material is precipitated from seawater only by evaporation to less than 10% of the initial volume. Finally, a bed with freshwater fossils was also encountered, indicating the sea was entirely locked at the Straits of Gibraltar, disconnecting it from its water supply from the Atlantic Ocean. So the entire Mediterranean basin dried up for a while, and then the Straits were reopened and the sea filled up again.

This is perhaps the most dramatic sea level fluctuation during the late geological periods, dwarfing in comparison the sea level changes during the glacial periods. The hydrological consequences must have been enormous:

The water table in the surrounding countries must have dropped by hundreds to thousands of meters.

The paleohydraulic gradient changed continuously, a fact that has to be taken into account in hydraulic calculations of the ancient groundwater flow velocities.

The depth of the aerated zone increased accordingly, influencing for example the noble gas recorded intake temperature (sections 13.7 and 13.8).

The surface waters incised the landscape, forming deep channels that were filled with new terrestrial and marine sediments, as the seawater rose again. These deep paleochannels serve, according to the nature of the filling sediments, either as potentially preferred flow paths, and preferred underground drainage, systems or as hydraulic barriers.

The seashore receded significantly, lowering the relative amount of arriving sea spray.

The sea itself turned saltier, and the former offshore terrains that became exposed, were covered with evaporitic salts. These two features must have increased the salt transport to the land (counteracting the effect of the receding shore) and changed the chemical composition of the sea spray.

As a result of the mentioned processes, slightly saline groundwaters must have formed in the surrounding continents during the regression of the Mediterranean Sea and during its refilling. The increase in the salinity of those groundwaters must have stemmed from the mentioned increase in the amount of arriving sea spray, and of an expected drop in the amount of rain as a result of the drying up of the Mediterranean Sea.

As the level of the recovering sea rose, the terminal base of drainage also rose. As a result, newly formed groundwater covered old groundwater. The older groundwater was submerged, and turned stagnant as it was buried in the zone of zero hydraulic potential (section 2.13, Fig. 2.14). Thus, one might expect that relatively saline old groundwater will be encountered in boreholes beneath the active through-flow zone, or base flow, in the coastal plains of the Mediterranean Sea. This is indeed intensively observed in Israel: boreholes that penetrate 30–100 m below sea level in various parts of the coastal plain encounter groundwaters with 500–700 mg Cl/l (Fig. 15.1). The composition of this more saline groundwater is characterized by less sodium and more calcium, as compared to seawater composition (i.e., in the direction of groundwater formed from sea spray of evaporitic nature). Overpumping of the shallow fresh coastal plain groundwater resulted in salinization to a degree that forced shutdown of wells.

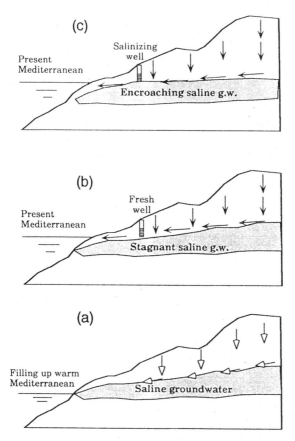

Fig. 15.1 Suggested formation of slightly saline groundwater, found in the Mediterranean coastal plane of Israel. (a) The filling up of the basin of the Mediterranean Sea as a result of the reopening of the Straits of Gibraltar: intensified sea spray of evaporitic nature prevailed, resulting in saltier rain and saline groundwater (open arrows). (b) The present Mediterranean Sea (open to the Atlantic Ocean), characterized by a higher sea level (bringing the ancient saline groundwater to below the terminal base of drainage, and thus making it stagnant), less sea spray, and formation of fresher groundwater (black arrows) overlying the ancient saline groundwater; a well producing fresh groundwater is marked. (c) Overpumping causes local lowering of the level of the freshwater table, resulting in lateral encroachment of remnants of the ancient saline groundwater, causing salinization of a coastal well.

15.6 Paleoclimate Changes

15.6.1 Changes in the Recharge Season: Possible Amplification of the Temperature Change Recorded by the Noble Gases

A relatively small drop in the average temperature during the Ice Age could cause a shift of the rainy season from summer rains to winter rains. Such a change in the recharge season would not affect the noble gas based paleo-temperatures in the case of groundwater recharged through a porous aerated zone of a thickness of greater than 15 m, that is enough to average out the seasonal and annual temperature variations. However, the change in the recharge season would result in an amplification of the temperature change recorded by noble gases, in the case of fast percolating water through a thin or conduit-dominated aerated zone. In such systems the noble gas concentrations reflect a temperature that is between that of the rainy season, or snowmelt, and the average annual temperature (Herzberg and Mazor, 1979).

Effects of this nature should be borne in mind when addressing an inconsistency between reconstructed oceanic and continental temperatures, as pointed out by Stute et al. (1995). The authors reported a temperature decrease of about 5°C during the last glacial maximum, as recorded by noble gases in groundwaters at several locations in temperate countries, whereas the corresponding temperature decrease of the ocean surface was only about 2°C, as indicated by oxygen isotope measurements on foraminifera and fauna abundances in deep sea sediments (section 13.8.3).

15.6.2 Changes in the Role of Snowmelt as a Recharge Agent

During colder climates the snow lines were lower, in extreme cases by up to 1000 m. This influenced:

The role of snowmelt in the recharge of a studied paleowater, relevant to the understanding of obtained noble gas based paleotemperatures, as discussed above.
The altitude of the area of recharge, relevant to the altitude effect, taken into account in stable isotope paleoclimate deductions (section 9.8).
The altitude to be applied as paleo-input in calculating noble gas temperatures. For example, a change of 200 m in the altitude of the recharge area implies a nearly 2.5% change in Xe solubility, equivalent to about 1°C in the noble gas temperature calculation (section 13.2).
The hydraulic gradient, to be applied for hydraulic groundwater age calculations.

15.6.3 Type of Plant Community that Grew in the Ancient Recharge Zone

The type of plant community that grew in the paleorecharge area is relevant to the $\delta^{13}C$ input value, applied to correct the measured ^{14}C concentration, in order to correct for water-rock interactions (sections 11.6 and 11.7). Plant communities in the ice ages were exposed to colder climate, generally drier air, and a distinctly lower concentration of CO_2 in the atmosphere. These parameters could cause the growth of plants that produced different (more positive?) $\delta^{13}C$ values in the soil air, as compared to the plants growing at present in the recharge area. This necessitates a different (smaller?) correction of the measured ^{14}C in order to calculate the initial ^{14}C concentration in a studied paleowater.

15.7 Selecting Sites for Noble Gas Based Paleotemperature Studies

In order to get noble gas based paleotemperatures that represent the average annual temperatures, study areas should be remote from past snowmelt recharge areas, and they should include only systems that received recharge through thick enough aerated zones, in which porous flow dominated. On the other hand, the seasonal variations that accompanied the climatic changes of the past are best retrieved from groundwaters that were recharged through thin aerated zones or karstic domains, and were located near former snowmelt recharge zones.

15.8 Stages of Data Processing

The data processing has to be done in stages. The data have to be screened for possible mixings of water from different aquifers, and only data from single aquifer samples, or values deduced for end members, should be applied in the following stages.

Calculation of an approximate water age, applying the present-day input functions, such as initial ^{14}C concentration in the recharged groundwater, and initial $\delta^{13}C$ value of soil CO_2; initial $^{36}Cl:Cl$ ratio in the recharged water; location and altitude of the present base of drainage, and the respective present hydraulic gradient, should be made.

With the crude age one can estimate the paleoenvironmental conditions in terms of glacial or interglacial period, sea level, distance of paleorecharge area from the seashore, season of major recharge, altitude of the former water table, former evapotranspiration intensity, or the type of plant community that prevailed at the studied paleorecharge site. These considerations should lead to the selection of the pertinent paleo-input functions.

One must then recalculate a more precise age for the studied groundwater, applying the deduced paleo-input functions. Final reconstruction of the paleo-climate conditions, based on the recalculated ages can then be done.

15.9 Summary Exercises

Exercise 15.1: An observation well, Curiosity 1, is located 3 km from the coast in a region of temperate climate. The well head is at an altitude of 15 masl, and water was sampled in a marine limestone aquifer from a depth of 105 m. A ^{14}C-based age of 17,000 ± 3000 years was deduced, and the δ^{13}C value was −8.1‰. Another observation well, Curiosity 2, has been drilled 20 m away to a depth of only 18 m, and it reached water of a recent ^{14}C age, and a δ^{13}C value of −8.3‰. Discuss the past and present hydrological setup of the study area.

Exercise 15.2: Following the conclusions stated in the answer to Exercise 15.1, is there a need to apply a correction to the deduced water age because the plant community could be different from that during a previous climate?

Exercise 15.3: Can there be some doubt on the validity of the ^{14}C-based age? What other parameters have to be measured?

Exercise 15.4: Is there another way to identify possible mixing of different waters as a result of corrosion of the casing of a deep well (or mixing along a fault plain in the case of a spring)?

Exercise 15.5: In light of the information discussed so far, is the Curiosity 1 well a good candidate for a study of noble gas based paleotemperature? Explain.

Exercise 15.6: What type of temperature will the atmospheric noble gases reflect?

16
DETECTING POLLUTION SOURCES

16.1 Scope of the Problem

Large-scale human operations carried out on the surface of the earth include ever-increasing intervention in groundwater systems. All uses of water result in an increase in dissolved components, all used water has to be disposed of somewhere, and all disposed water eventually reaches existing water reservoirs: lakes, rivers, groundwater, and the ocean. Hence, an intrinsic part of water use is water contamination—the addition of components that were not there before. Water quality control is, thus, an integral part of water production and consumption and, in the same way, water monitoring and tracing of pollutants is an integral part of the hydrochemist's field work.

Monitoring of contaminants is somewhat like troubleshooting, but each contamination case study also has a scientific aspect, because each pollution case serves as a large-scale tracing experiment. The tracing of pollutants complements the methods presented in section 6.5, which is devoted to the establishment of hydraulic interconnections, the basic premise that underlies most groundwater models.

The use of artificial tracers introduced into groundwater systems, for direct tracing of hydraulic interconnections, is limited to wells that are tens of meters apart, and in karstic systems artificial tracing may reach a few kilometers. For larger-scale experiments enormous quantities of tracers are needed, and these are costly and rarely authorized by the local health authorities. Pollution from a landfill, waste dump, sewage disposal, factory effluents, or pesticide storage, are all threatening. It is our challenge to use such cases as large-scale tracing experiments, and to use each catastrophe to improve our understanding of the hydraulic underground plumbing.

In essence, contamination monitoring and pollutant tracing are special cases of the general topic of water mixing; hence, the emphasis placed on the detection of mixing of different water types, identification of end members, and calculation of mixing percentages (sections 6.4, 6.6 and 6.7). The mixing in contaminated water systems is twofold:

Introduction of the contaminants into uncontaminated water.
Introduction of specific properties of the water, which serves as a carrier for the contaminants, that may differ from the characteristics of the threatened groundwater system.

Examples of these include irrigation water that is partially evaporated and becomes enriched in deuterium and ^{18}O, providing a signature that differs from the local groundwater into which the irrigation water carries fertilizers and pesticides. Water used by industry often comes from rivers or surface reservoirs with atmospheric tritium concentrations and enrichment in the heavy isotopes of hydrogen and oxygen, whereas threatened groundwater may be devoid of measurable tritium and have a light isotopic signature.

In the deciphering of contamination problems all the tools of the hydrochemist come into play: many parameters have to be measured, repeated sampling is needed to understand the dynamics of the systems, and large amounts of data have to be processed. Based on the observations, a conceptual model is constructed, and this model has to be checked by quantitative (mathematical) modeling and by further checking of predicted consequences.

16.2 Detection and Monitoring of Pollutants: Some Basic Rules

16.2.1 Prepollution Database

Let's begin the discussion with a question: 240 $mgNO_3/l$ have been detected in a farm well, much above the permissible concentration for domestic water. What should be done?

Two steps are needed: (1) notify the local population of the finding; and (2) search for the source of the contaminant.

The first step in the search is to establish the time when the pollutant arrived, and, in a case like nitrate, to find out whether it is from a local natural source or a man-made phenomenon?

When a pollutant is detected, all contamination sources should be considered. Knowledge of the pollutant arrival time may eliminate some suspected sources that arrived at later dates, and helps narrow the list of possible sources of contamination. Hence the need for historic data (section 7.5). This information may be gathered from archives at local authorities, obtained from the well owner, or found in published reports. The recommendation to collect historical data as a basic part of every pollution investigation sounds trivial, but in reality it is only partially followed. The need for a prepollution database is a major incentive for regional hydrochemical surveys that are performed periodically.

16.2.2 Identification of Specific Labels in Potential Contamination Sources

Work on potential pollution sources requires a thorough knowledge of the composition of potential pollutants. Maximum parameters should be measured on every industrial effluent that leaves a factory, and the fluid that is formed in each landfill. The parameters to be measured include temperature, pH, major dissolved ions, trace elements (mainly metals and metal compounds), organic compounds, and isotopic compositions of hydrogen, oxygen, carbon, sulfur, and nitrogen. To perform such analyses, special laboratories have to be contacted. In certain cases the main product may serve as the label, and in other cases the labels are supplied by accompanying compounds, which in themselves may not be poisonous.

16.2.3 Movement of Pollutants in the Aerated Zone

Most pollutants are released on the ground and have to pass through the aerated zone before they arrive at the saturated groundwater zone. The nature of the aerated zone must be well understood in each case. Aerated zones with dominating flow in pores may retard wastewaters and cause biological decomposition of part of the toxic components or absorption of toxic ions on clay particles. These protections are absent in areas of conduit-dominated recharge. Hence, knowledge of the flow regime in the aerated zone can help in assessing the influence of potential contaminating agents.

The mode of fluid flow in the aerated zone may be changed by the hydraulic aspects of fluid waste disposal. Constant release of large amounts of fluids may cause a local rise in the hydraulic head. Coupled with chemical fluid-rock interactions, this may form new high-conducting conduits that can lead the contaminants directly into the saturated zone. In other cases, fine particles that come with the contaminating fluids can clog pores in the aerated zone and reduce through-flow.

16.2.4 Importance of Very Sensitive Monitoring Techniques

The propagation of pollutants takes time, and their arrival in springs and wells is often gradual. In ideal situations the arrival of contaminating fluids can be detected by changes in the concentration of one or several parameters that are measurable but not toxic. In such cases preventive measures can be started in time to prevent toxic contamination. Hence the emphasis on the most sensitive measuring techniques, either of in situ measuring probes, or of laboratory techniques conducted on collected water samples.

16.2.5 Importance of Extensive Hydrochemical Studies

Hydrochemical surveys that encompass many wells and springs in extended areas may locate water sources of exceptional compositions, possibly caused by man-made contamination. Such surveys may detect contamination long before any complaints are expressed by water consumers. The next step in dealing with water sources that reveal suspected contamination is comparison of the composition with historical data, as discussed above.

16.3 Groundwater Pollution Case Studies

A large number of groundwater pollution cases are never published. This is especially true for severe cases. Yet, an ever-growing number of case studies are being published. Every hydrochemist should read as many of these papers as possible, as each provides examples of possible scenarios and possible modes of diagnosis and remedy. A selection of pollution case studies is reviewed in the following sections.

16.3.1 Nitrate Contamination Detected by Depth Profiles

Pickens et al. (1978) worked on nitrate contamination in wells drilled in a shallow sandy aquifer. An NO_3 plume was detected in carefully sampled depth profiles (section 7.3), as shown in Fig. 16.1. The contamination was attributed to extensive use of nitrate fertilizers.

Discussion:

The arrival of the pollutant was detected at a very early stage—12 ppm of NO_3 being alarming, but not threatening.

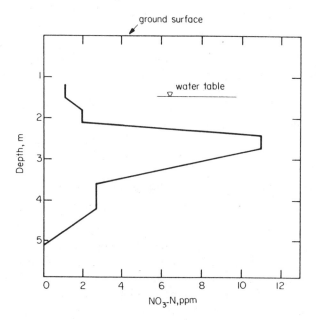

Fig. 16.1 NO_3 (expressed in ppm N) in a depth profile in a piezometer (from Pickens et al., 1971). A contaminated water plume is seen at a depth of 3 m.

The nitrate signal would not have been seen on a pumped sample, as pumping would lower the NO_3 concentration by its averaging effect, hence the advantage of depth profiles was demonstrated in this case.

The nitrate peak indicates higher water conductance in the 2.5–3.0 m zone—a direct contribution to the understanding of local groundwater flow.

16.3.2 Oxidation and Reduction Zones in Groundwater Profiles and Their Bearing on Nitrate Contamination

Anderson et al. (1980) worked on nitrate distribution in the Karup Basin, Denmark. Results of depth profiles in two boreholes are shown in Fig. 16.2.

Discussion. Nitrate is seen to decrease at a water depth of 8 m (15 m below the surface), whereas iron is low at this depth and rises beneath. This indicates the existence of an oxidizing zone in the upper 8 m of water, and a reducing zone below. The nitrate was reduced by ferrous iron to N_2, which was expelled into the atmosphere. The tritium data were most informative: values up to 200 TU were detected in the NO_3 contaminated upper water zone, indicating post-1963 (man-made tritium peak) recharge (the study was conducted in 1976).

Detecting Pollution Sources

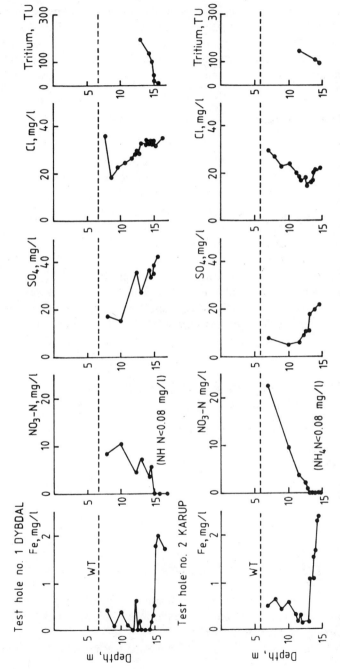

Fig. 16.2 Depth profiles of Fe, NO_3, SO_4, Cl, and tritium in two test holes, Karup Basin, Denmark. A sudden change is observed at a depth of 15 m (8 m below water table), explained as indicating the occurrence of an oxidizing regime above 15 m and a reducing regime below (see text).

Thus, fast-arriving recharge was demonstrated, supporting the hypothesis of contamination by locally used fertilizers. The investigators determined the soil's capacity to contribute ferrous iron to the reaction. This value is needed to assess how long one may rely on this mode of nitrate decomposition, and to predict when nitrification will become a problem if the use of excess fertilizers continues.

16.3.3 Water Salinization Due to Removal of Vegetation

Borman and Likens (1970) reported changes in the composition of water flowing from a watershed, caused by removal of vegetation at the Brook Experimental Forest, Hubbard, New Hampshire (Fig. 16.3).

Discussion. The time of experimental removal of vegetation is indicated in Fig. 16.3. A distinct rise in nitrate and calcium was noticed, while no rise was noticed in an adjacent undisturbed watershed. Thus, the increase in nitrate was clearly related to the removal of vegetation, which lowered the denitrification effect. The increase in nitrate and calcium was abrupt, and occurred 5 months after the vegetation clearing, demonstrating a piston flow recharge. Studies of this nature are relevant to the influence of deforestation and vegetation stripping by coal mining. Such operations may deteriorate the quality of local surface and groundwater. Thus, such operations should be accompanied by hydrochemical monitoring before, during, and after operations.

16.3.4 Monitoring of Nitrate and Chloride in an Irrigated Farmland

Saffinga and Keeney (1977) monitored Cl and NO_3 in shallow wells near a farm in central Wisconsin. In Fig. 16.4, the results from a well in an area not irrigated by the farm (well 7) are compared with the results from wells located in the irrigated farmland (wells 8 and 9, the latter was situated in a less cultivated part). The farming intensity dropped after 1972.

Discussion:

The NO_3 and Cl concentrations were correlated to irrigation intensity, since no increase was observed in well 7, located in a nonirrigated area, and as seen by the drop in Cl and NO_3 concentrations with the decrease in irrigation between 1972 and 1975.

The Cl and NO_3 covaried, indicating a common source, which was found to be two fertilizers applied in constant proportions: a nitrogen-rich fertilizer and a potassium chloride fertilizer.

The recovery of the water quality in 1975 indicated that no salts were stored in the ground.

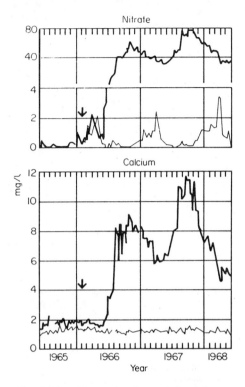

Fig. 16.3 Change in nitrate and calcium concentration in water flowing from a watershed cleared of vegetation (thick line) and an adjacent watershed with preserved vegetation (thin line). The arrow marks the date of vegetation clearance. Nitrate and calcium increased, with a 5-month delay, indicating piston flow recharge (Bormann and Likens, 1970).

The rapid recovery also indicated that the irrigation water reached the local groundwater rapidly, changes being observed from month to month (Fig. 16.4).

16.3.5 Pesticide Contamination

Lewallen (1971) studied a farmer's well in Florida, USA. High DDT, DDE, and toxaphane concentrations were noticed shortly after well completion. The detected source was backfill soil, brought from a location where pesticide materials were dumped. The soil of the backfill, and sediment accumulated at the bottom of the poorly cased well, contained high pesticide concentrations. In the second half of 1967 the well was cleaned and a proper casing was installed, causing an immediate drop in pesticide concentration (Fig. 16.5).

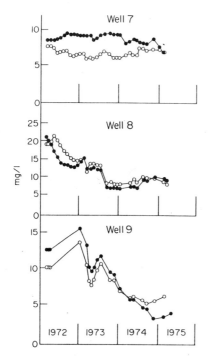

Fig. 16.4 Nitrate (dots) and chloride (circles) in groundwater of a well (no. 7) in a nonirrigated area, and wells (nos. 8 and 9) located in an irrigated area, Hancock Experimental Farm. Irrigation caused an increase in nitrate and chloride, caused by nitrogen and potassium fertilizers, the latter containing Cl (KCl). The quality of water was restored because of a decrease in farming activities (from Saffinga and Keeney, 1977).

Discussion:

- The contamination decreased significantly after cleaning and casing of the well, although no change was introduced to the adjacent backfill. Hence, the pesticides were fixed on the soil.
- No conduit-controlled flow of water occurred between the backfill site and the well area. Possible flow through a granular medium retarded pesticide contaminants by fixation to the soil.
- Although the quality of the water in the particular well was restored, much potential danger remains in the region and the surface and groundwater zones should be carefully managed and monitored.

16.3.6 *Large-Scale Industrial Fluid Waste Injection*

Goolsby (1971) described a thorough monitoring project of a large-scale industrial waste injection project in Florida. The task was to inject fluid waste at a

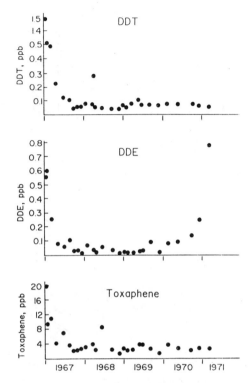

Fig. 16.5 Pesticide compounds in a well in Florida. Contamination was caused by pesticide-contaminated soil in the backfill and in sediment washed into the well. Cleaning of the well and repair of the casing removed the contamination and the well recovered.

rate of up to 500 m^3/h and about 3×10^6 m^3/year. The fluid composition was described as an aqueous solution of organic acids, nitric acid aminos, alcohols, ketons, and inorganic salts.

Discussion. The injection site was selected because of geological and hydrological considerations (Fig. 16.6):

A free surface sand and gravel aquifer was found to overlay two major confined aquifers of Floridian limestone, the three being separated by distinct clay layers.

The water in the upper limestone aquifer contained 425 mg Cl/l, whereas the lower limestone aquifer contained 7900 mg Cl/l. Thus, the two aquifers did not communicate hydraulically.

The water of the lower limestone aquifer was nonpotable.

In no place was the saline lower aquifer water seen to ascent to the surface.

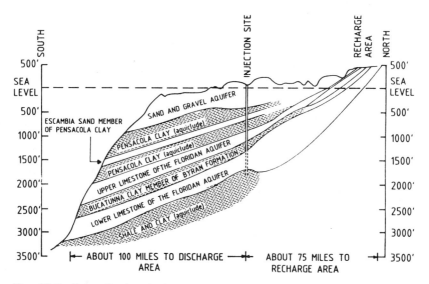

Fig. 16.6 Generalized geological cross section through southern Alabama and northwestern Florida (after Goolsby, 1971). An injection well penetrated the confined lower limestone aquifer.

The recharge area of the lower aquifer was about 120 km north.

The nearest possible discharge area of the lower aquifer, the ocean, was about 160 km south of the selected injection area. Thus, the injection site was remote from recharge and discharge points, and the injected aquifer was saline and isolated from the overlying aquifer.

Injection and Monitoring Systems. An injection well (A), 500 m deep, was drilled into the lower confined limestone aquifer, and later a second well (B) was drilled 400 m west of the first one. A deep monitoring well, 400 m south of A, was drilled into the receiving aquifer. A shallow monitoring well was drilled into the upper limestone aquifer, 30 m from injection well A. Hence, changes in the chemical composition could be monitored both in the receiving and overlying aquifers.

Water levels recorded in the two injection and two observation wells are given in Fig. 16.7, along with the injection rates. The following observations were made:

The deep monitoring well responded with an increase in water level as a function of the rate (and pressure) of injection.

The shallow monitoring well revealed no change in water level, indicating the upper limestone aquifer was not affected.

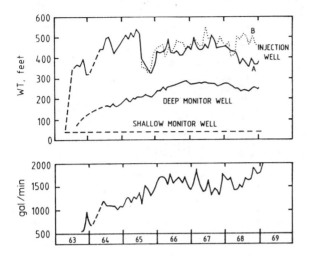

Fig. 16.7 Piezometric water levels in waste injection and monitoring wells, and injection rates in Florida. (From Goolsby, 1971.)

The calcium concentration in the deep monitoring well is given in Fig. 16.8. The following patterns were observed:

Although the waste fluid was low in calcium (about 3 ppm), it was calcium-tagged in the aquifer. This was actually expected because of the reaction of the acid waste (pH = 5, and later pH = 3) with the aquifer limestone.

The first front of injected waste arrived 10 months after commencement of the operation. The arrival of the fluid in the 400-m distant monitoring well was noticed by a marked increase in calcium (Fig. 16.8).

At the beginning of 1966 the calcium concentration decreased in the deep monitoring well, a feature not well understood.

In 1968 a previous neutralizing process was stopped, and the waste was injected with a pH = 3, resulting in a sharp increase in dissolved calcium in the monitoring well (Fig. 16.8). This indicated that limestone was dissolved at a high rate. The resulting "karstification" probably enhanced the movement of the waste in the aquifer.

Additional chemical data are given in Table 16.1. As mentioned, the composition of injected waste was changed in the middle of 1968, when neutralization was dropped. As a result, the pH went down to 3.3 and the nitrate concentration went up. The deep monitoring well responded in the following way (Table 16.1): the October 1967 and May 1968 analyses showed high sodium and chlorine concentrations, revealing dominance of the original saline aquifer water.

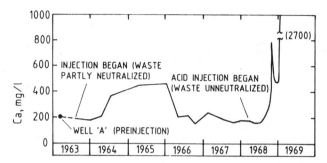

Fig. 16.8 Calcium concentration, deep monitoring well, 1963–1964, Waste Injection Projection, Florida. (From Goolsby, 1971.)

The percent wastewater was calculated by the investigator's from the chlorine concentration in the monitoring well (5800 mg/l) and the initial chlorine concentration in the lower limestone aquifer (7900 mg/l):

$$\text{percentage injected water: } \frac{7900 - 5800}{7900} \times 100 = 26\%$$

In other words, up to May 1968, the chlorine-rich aquifer water was diluted by 26% with injected chlorine-devoid water. The newer, nonneutralized waste with pH = 3.3 caused a remarkable change in the composition of the fluid at the deep monitoring well: chlorine dropped to 160 mg/l, indicating the deep monitoring well was dominated by 98% injected water:

$$\frac{7900 - 160}{7900} \times 100 = 98\%$$

The rapid increase from 26% to 98% waste fluid in the deep monitoring well indicated that the nonneutralized waste dissolved the aquifer carbonates and

Table 16.1 Composition (mg/l) of waste and of groundwater (Goolsby, 1971)

	Injected wastes		Deep monitor well		
	Nov. 1967	Apr.–Dec. 1968 (Average)	Oct. 1967	May 1968	Jan. 1969
Ca	3.4	<20	168	165	2350
No_3	2420	5070	2	0	5760
Na	610	<1000	3720	3720	635
Cl	82	200	5700	5900	161
COD	–	22980	–	2355	22100
pH	5.2	5.3	6.80	7.00	4.75

Detecting Pollution Sources

developed a highly conductive karstic system, which was very alarming in terms of waste disposal in the aquifer. During all this time no chemicals arrived in the upper monitoring well.

16.3.7 Short-Circuiting of Aquifers by an Abandoned Well

Jorgensen (1968) investigated a case of water quality deterioration in a municipal well at Avon, South Dakota. The city wells pumped the free water table aquifer of the Codell sandstone (Fig. 16.9). Well A had a significantly inferior quality. Pumping tests supplied the first clue: during a pumping test a contribution of 48% from another aquifer was detected. Local intrusion of water from the lower confined Dakota sandstone aquifer (Fig. 16.9) was suspected, as its water was known to be saline and of a piezometric head that exceeded the water level in the phreatic Codell sandstone aquifer. Chemical compositions and temperatures for the undisturbed phreatic aquifer, the confined aquifer, and well A are given in Table 16.2. A glance at the table reveals that the properties of the pumping test sample of the deteriorated well are indeed between the properties of the water in the Dakota sandstone aquifer and the Codell sandstone aquifer. The percentage (x) of saline and warm Dakota sandstone water in the pumping test sample can be calculated (section 6.7), applying several parameters:

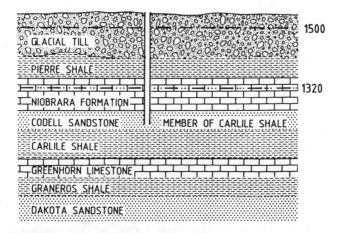

Fig. 16.9 Geologic cross section, Avon, South Dakota (from Jorgensen, 1968). An old corroded well short-circuited the deep pressurized Codell aquifer and the shallow Dakota aquifer, contaminating the latter.

Table 16.2 Dissolved ions (meq/l) and temperature in wells near Avon, South Dakota, USA (Jorgensen, 1968)

	Dakota sandstone water	Codell sandstone water	Pumping test well A water	% Dakota water in A
Temperature	18°C	11°C	14°C	43
Ca	15.2	7.0	11.0	48
HCO_3	2.7	4.9	3.6	57
Mg	4.7	3.5	2.3	
Na	4.0	19.6	18.7	
K	0.51	1.04	1.05	
CO_3	0.0	0.0	0.0	
SO_4	16.8	21.6	25.2	
Cl	2.6	3.4	2.8	

By temperature: $14 = 18x + 11(1 - x)$

$x = 0.43$, or 43%

By calcium: $11 = 15.24x + 7(1 - x)$

$x = 0.48$, or 48%

By bicarbonate: $3.64 = 2.69x + 4.91(1 - x)$

$x = 0.57$, or 57%

These calculated mixing ratios agree well with the value of 48% contribution from the deep aquifer, determined from the pumping test. A later inquiry revealed that in 1870 a city well was constructed only 6 m away from the deteriorated well. That well reached the confined Dakota sandstone aquifer and water rose in it above the piezometric level of the phreatic aquifer. That well was capped 50 years ago. It, thus, seemed that the casing of the old well had failed, short-circuiting the two aquifers and creating a saline plume in the upper aquifer, locally deteriorating the water quality. The agreement between mixing percentages calculated by the pumping test, temperature, Ca, and HCO_3 supported the hydraulic short-circuiting hypothesis.

Discussion. What should be done? The old well should be plugged in order to restore the separation between the two aquifers. Following the discussed mode of calculating the percentage of saline water in well A, the data in Table 16.2 can be used to calculate the percentage of saline water in well A based on the other parameters: Mg, K, CO_3, SO_4, and Cl. It turns out that these parameters do not play the game: the mixing percentages deduced from Na, K, and Cl differ from each other and from the values obtained by the pumping test, temperature, Ca, and HCO_3.

What could cause this disagreement? Is the whole short-circuiting hypothesis erroneous? The answer is that because of the mixing of two such different waters, it is possible that secondary interactions took place, affecting some of the dissolved ions. An example for such secondary reactions in a disposed slaughterhouse effluent is discussed in section 16.3.15.

16.3.8 Contamination by Released Oil Brines

Fryberger (1975) described a case study of groundwater contamination by displaced oil production brines. In 1967 a farmer lost his rice crop, near Garland City, southwest Arkansas, USA, because his well water became saline. An adjacent oil field came into production in 1955, with an increasing amount of separated brine that had been diverted into a poorly lined evaporation pond that leaked into the local aquifer. The brine was subsequently recharged into the ground through an abandoned production well, but soon the casing became corroded and the brine again reached the local aquifer. Eventually a new recharge well was constructed.

Monitoring Project. To monitor the brine contamination, 36 observation wells were drilled to different depths. Chlorine, regarded as best representing the brine, was measured, and the data served to produce a chlorine concentration contour map (Fig. 16.10). The spread of the lines indicates the water in the aquifer flowed southward. A section showing chlorine distribution (Fig. 16.11) also indicated a southward flow. To get a more complete picture of the brine movement, a schematic compositional cross section was prepared (Fig. 16.12). The results were plotted as a function of the distance of each observation well from the contamination center. The concentration of Cl, Br, F, Pb, Ca, Mn, Ni, and Al decreased in a pattern explainable by mixing, in various proportions, of contaminating brine and local groundwater. However, several dissolved ions revealed different patterns, indicating secondary processes occurred:

- Iodine, iron, and zinc increased in wells over the concentrations in the brine or local groundwater, represented by the private (PVT) well. The researchers hypothesized that chemical displacement of ions coating the sand grains took place. This explanation is plausible for the iron alone. Hence, another explanation was offered—that the brine disposed in earlier stages had a different composition.
- Barium is seen in Fig. 16.12 to drop faster than chlorine in wells TW 11 and TW 20, indicating effective precipitation en route.

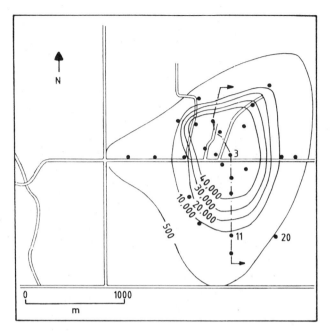

Fig. 16.10 Isochloride contours at the bottom of the alluvium aquifer, Gerald City. Flow to the south is seen. (From Fryberger, 1975.)

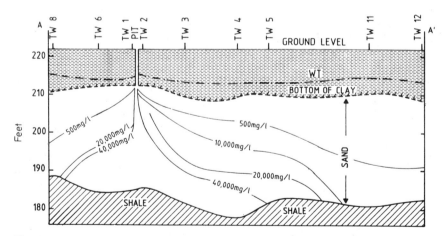

Fig. 16.11 A chloride concentration cross section, Gerald City. Leakage from the brine evaporation pit is recognizable. The heavy brine accumulated above the shale aquiclude. (From Fryberger, 1975.)

Detecting Pollution Sources

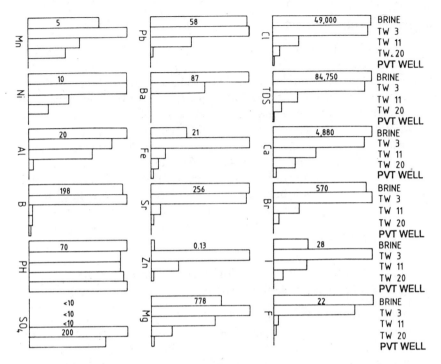

Fig. 16.12 Composition histograms (mg/l) of the brine pit of Gerald City, and monitoring wells, arranged by increasing distance from the pit. A contamination gradient is observed from the brine pit toward the freshwater well PVT. (From Fryberger, 1975.)

SO_4 is high in the private well and TW 20, indicating it is high in the local groundwater. The observed drop in SO_4 concentration in TW 11 and TW 3 is probably due to consumption by barium precipitation.

Discussion. The term *conservative* is applied to dissolved ions that, except for mixing, take part in no other process. In contrast, *nonconservative ions* reveal changes in concentrations caused by mixing of two water types and, in addition, by water-rock interactions. Looking at Fig. 16.12, which ions are conservative and which are nonconservative? Cl, Br, Ca, Ni behaved as conservative ions. In contrast, secondary reactions changed the composition of the water by exchange reaction with rocks, for example, with Fe and possibly Zn.

Conclusion. Different ions travel at different velocities, the conservative ions moving fastest. The conservative ions are most helpful for the understanding of groundwater flow.

16.3.9 Contamination by Repressuring of an Oil Field

Wilmoth (1972) reported that a domestic well in Kanawha County, West Virginia, USA, contained only 32 mgCl/l until 1967. This excellent water quality deteriorated because of brine injected into an adjacent oil well. Within 1 year the chlorine concentration rose to 1140 mg/l (Fig. 16.13). Operations were stopped and the chlorine concentrations decreased slowly, reaching a value of 450 mgCl/l early in 1971, and the well could be used once more. The decrease in chlorine was caused by natural flushing of the water flowing in the aquifer.

A second case described by Wilmoth (1972) in the same country related to a freshwater aquifer near Wallace that contained 100 mg Cl/l. In late 1967 injection of brines, as part of subsurface operations in an adjacent oil field, caused an increase to 2950 mg/l chlorine concentrations within 2 months. Operations were stopped and the water quality was soon restored, attaining concentrations of 190 mg/l after 2 1/2 years (Fig. 16.14).

16.3.10 Leakage from a Cooling Pond Monitored by Temperature Measurements

Andrews and Anderson (1978) applied temperature measurements to trace leakage from a cooling pond. The pond served two 500-MW electric power plants of the Columbia Generating Station, located on the floodplain of the Wisconsin River, near Portage, Wisconsin, USA. The cooling pond was constructed of 5-m high dikes, covering an area of 200 ha. The bottom, a silty sand, was partially lined with bentonite. The area is a discharge zone of marshes and wetlands. The geology of the area was studied with 100 borings, the water table was monitored by 80 small-diameter observation wells, and heat profiles were

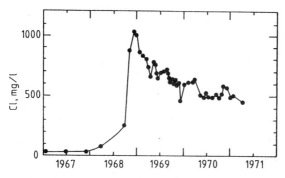

Fig. 16.13 Changes in chloride in a contaminated well adjacent to an oil field, Kanawha County (following Wilmoth, 1972). Fast response to brine injection in a nearby well was observed, as well as fast recovery after brine injection was stopped.

Detecting Pollution Sources

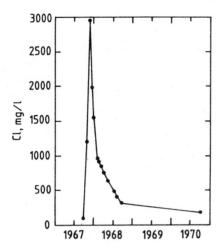

Fig. 16.14 Changes in chloride in a contaminated well adjacent to an oil field, Kanawha County (after Wilmoth, 1972). Injection of brines into nearby wells caused a dramatic increase in chlorine, which was cured in 2 years.

measured in 19 observation wells. Figure 16.15 provides an example of temperature data obtained in the extensive monitoring operation. The summer temperature peak reached well B (60 m from the pond) 60 days after it reached well A (3 m from the pond). Thus, the water leaking from the pond traveled 60 m in 60 days, or 1 m/day. A network of 47 boreholes of 10-m depth were equipped with thermistors for extensive monitoring.

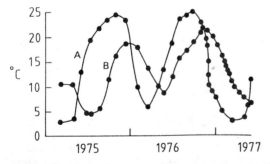

Fig. 16.15 Temperature variations in well A (4 m depth, 3 m west of the dike) and well B (2.5 m deep, 60 m west of dike) (after Andrews and Anderson, 1978). The summer temperature peak reached well B 60 days after it reached well A, indicating a flow velocity of 1 m/day.

16.3.11 Contamination by a Coal Ash Pit

Andrews and Anderson (1978) studied the impact of a coal ash pit that served two 500-MW electric power generating units near Portage, Wisconsin, USA. Water in the ash pit differed in its composition from the local groundwater. Operation of the power plant began in May 1975, monitored by four wells that were constructed south and north of the pit dikes. Selected data are reported in Table 16.3.

Discussion. From Table 16.3, it emerges that the data for 1972–1973 predated the operation of the power plant and reflect the composition of the nonpolluted local groundwater. The variations in composition between wells A, B, C, and D reflect natural fluctuations in the local groundwater composition. For example, calcium occurred in the range of 13.9 to 31.6 mg/l, and SO_4 varied from 0.8 mg/l (well B) to 9.7 mg/l (well D). Bearing these variations in mind, it seems that:

Ash pit water reached no well, considering potassium and magnesium—the relatively high values observed exceeded the values in the ash pit water and must be due to some other reason (e.g., secondary water-rock interactions).
Sodium indicates contamination in well A alone.
Ca, SO_4, and Cl indicate pollution in wells A and C.

The conclusion from all observations is that the water reached wells A and C, located 1 m west and north of the pit, and no contaminating water reached wells B and D, located 75 m west and 25 m north. The contamination was restricted to the immediate surroundings only.

16.3.12 Fly Ash (Coal) Contamination

Cherkauer (1980) monitored groundwater near a fly ash site, operating 8 years, for the Port Washington Power Plant in southeastern Wisconsin, USA. The maps in Fig. 16.16 show the potentiometric surface contours and direction of flow, TDS, SO_4, Ca, and HCO_3, along with the positions of the monitoring piezometers. The ash leachate contained various metals that were not observed in the monitoring wells, except some iron.

16.3.13 Monitoring Leachate of a Sanitary Landfill

Fritz et al. (1976) studied possible movement of leachate from a sanitary landfill in Frankfurt am Main, Germany. In a preliminary search for suitable tracers it was found that the leachate leaving the waste disposal site was enriched in deuterium and ^{18}O (possibly due to evaporation). The leachate was enriched by 1.5–2.0‰ of $\delta^{18}O$, whereas the analytical resolution was ±0.1‰. The site

Table 16.3 Water composition (mg/l) in ash pit and monitoring wells near Portage (from Andrews and Anderson, 1978)

	Ash pit water	West of the dike						North of the dike						
		A – 1 m West			B – 75 m West			Ash pit water	C – 1 m North			D – 25 m North		
		1972-3	1975-6	1977	1972-3	1975-6	1977		1972-3	1975-6	1977	1972-3	1975-6	1977
K^+	2.3	1.2	9.2	6.9	1.2	0.75	0.5	2.4	1.7	0.75	1.5	2.0	0.75	0.5
Na^+	10.2	2.2	8.5	13.2	3.6	3.5	4.0	10.3	4.3	2.6	5.1	5.2	3.1	3.7
Ca^{2+}	77.8	13.9	16.9	21.6	31.6	28.8	24.3	54.7	18.8	23.3	37.0	22.0	22.6	21.5
Mg^{2+}	10.6	11.1	9.8	14.7	20.3	22.0	16.9	4.3	17.3	22.5	33.8	18.0	18.8	25.8
SO_4^{2-}	44.6	5.4	18.5	10.5	0.8	0.2	0.1	36.0	7.7	9.6	22.0	9.7	9.3	14.0
Cl^-	10.6	3.4	4.5	6.5	4.49	2.64	5.0	10.5	2.8	4.2	8.5	7.2	3.2	3.5

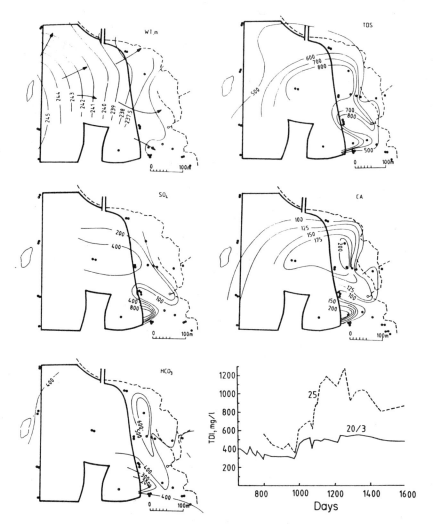

Fig. 16.16 Data of a fly ash site monitoring operation, southeastern Wisconsin (after Cherkauer, 1980). Water table contours (upper left diagram), September 1978, masl: dots mark locations of piezometers; heavy line marks the boundary of the filled area; dashed line marks surface drainage. Contours of dissolved ions are in mg/l. Right bottom: TDI changes in contaminated well (25) and uncontaminated control well (20/3).

operated from 1925–1968, and about 1.8×10^7 m^3 of garbage were deposited there. The underlying phreatic aquifer, built of quaternary sand and gravel, was biologically polluted. Samples were collected in test wells that penetrated the aquifer, at depths of 8–11 m. Samples were also collected from drainage ditches in the landfill. Partial chemical data are given in Table 16.4, along with chemical oxygen demand (COD), reflecting the concentration of organic matter in the polluted fluid. The locations of the test wells and piezometric contours are given in Fig. 16.17 and 16.18, respectively.

Discussion. Looking at the water table elevations (Fig. 16.18) and the analytical results (Table 16.4), which wells may be regarded as representing uncontaminated groundwater? Well 401 is upstream (Fig. 16.18), and on this basis alone should be uncontaminated. This is confirmed by the observation that the water in this well is lowest in Cl, NH_4, COD, and TDS. Wells 404, 405, 406, 415, 416, and 417 are closest to the landfill, and in the groundwater flow direction (Fig. 16.18). Therefore, they should to be suspected of contamination. The data reveal that these wells, and additional wells that are located downflow, are indeed contaminated.

Of special interest (and warning) is the observation that different parameters assign a different degree of contamination to individual wells. For example, well 404 is most contaminated by Cl and COD but slightly less by NH_4. To make this point clear, the wells are listed in Table 16.5 in decreasing order of concentration of each contaminant. The relative order of the wells in the Cl column is seen to differ from the arrangements by SO_4, NH_4, and COD. Yet, an

Table 16.4 Composition of surface and groundwater near landfill site, Frankfurt am Main (Fritz et al., 1976)

Station No.	Cl^- (ppm)	SO_4^{2-} (ppm)	NH_4^+ (ppm)	COD (ppm)	TDS (by evaporation) (ppm)	Comments
401	19.3	122.5	0.16	10	256	Groundwater
404	2682	867.3	244	1310	9250	Groundwater
405	1806	89.5	1404	1070	8670	Groundwater
406	1725	990.5	603	1310	7800	Groundwater
407	987.7	285.8	623	460	6700	Groundwater
408	1670	117.1	198	880	5700	Draws lake water
409	26.0	48.1	0.16	8	233	Groundwater
410	73.1	131.3	1.55	4	493	Groundwater
411	168.9	224.7	0.74	46	722	Groundwater
412	43.1	82.4	1.78	29	344	Groundwater
413	14.6	45.6	0.06	8	348	Groundwater
414	11.7	40.5	3.09	24	287	Groundwater
415	1883	137.2	119	960	6950	Groundwater
416	1610	313.4	114	810	6300	Groundwater
417	719.3	217.1	3.19	305	2665	Groundwater
418	40.4	35.6	0.04	25	551	Groundwater
20	1636	214.3	104	820	5400	Lake water
21	1713	211.5	104	765	5400	Lake water
26	61.2	126.1	73.1	375	2011	Lake fed by ditch
28	920.5	127.9	135	520	3350	Ditch fed by groundwater

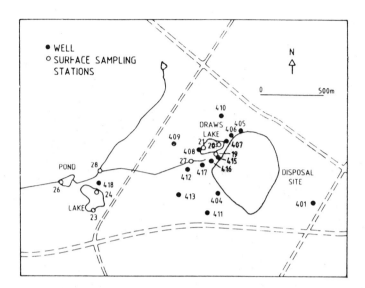

Fig. 16.17 Sanitary landfill, Frankfurt am Main. (From Fritz et al., 1976.)

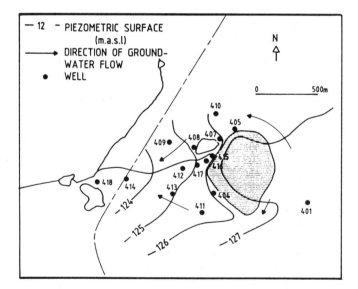

Fig. 16.18 Water levels, masl, in a shallow aquifer at a sanitary landfill, Frankfurt am Main (Fritz et al., 1976).

Table 16.5 Monitoring wells arranged in descending order by the various tracers (after Table 16.3)

Cl^-	SO_4^{2-}	NH_4^+	COD
404	406	405	404
415	404	407	406
405	416	406	405
406	407	404	415
408	411	408	408
416	417	415	416
407	415	416	407
417	410	417	417
411	**401**	414	411
410	408	412	412
412	405	410	418
418	412	411	414
409	409	**401**	**401**
401	413	409	409
413	414	413	413
414	418	418	410

overall picture is obtained: wells 404, 405, 406, 407, and 408 are clearly contaminated. There are several reasons that could explain the observed differences:

The sanitary landfill is not homogeneous, and different parts seem to contain different types of garbage.

The uncontaminated groundwater might have had slight differences of composition.

Some of the ions could be retarded by soil components or exchanged with soil and rock, and these changes might differ locally.

Where was the contaminating groundwater front? It was situated between the most distant wells that reveal contamination and the closest wells that reveal no contamination. Well 401 has been selected as representing the uncontaminated groundwater, as discussed above. Well 401 is emphasized in Table 16.5, and the wells listed below well 401 may be regarded as uncontaminated. By this approach it is seen that only two wells are uncontaminated, taking Cl as a tracer, whereas 7 wells are uncontaminated using SO_4 as a tracer.

Chlorine and COD were the most conservative (and therefore most sensitive) tracers in the Frankfurt am Main study. The importance of analyzing many parameters is hereby demonstrated. Later measurements may be limited to the established conservative, most sensitive tracers.

Figure 16.17 shows that the landfill is situated near two lakes and a pond, which have a possible influence on local groundwater and need to be understood and taken into account in the interpretation of the pollution monitoring results. δD and $\delta^{18}O$ turned out to be most useful in this respect. The $\delta^{18}O$ data, given in Table 16.6, reveal three groups of values: $-8.6‰$ to $-8.0‰$, $-6.7‰$ to $-6.3‰$, and $-3.0‰$ to $-2.0‰$. Can sense be made of these three groups, denoted by $\delta^{18}O$ (and δD)? Consult the map given in Fig. 16.17 and the following picture emerges:

The first group, which includes well 401, is of uncontaminated groundwater.
The second group is of polluted groundwater.
The third group, including well 408, reflects recharge from Draws Lake.

A second pattern, noticeable in the isotopic data of Table 16.6, is a systematic change to more negative values, seen in those wells that have been shown by their dissolved ions to be contaminated (Table 16.5). This demonstrates the progressive increase of water from the landfill.

The $\delta^{18}O$ data has been used by the researchers to calculate the percentage of leachate water in each well, summed up in a map of the pollution plume (Fig. 16.19). The correlation between the isotopic data and dissolved ions has been checked in Fig. 16.20. Good positive correlations are observed between Cl, COD, and $\delta^{18}O$, and almost no correlation is seen with NH_4 and SO_4. This supports the conclusion that Cl and COD (as well as $\delta^{18}O$ and δD) were conservative tracers.

16.3.14 Contamination by Highway Deicing Salt

Dennis (1973) studied the impact of a stockpile of 50,000 tons of salt used for highway deicing at Mars Hill, Indianapolis, Indiana. The salt was stored in 1966 and removed in 1968 because the local aquifer turned saline (several thousand mgCl/l). The aquifer recovered gradually. In 1976 the water of most wells was potable, but traces of the contamination were still reflected in an isochlor contour map (Fig. 16.21).

This accident supplied interesting hydraulic information: it established the direction of flow at Mars Hill, and the observation that the wells recovered in 6 years indicates the aquifer is highly dynamic.

A second case of contamination by road salt piles has been described for Monroe County, West Virginia (Wilmoth, 1972). Chloride concentrations in water of a 500-m distant well (25 m deep) increased from 185 mg/l to 1000 mg/l in 5 years. The chlorine concentration rose to 7200 mg/l, in response to an enlargement of the salt storage area. The salt piles were removed in 1970,

Detecting Pollution Sources

Table 16.6 Isotopic composition of surface and groundwater near the landfill site, Frankfurt am Main (Fritz et al., 1976)

Sampling point	18_O(‰) Nov 1972	18_O(‰) April 1971	18_O(‰) June 1975	δD(‰) April 1973	δD(‰) June 1975	Comments
401	−8.6	−8.2	−	−	−58	Groundwater
409	−8.3	−8.3	−8.5	−	−57	Groundwater
410	−8.5	−8.6	−8.4	−55	−59	Groundwater
411	−8.2	−8.6	−8.6	−	−59	Groundwater
412	−8.3	−8.4	−8.5	−	−58	Groundwater
413	−8.2	−8.5	−8.6	−55	−58	Groundwater
414	−8.1	−8.3	−8.5	−	55	Groundwater
417	−8.0	−7.8	8.0	43	−58	Groundwater
418	−8.4	−	−8.4	−	−42	Groundwater
404	−6.3	−6.5	−6.2	−	−45	Groundwater
405	−6.7	−6.8	−7.3	−46	−47	Groundwater
406	−6.4	−7.1	−7.4	−	−43	Groundwater
407	−6.5	−7.0	−6.9	−	−39	Groundwater
415	−6.4	−6.6	−6.9	−32	−48	Groundwater
416	−6.4	−6.7	−7.3	−	−33	Groundwater
408	−2.9	−4.0	−4.4	−31	−	Draws lake water
20	−2.0	−3.5	−	−	−37	Lake water
21	−3.1	−3.7	−4.5	−24	−	Lake water
21	−3.0	−4.0	−	−	−46	Lake water
23/24	−	−3.2	−5.3	−29	−55	Lake fed by ditch
26	−7.2	−7.0	−7.8	−41	−	Ditch fed by groundwater
26	−7.1	−7.2	−	−		

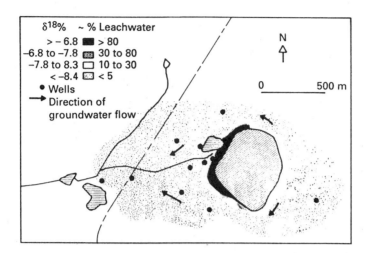

Fig. 16.19 Pollution plume, outlined by $\delta^{18}O$, landfill site, Frankfurt am Main. (From Fritz et al., 1976.)

and within 2 months the chlorine concentration decreased to 188 mg/l (Fig. 16.22). The extremely fast recovery indicates that the studied aquifer has a large through-flow.

16.3.15 Slaughterhouse Waste and Secondary Ion Exchanges in Soil

Mazor et al. (1981) studied contamination of groundwater by a slaughterhouse at Lobastse, Botswana. Sodium chloride, used to wash and preserve skins, was a major contaminant released with the slaughterhouse effluent. Preliminary studies revealed chlorine and tritium as the most conservative tracers in this case. The slaughterhouse used water from a dam, and this water had tritium concentrations that were significantly higher than in the local groundwater. The location of the slaughterhouse, and chlorine and tritium concentrations are shown in Figs. 16.23 and 16.24, respectively. Significant contamination is seen in wells close to the slaughterhouse and along the Peleng riverbed, into which the effluent was released. Other wells reveal low chlorine and tritium values, indicating they remained uncontaminated. Contamination in the wells fluctuated with time (Fig.16.25), reflecting the seasonal nature of the slaughterhouse output and demonstrating the immediate affect of contamination.

Composition plots (Fig. 16.26) revealed an interesting pattern: chlorine in the contaminated wells was positively correlated to TDI, as expected from the

Detecting Pollution Sources

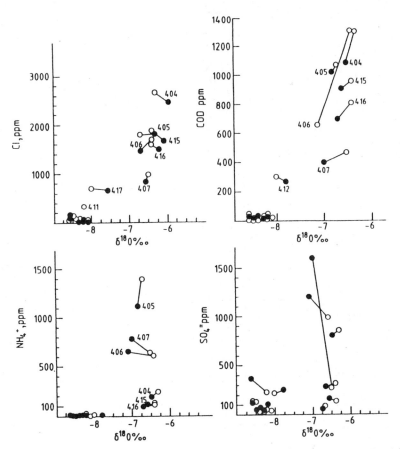

Fig. 16.20 Cl, COD, NH$_4$, and SO$_4$ as a function of δ^{18}O, landfill site, Frankfurt am Main; circles: November 1972; dots: April 1973. (From Fritz et al., 1976.)

NaCl disposal. In contrast, sodium revealed only a small increase in the contaminated wells and no correlation to TDI. This was compensated by increases in calcium and magnesium in the contaminated wells (Fig. 16.26). The explanation offered was that sodium was exchanged for calcium and magnesium upon interaction of the effluent with calcium and magnesium carbonates or with clays in the local soil.

The observation of sodium exchanging for calcium and magnesium is of general importance—it is, for example, observed in certain cases of seawater intrusion (Mazor, 1978).

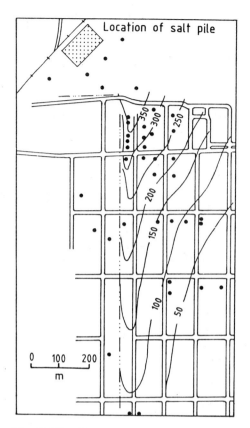

Fig. 16.21 Mars Hill, 1973: location and isochloride contours (mg/l) (from Dennis, 1973). Groundwater has almost recovered, but the salinity distribution still reflected the southward direction of groundwater flow.

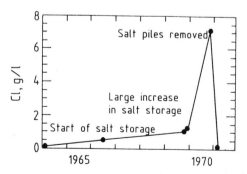

Fig. 16.22 Chloride in shallow groundwater, Monroe County (from Wilmoth, 1972). A salt pile caused severe contamination in a 500-m distant well. The system recovered within 2 months after the salt pile was removed.

Detecting Pollution Sources

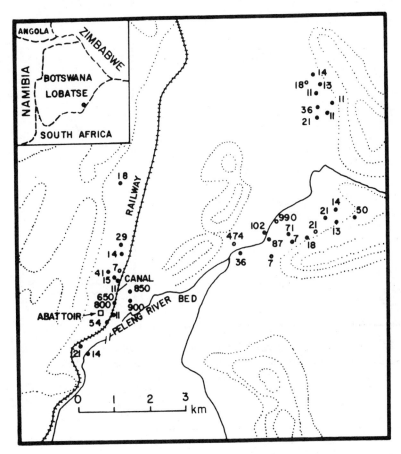

Fig. 16.23 Chloride concentrations (mg/l) observed in wells (dots), Lobatse Botswana. High concentrations were observed near the slaughterhouse and close to the Peleng riverbed. (From Mazor et al., 1981.)

The observations at Lobatse were, thus, another case in which pollution served as a large-scale tracing experiment, checking a natural phenomenon. Because of its special interest, the exchange was checked in laboratory experiments in which the sewage effluent was stored with Lobatase soil samples for 170 days (Table 16.7). The results are plotted in Fig. 16.27, revealing a drop in sodium and gain in calcium and magnesium. Such exchanges are common and show that a contaminant released into the ground may cause pollution of groundwater by another contaminant.

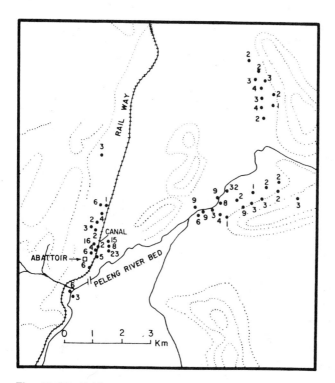

Fig. 16.24 Tritium concentrations (TU) observed in wells, Lobatse, Botswana. High values were observed in wells (dots) near the slaughterhouse and near the Peleng riverbed. (From Mazor et al., 1981.)

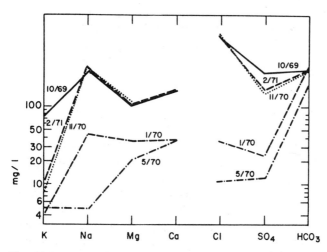

Fig. 16.25 Variations in dissolved ions in well P209, located on the slaughterhouse grounds, at Lobatse, Botswana, reflecting the seasonal nature of the release of effluent.

Detecting Pollution Sources

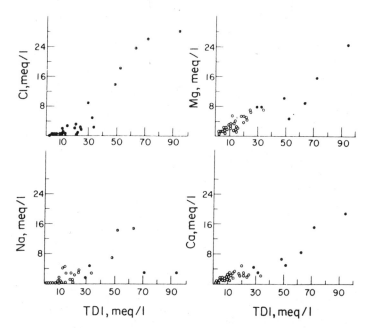

Fig. 16.26 Composition diagrams, wells at Lobatse, Botswana; uncontaminated wells (circles) and contaminated wells (dots). Sodium does not reveal an increase with TDI, but calcium and magnesium do, indicating sodium was exchanged for calcium and magnesium. (From Mazor et al., 1981.)

A second example of exchange is seen for the Lobatase case (Fig. 16.26). The SO_4 is seen to increase along with other ions. The explanation: the NaCl effluent enhanced dissolution of secondary gypsum in the soil and the aquifer rocks.

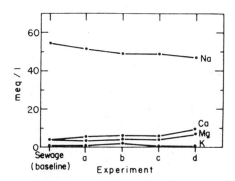

Fig. 16.27 Laboratory experiments: compositional changes in NaCl-rich industrial and domestic sewage after 170 days contact with soils of Lobatse, Botswana (Table 16.7). Sodium was exchanged for a calcium and magnesium. (From Mazor et al., 1981.)

Table 16.7 Chemical changes in NaCl-rich industrial and domestic sewage stored in the laboratory with soils of Lobatse, Botswana (Mazor et al., 1981)

Experiment	Description	K^+	Na^+	Mg^{2+}	Ca^{2+}	Cl^-	SO_4^{2-}	HCO_3^-
Base line	Sewage effluent	1.2	52.6	4.7	3.6	59.2	4.4	4.2
	Sewage effluent stored 170 days ('blank')	1.0	54.4	4.0	3.9	60.3	5.5	2.2
a	400 g soil (from well 1931 area), with 500 cm^3 sewage effluent, stored 170 days	1.1	51.1	3.4	5.5	58.1	4.3	0.33
b	400 g soil (top soil, near Geol. Survey) with 500 cm^3 sewage effluent, stored 170 days	2.0	47.9	4.0	6.2	58.9	5.0	0.46
c	400 g soil (from 1.8 m depth, near Geol. Survey) with 500 cm^3 sewage effluent, stored 170 days	0.51	48.9	3.9	6.2	58.3	2.9	0.24
d	400 g red soil (from well 1379) with 500 cm^3 sewage effluent, stored 170 days	0.46	46.8	6.9	9.4	56.2	4.0	0.33

16.4 Summary of Case Studies

Detection of sources of contamination and monitoring of waste disposal of all kinds are tasks that have to be handled with the highest standards and expertise. No case is the same as any previous cases, so experience plays a major role. The case studies described in this chapter are a small assortment—just to convey the nature of the job. A recommendation is to read and get familiar with as many case studies as possible. The ones published in professional journals have the advantage of being well presented and concentrate on the highlights. But most pollution research is summed up in bulky reports that can be obtained from relevant agencies. Such reports provide a more realistic picture of the difficulties that have to be overcome, the huge amount of data accumulated, and the countless number of possible modes of data processing and problem solutions.

Pollution detection and monitoring projects should be planned and conducted within the guidelines that were found useful for all hydrochemical studies (chapter 7). To these should be added the requirement for specialized laboratories that analyze the specific compounds present in unusual wastes.

16.5 Summary Exercises

Exercise 16.1:   Read again the nitrate contamination case study (section 16.3.1) and suggest how to proceed? What should be done next?

Exercise 16.2:   Section 16.3.9 deals with a case of groundwater contamination by injected oil brine. Figure 16.13 reveals that the chlorine concentration decreased quickly after the brine injection ceased. Is this full proof for complete restoration?

Exercise 16.3:   Looking at the composition of the water in the ash pit (Table 16.3), and bearing in mind the discussed observations on the rate of water movement (section 16.3.11), can the monitoring operations be stopped? Are there any other suggestions?

Exercise 16.4:   Study the data plotted in Fig. 16.16. What is your expert opinion?

17
SUSTAINABLE DEVELOPMENT OF GROUNDWATER

17.1 The Concept

Sustainable development means managing a resource in a way that it is profitable without that resource being depleted. In regards to groundwater, sustainable development means:

Pumping that is limited to the volume of recharged water. In other words, it is groundwater exploitation that nearly equals, but never surpasses, the amount of groundwater that would be drained naturally into the ocean.
Maintaining the natural recharge mechanisms and enhancing them.
Preserving groundwater quality—avoiding deterioration by arrival of anthropogenic pollution or by pumping-induced intermixing with low-quality groundwater.
Avoiding overpumping-induced cessation of groundwater drainage to the ocean, as it also means loss of drainage of salts and pollutants.

17.2 Man and Groundwater: A Historical Perspective

For hundreds and thousands of years man was limited in his activity to the immediate vicinity of available water sources, thus restricting his territorial distribution. The invention of water carrying vessels increased the radius of travel a little. With the passage to animal husbandry and farming in the Neolithic period, some 7000 years ago, the demand for water increased. This was the background for the development of the large river civilizations in the Middle East, based on irrigation canals. In the vast semiarid areas that have no perennial rivers, wells were dug, and the Bible contains testimonies to their importance to the well-being of whoever had control of them. The city of Jerusalem is situated in the mountains in a region that is devoid of rivers or lakes, and from its founding, some 3000 years ago, it was dependent on the collection of rainwater and water brought in from springs (the local groundwater table was too deep to be tapped by wells in those days).

In Europe, America, and other areas cities grew most commonly on the banks of large rivers, and until recently most urban water supplies came from dammed surface reservoirs. To overcome years of drought and cope with pollution of the surface waters, these cities incorporate in their supply systems an ever-growing number of wells, and their dependence on the groundwater resource is constantly growing.

Groundwater exploitation has come a long way since the early days, when man derived water from shallow wells using a bucket and a rope, to the present use of powerful deeply submerged pumps. In ancient times wells were dug to depths of 1–2 m below the water table, and therefore water could only be "skimmed". The wells were left to recover overnight. Thus, in ancient times, the exploitation did not cause a lowering of the natural water table. Today, groundwater is often overpumped, resulting in significant lowering of the local water table, on the order of up to tens of meters, making room for encroachment of saline water from the sea or from stagnant deeper aquifers. In addition, modern man heavily pollutes his environment to a degree the endangers groundwater.

The demand for high-quality groundwater will steadily increase in the future, because of population growth and because of an expected rise in the living standard of the world's population. This increased economic activity brings with it the danger of increased pollution of the groundwater. Pessimists see the situation as grim, and optimists see the tremendous challenge of proper management of the groundwater resource. It is in this niche—the challenge to optimally manage groundwater systems—that the hydrochemist is called to play a key role.

17.3 The Concern over Groundwater Quality: A Lever for the Demand of Clean Nationwide Management

The atmospheric system rapidly recovers after cessation of contamination, and the seashores rapidly recover whenever sea pollution is stopped. In contrast, certain types of nondegradable pollutants accumulate in groundwater systems, spoiling them for long periods of time. Some contamination of groundwater systems is irreversible in human time scales. In addition, cleaning up of ruined aquifers is an unrealistic mission, technologically and financially, a lesson learned from large efforts invested in small-scale remedial operations all over the world. Because of this vulnerability of groundwater, it is important for the future of this resource that man switchs to clean management of all his enterprises.

Environmental activities are so far mainly directed at cleaning pollution crises after they happen. The main current approach is to treat the symptoms—cleaning up contaminated sites, instead of fighting the diseases—the making of pollution. It seems that future development can be reasonable only if pollution is considerably restricted. Every factory needs to treat its own waste, or switch to clean production processes; every municipality needs to treat its own sewage effluents; and the issuing of licenses for new factories should be conditioned on clean management that ensures that no pollutants that are harmful to groundwater are released. The outcry for clean management has to be lead by groundwater specialists acquainted with the quality aspects of groundwater systems.

17.4 The Value of the Environment: An Essential Parameter Is Economic Calculations

Clean management is often put off by economic considerations: If a certain chemical factory has to be fully clean, the products will be too expensive; if sewage treatment installations are required to be absolutely leakproof and covered by an appropriate monitoring network, they will not be profitable; local landfills instead of centralized landfills may have environmental advantages, but may be more expensive to operate. This is poor and short-sighted thinking. In the name of short-term interests, groundwater systems are allowed to deteriorate, especially so under circumstances in which observable harm is delayed for a while.

The value of the environment has to be an integral part of the equations applied to profitability calculations. The value of the respective aquifer has to be taken into account in economic considerations. This approach will no doubt reveal that cleanly managed enterprises are profitable, and all the rest are in the long run economic disasters. Evaluating the value of the relevant aquifers necessitates delineation of their extent, and this is in the realm of the hydrologist's and hydrochemist's field of study.

17.5 Recharge and Its Controls

Upon reaching the ground, precipitation is divided into runoff and infiltrating water. The latter is divided into water that is returned to the atmosphere by evapotranspiration, and recharge that reaches the saturated groundwater zone.

Agricultural activities, mining, dams, canals, roads, and buildings disturb the natural balance between these parameters of the water cycle. In semiarid and arid regions recharge of groundwater is a necessity, and it has to be preserved and even enhanced. In contrast, in regions with high precipitation and in certain urban areas in flatlands, recharge may be in excess and has to be restricted.

The processes involved in recharge control are numerous, with a delicate interplay between the natural processes and anthropogenic interventions. The following sections describe the most important processes.

17.5.1 Permeability of the Ground Surface

Sand dunes are highly permeable and water readily infiltrates in them, resulting in minimal water loss by evapotranspiration, which is reflected in a very low salinity, and an isotopic composition that is as light as the composition of the local precipitation water. Soils have varying permeabilities, according to their clay content. The soil's ability to take in water may be significantly increased by desiccation cracks, animal holes, or plant roots (Fig. 2.3). Man enhances water intake by plowing. Anthropogenic pollution moves with the infiltrating water. A profound understanding of the listed processes is essential in order to decide on pollution prevention measures.

17.5.2 Landscape Surface Morphology

Runoff is controlled by the landscape relief and the riverbed network. All earth-moving construction interferes with the natural setup, in some places arresting water and causing increased water intake and in other cases enhancing runoff. Agricultural terraces, plowing along elevation contour lines, and dams are the most common means by which man slows down runoff and enhances recharge. In contrast, drainage trenches are intended to reduce water intake and remove part of it as runoff. Man interferes with the water cycle by constructing mines, quarries, roads, railroads, and irrigation and transportation canals. Thus, every construction project must be accompanied by careful planning to minimize damage to the groundwater system. In addition, monitoring to determine the hydrological outcomes should be an integral part of every construction project. Monitoring should include composition analyses and water table measurements. The results may be crucial in the identification of groundwater damaging obstacles.

17.5.3 Irrigation

Irrigation is a most noticeable man-made contribution to recharge. Irrigation is harmless, or even positive, as long as the following conditions are met:

The soil and rocks in the aerated zone are of good permeability, making room for efficient drainage to below the root zone.
The local through-flow groundwater system is well drained to the sea.

Under such conditions irrigation can go on uninterrupted for centuries. However, under poor drainage conditions irrigation can be harmful. Irrigation water is exposed to intensive evapotranspiration that returns pure water back to the atmosphere but leaves behind all the dissolved compounds. Thus, the recharged irrigation water is enriched in dissolved components, compared to the supplied water. In many regions the water for irrigation is supplied from local wells. Thus, in cases of poor drainage to the sea, water may be cycled between the irrigated fields and the local pumped aquifer, causing salts to accumulate and deteriorate local groundwater quality.

Groundwater deterioration in poorly drained terrains is enhanced by accumulation of fertilizers and pesticides, washed into the local groundwater system. This process is further aggravated by irrigation with treated sewage water, which contains residual dissolved ions and organic materials. To what extent these processes take place can be established by monitoring to detect changes in the chemical composition of the groundwater (appearance of compounds typifying the irrigation activity) and its isotopic composition (arrival of highly evapotranspired water).

17.6 Overpumping Policy: The Through-Flow Systems

What effects a significant lowering of the water table by intensive pumping will have, is a major question in groundwater exploitation policy. A number of processes warrant consideration in each case of recommended overpumping. The role of each of these processes has to be estimated in each case, based on knowledge of the local system and ongoing monitoring.

17.6.1 Seawater Encroachment

Salinization is often observed in heavily pumped coastal wells. This observation is commonly based solely on chlorine concentration data, and is almost always attributed to seawater encroachment, which is made possible because of overpumping. The real cause of salinization has to be investigated in each case with the aid of complete chemical and isotopic analyzes.

Seawater intrusion is usually limited to a distance of 1–2 km from the sea, but it varies. The depth at which seawater-induced salinization occurs increases with the distance of the well from the sea. The invading seawater may have direct hydraulic interconnection with the open sea, or it may originate from a stagnant compartment of seawater that was trapped during past hydrological conditions. The latter case will be recognizable because of the old age of the saline water and by deviations from the present seawater composition, reflecting seawater-rock interactions.

17.6.2 Encroachment of Stagnant Saline Groundwater

Saline groundwater occasionally invades inland wells of high-quality groundwater because of overpumping. An example of the occurrence of such saline paleowater, and its encroachment into overpumped wells, has been described for the coastal plain of the Mediterranean Sea (section 15.5.3). The occurrence of shallow stagnant groundwater has been little investigated, but it might become better known thanks to the increasing number of observations of salinization in inland wells.

17.6.3 Cessation of Drainage to the Ocean

The bulk of the through-flowing groundwater is drained by lateral base flow to the ocean (section 2.13.1; Figs. 2.14–2.16). Under natural steady-state conditions all the recharged water is discharged, guaranteeing that all the dissolved components are drained as well. The groundwater cycle functions in the following mode: distilled water is pumped by the sun into the clouds and precipitated on the continent, where part of it is recharged underground. Groundwater becomes enriched in dissolved salts from sea spray and from interaction with rocks along its flow course, but ultimately it is drained back into the ocean, the sea acting as a major sink for dissolved compounds. This steady state can be disturbed by lowering of the water table to an extent where the lateral base flow to the ocean is interrupted. From this moment on the salts accumulate in the groundwater system instead of being drained into the ocean. The same holds true for contaminants, they accumulate in the groundwater system as well.

The need to preserve some degree of seaward drainage has to be taken into account in groundwater management considerations. At the least, periodic drainage to the sea has to be maintained.

17.7 Optimization of Pumping Rates in the Through-Flow System

Authorities who operate wells are constantly tempted to pump as much as possible, either to supply ever-increasing demand or to save the drilling costs of more wells at more distant locations. Overpumping occurs whenever the amount of pumped water is greater than the amount of recharged water that reaches the respective part of the groundwater system. Overpumping is evidenced by a significant lowering of the local or regional water table. Study of the compositional changes of the groundwater, as a function of the lowering of the water table, can reveal the nature of the secondary processes that occur. The following points have to be taken into account in the pumping rate considerations:

- Overpumping is a mining operation—it provides a one-time, nonrenewed amount of water (once the decided minimum water table has been reached, all subsequent pumping is limited to the amount of recharge).
- In undisturbed systems pollutants reach the base flow zone and do not penetrate into deeper (below sea level) stagnant parts of the system. Thus, the deeper stagnant parts below through-flow systems may serve as a strategic reserve of clean water. Lowering of the water table may ruin this reserve because it lets pollutants reach the deeper parts of the system.
- Overpumping makes room for the encroachment of seawater (close to the coast) and deeper saline groundwaters (in certain regions).
- Overpumping reduces the seaward drainage of dissolved compounds and pollutants.
- A positive outcome of water table lowering is an increase in the storage capacity of the system, to be used for occasions of extra-large recharge following rainy periods or times of reduced water demand.

17.7.1 Setting of "Red Lines" and Monitoring

A clear pumping policy has to be formulated for every region in accordance with its hydrological structure. It seems useful to sum up the accepted policy in terms of "red lines" of maximum tolerable lowering of the water table and maximum tolerable concentrations of water quality key parameters. These parameters have to be monitored routinely. Electric conductivity and water table may be monitored continuously by instruments placed in observation wells or pumping wells. In the latter case, the respective dynamic water table will be measured. The monitoring network has to be organized to issue real-time warnings to the well operating agency.

17.8 Optimal Utilization of the Groundwater Resource

A basic rule of economic geology states that an ore deposit should always be exploited so that its highest potential value is harvested. This rule is applicable for groundwater—the many different qualities and modes of emergence call for careful marketing evaluation and rational planning of the exploitation of this basic commodity.

17.8.1 Pricing by Quality

Groundwater is used for a large number of different purposes, and each consumer has his specific minimum quality demand. Yet, water is often supplied with a higher quality than needed, for example, to certain types of industry or agriculture. If the price of water is set according to its quality, some consumers may prefer lower quality water, relieving the pressure on the high-quality resource. This policy implementation is based on hydrochemical expertise.

17.8.2 Groundwater-Fed Sources as Assets of Nature

Cold and warm springs, oases, waterfalls, and wetlands are a few examples of groundwater-fed assets of nature. In the hasty drive to exploit groundwater, these presents of nature are often destroyed. However, these features should be protected and preserved as remnants of the unspoiled systems of the earth. Besides their aesthetic merit, these assets have a high economic potential, in the frame of national parks, recreation areas, or central attractions in resorts. Spotting the best candidates for conservation, and securing their preservation, are in the realm of the hydrochemists.

17.8.3 Mineral Water: Its Double Meaning

In the old days it was a custom to travel to specific localities in order to drink "healthy water." This water was rich in dissolved ions, or salts, and it was called "mineral water." Mineral water often had a bad taste, but people drank it as a medicine.

In recent years the custom of drinking bottled water has spread, mainly because on of the deterioration of tap water in a growing number of localities. Bottled water is expected to have a good taste, which means a low concentration of dissolved ions. Yet, this water is often called mineral water.

The bottled water industry is enormous. It demands prime quality water, but offers the highest added value. The name "spring water" is much used to promote sales, but water from wells is just as good. The prime concerns in the supply of water for bottling is low ion content, no odor, good taste, and no biological or other contaminants.

17.8.4 Spas

Spas are in great demand all over the world. There is an enormous range of water properties by which the various spas are distinguished and advertised, including high salt content, sulfuric smell, radioactivity (high concentrations of radium and radon), and elevated temperature. The therapeutic activities include bathing, drinking, and gas and steam inhalation.

The long established spas each have a specific list of illnesses that are cured by its waters. Almost by convention, physicians assess the medical value of a new candidate water source not by the results of its compositional analysis, but by its likeness to other well-known spas. Into this niche steps the hydrochemist. He has to assess the chemical and physical descriptions of the world's leading spas and find the best match to the new spa. The list of curable maladies can then be obtained.

In principle, water with all the desired compositional combinations can be prepared artificially, but the spas' trade is based solely on its natural sources. Similarly, naturally heated water is preferred by most clients. Hence, the high economic value of springs and wells of special composition and elevated temperature. These waters are often unusable for other purposes.

17.8.5 Stagnant Systems: A Strategic Reserve

Stagnant groundwater systems, especially the pressurized and hydraulically isolated ones (sections 2.13 and 2.14), call for a management policy that is entirely different from that developed for through-flow systems (section 17.6). The rock compartments that host stagnant groundwater are isolated from the surface and have no drainage. These stagnant groundwaters have the following characteristics:

They constitute a nonrenewable resource.
They are protected from anthropogenic pollution.
They possess an excellent storage capability—for durations reflected in the old age of the waters (10^4–10^6 years).

Many of the known artesian groundwater systems are used for irrigation, or occasionally for space heating. Being regarded as a cheap commodity, they are often wasted. However, in light of their special qualities, pressurized stagnant groundwaters are a precious resource that should be developed with concern and planning.

Pressurized stagnant aquifers with good-quality water should be preserved for times of need: severe droughts, large-scale pollution accidents, or nuclear catastrophes. In times of crises, a few weeks may pass before it is clear which of the ordinary wells are contaminated. In this critical period stagnant aquifers can supply pollution-free water. Hence, selected high-quality pressurized aquifers should be drilled and equipped for emergency water extraction.

Pressurized stagnant aquifers with slightly saline water should be preserved for similar reasons, because their water may be useful for large-scale decontamination activities, for domestic and industrial needs in times of shortage, and for future desalinization, especially in remote inland regions.

Sustainable Development of Groundwater

The practice of exploiting artesian systems has produced a large number of depressurized stagnant aquifers that are at present of no use. Such aquifers, which supplied high-quality water, can be reused by reinjecting fresh water to create pollution-immune underground water reservoirs.

The different management strategies recommended for through-flow systems and stagnant systems demonstrate the importance of correctly identifying the hydrological regime of every exploited groundwater system.

17.8.6 Stagnant Systems as Potential Waste Repositories

Abandoned depressurized aquifers, that supplied low-quality water, may serve as potential sites for injection of toxic fluids. The threat of developing uncontrollable hydrofractures during injection is expected to be minimal, as long as the operational pressure does not surpass the initial natural pressure.

18
GROUNDWATER ASPECTS OF URBAN AND STATEWIDE DEVELOPMENT PLANNING AND MANAGEMENT

The topic of the present chapter may seem strange in a textbook dealing with chemical and isotopic hydrology. But the success of residential, industrial, and agricultural projects is today deeply anchored in the proper approach to a long list of groundwater-related aspects. It is the hydrochemist who is best equipped to formulate the necessary questions and to suggest the possible answers. This is well in the frame of the applied approach of chemical and isotopic hydrology. Advice to planners can be given on the basis of general knowledge of groundwater behavior, but in actual projects specific research has to be conducted in order to understand the specific groundwater systems.

In the constant battle for clean environmental management, the hydrochemist has a key role. And the key to fulfilling this role is to thoroughly understand the respective hydrological systems, a task that can be achieved with the tools discussed in the previous chapters.

18.1 Surface Coverage by Roads and Buildings

A significant quantity of rain and snow water comes off of the roof of every house. Prior to erection of the building, this water reached the ground and contributed to the local recharge (section 17.5). Thus, buildings make the ground impermeable. In ancient times this water was collected in cisterns and used for the household, but this wise practice has to a large extent been abandoned. Similarly, paved roads cover the earth's surface with an impermeable layer. In extreme cases the pavement entirely closes in the trunks of the trees that grow along the streets, leaving no space for water infiltration. The impermeable coverage of cities may exceed 90% of their area. The result: the region is deprived of its former recharge and the local groundwater table drops, especially if pumping from local wells continues.

18.2 Street Runoff Mitigation

The rain and snow water reaching the urbanized areas is transformed into street runoff. Unless it is specially managed, street runoff causes a number of problems:

Flooding of streets at peak rains.
Flooding of basements and first floors.
Flooding of the existing sewage system.

The common way to cope with street runoff is to construct a special drainage system; the water being evacuated to the sea, a lake, or a river. In this way the comfort of the citizens is maintained, but the region is deprived of its original recharge (previous section). The latter effect is of importance in semiarid and arid regions, since in these regions water supply is mostly from local wells.

There is another solution. Water from roofs can be diverted to adjacent open areas or gardens. The streets and pavements should be built with local slopes toward trees growing along the streets, and the streets should slope toward adjacent small and large public gardens and other open spaces. In this way a significant portion of the rain and snow water can be diverted back into local recharge, and the negative effects of street runoff will be eased. The rest of the runoff should be recharged into the ground by spreading over special infiltration fields or via reinjection boreholes.

18.3 Raising of the Water Table Due to Sewage Infiltration

Many of the world's residential areas have no central sewage systems, and the domestic sewage goes to septic tanks or to toilet pits. The resulting effects are:

A general rise in the local water table, sometimes to an extent that building foundations are damaged and basements are constantly damp.
The local groundwater quality is deteriorated to an extent that it is unhealthy and nonpotable.

The remedy in such cases is clear: construction of central sewage collection and treatment installations, and acquiring a supply of water from wells that are outside the residential area.

The ultimate goal is to reach the state described in the previous section, namely, a well-organized street runoff mitigation system beside a central sewage system. The moment this stage is reached, groundwater extraction from local wells can be renewed.

Monitoring of the groundwater condition is essential until all the remedial steps are undertaken. There are different levels on which the monitoring can be performed. But, means being limited, any monitoring is better than none. The results of the monitoring are crucial in order to avoid immediate health disasters, make the authorities and the public aware of the urgent need for the remedial measures, and prepare the database needed for the optimal planning of the remedial measures.

18.4 Location of Urban Landfills, Sewage Treatment Plants, and Industrial Zones

Landfills, sewage treatment plants, and factories can be planned and managed with high environmental standards. The location of such undertakings has to ensure the safety of the groundwater. For this reason, the following surveys and studies are essential:

The groundwater systems of the planned area and its immediate surroundings have to be identified in terms of through-flow systems (may belong to several local drainage basins) and stagnant systems (unconfined and confined).

Each groundwater system has to be characterized in terms of its depth and aerial extension, water composition, isotopic composition, concentration of water age indicators, and arrival of specified pollutants.

The flow direction in each of the local through-flow systems has to be established.

The thickness of the aerated zone and the nature of its soil and rocks have to be established in terms of the lithologies, prevalence of conduit flow or porous flow, and absorption capacity.

Organization of an updated data bank, including as much old (historical) data as possible.

Based on the outcome of these detailed studies, landfills, sewage treatment installations, and factories should be placed according to the following considerations:

At the downflow end, to ensure that arriving contaminants do not reach local wells.

Above thick parts of the aerated zone, to ensure maximum delay of infiltrating fluids.

Above sections of soils and rocks of the aerated zone that are richest in clay and other absorbing rocks.

Away from operating wells or surface water supply installations (lakes, dams, or rivers).

18.5 Irrigation Optimization

Irrigation is a major water-consuming activity that results in significant release of water into the ground. The outcomes are crucial to the groundwater, both quantitatively and qualitatively. Irrigation effects groundwater in the following ways:

Addition of water recharged into the unconfined through-flow system.

The recharged irrigation water is enriched in dissolved ions as a result of evapotranspiration activity, a feature also reflected in enrichment in the heavier hydrogen and oxygen isotopes.

The irrigation recharge water is often enriched in fertilizers, pesticides, nitrate, and other compounds that are formed in the soil zone.

Irrigation water transports pollutants from the surface down into the through-flow system.

The impact of irrigation on groundwater can be mitigated in the following ways:

Irrigate at night, in order to significantly reduce evaporation, thus reducing the quantity of consumed water, and reducing evaporation-induced water salinization.

Irrigation by methods that reduce evaporation (i.e., avoiding sprinkling and flow in ditches), but do not increase local soil salinization.

Installation of sensors to monitor the water content in the soil root zone, to automatically control irrigation to amounts that are just enough to saturate the root zone or slightly beneath it.

Using closed greenhouses.

Avoiding crops with extra high water demands.

Development and installation of sensors for fertilizer compounds in order to restrict the use of fertilizers in the root zone, with no leftovers to be washed into the groundwater system or to accumulate above it.

Development and installation of sensors to determine the concentration of specific pesticides on plants in order to optimize their application.

18.6 Agricultural and Urban Use of Treated Effluents

Sewage effluents can be reclaimed in quality grades that are perfectly suited to irrigation purposes. In an increasing number of countries, water shortage makes irrigation with treated effluents attractive. This practice is feasible, provided a number of precautions are taken:

The quality of the treated effluent must be routinely monitored, and the use of contaminated batches must be avoided.

The local groundwater quality must be routinely monitored to ensure its quality is not lowered.

The soil condition must be routinely monitored to ensure that its physical properties are not harmed, for example, by too high doses of sodium that may cause particle dispersion.

All the precautions recommended in the previous section for irrigation are implemented.

Treated effluents are also suitable for public gardens, golf courses, playgrounds, forests, and other public open areas.

18.7 Groundwater Vulnerability Maps: A Questionable Approach

Maps that delineate the vulnerability of various regions to their groundwater deterioration, as a consequence of planned human activity, have become a fashionable tool tailored for planners. The basic idea sounds good: hydrologists take existing geological maps and translate them into groundwater vulnerability maps. The most common type of map is of the time predicted for the arrival of pollutants in the local groundwater. The criteria applied to prepare such maps include:

Rock types, classified by their permeability: the less permeable, the longer is the predicted time span until pollution will arrive.

Thickness of the aerated zone; the thicker it is the longer are the pollutants predicted to be retarded.

The nature of the infiltration system: if porous, the movement of pollutants is assumed to be slow, if conduit controlled (karstic), pollutants will arrive fast.

There are several major problems with the described approach. First, the geological maps are prepared by stratigraphic units, and the lithology has to be inferred. In any case, stratigraphic units include, in most cases, alternations of various rock types, often with a wide range of permeabilities. It is almost impossible to translate a geological map into permeability units.

Second, it is almost impossible to quantify the terms "fast arrival" and "slow arrival" of contaminants. In some vulnerability maps the legend is divided into intervals of arrival time (e.g., within 3 years, within 20 years, and later than 20 years). Third, the nature of the infiltration setup (depth to water table, porous or fissured medium) varies on a small scale, and there are no hints to it in geological maps. The information has to be derived from other sources, which are rarely at hand.

Planning and Management   371

A major problem with the preparation of groundwater vulnerability maps is anchored in the basic question of whether the time of arrival of contaminants is at all a base to assess vulnerability. Is fast arrival better or worse than slow arrival? Fast arrival may in fact be advantageous, because the outcry against the pollution act will be immediate, and dilution with through-flowing groundwater will be relatively efficient. Late arrival postpones the protest, and in the meantime pollutants can accumulate in the aerated zone to a degree that can be built up causing unreversible damage.

The issue of vulnerability maps deserves serious consideration. The topic is presented here in order to draw attention to the problems, and to warn against uncriticized use of such maps by planners. The warnings and solutions have to be provided by hydrochemists.

18.8 Monitoring Networks: Basic Principles

The need for monitoring is emphasized throughout this chapter. Monitoring deals with repeated measurements of properties of a system in order to get a description of the system at a certain initial point in time, and to follow ongoing changes in the system. The extent and detail of the monitoring will vary according to the problems involved and the budgets that are available. The following are some general considerations:

There are basically two types of monitoring wells—observation wells that are never pumped, and producing wells. The advantage of the first type is that they reflect the regional situation and, in terms of water table, they reflect the static water table. The second type reflects the water that is supplied to consumers, and in terms of water table, indicate the dynamic water table (i.e., the depth of the pumping cone). Monitoring of both types of wells is needed, and the information gained is complementary.

The number of measurable parameters is huge and therefore costly. Thus, priority of measurements has to be worked out for every monitored well. For example, at a well situated near a pesticide storage facility (at the downflow direction) the entire list of stored and handled pesticides should be analyzed. The list of measurements will be completely different for a well that is remote from such materials but adjacent to a dairy. In this well, specific organic and biologic parameters should be selected.

It is highly recommended that every well in a monitoring network be thoroughly measured at least once, in order to obtain the base data needed for a comprehensive understanding of the respective groundwater system, and in order to understand the position of the well in that system. This is, in turn, necessary in order to establish the list of routine measurements, and to be able to properly interpret observed changes in the well's properties. The list of measurements to be included in such a one-time thorough investigation includes the water table height, temperature, common ions, trace elements, stable isotopes of hydrogen and oxygen, concentration of radioactive isotopes, concentration of various age indicators, and a long list of potential contaminants. Actually, all the parameters included in the list of drinking water standards should be analyzed at least once in every well.

The monitoring network should cover every potential source of major pollution (e.g., sewage treatment plants, landfills, factories, farms), every distinct through-flow and stagnant groundwater system, and special areas (e.g., adjacent to the seashore where there is a danger of seawater intrusion).

At a given source of potential pollution, the monitoring well has to be selected so that it is nearby, in the groundwater downflow direction. In many cases more than one well will be required to adequately cover the potential pollution source.

In pumping wells, it is important to find out at what depth they sample, and whether waters from several aquifers are mixed. Wells that pump from a single water system are preferable. In observation wells it is highly recommended that a depth profile of electric conductivity and temperature be determined.

The monitoring frequency has to be defined according to the relevant problem. For example, determining recharge efficiency requires measurements at the end of the dry season and at the end of the wet season. An uncontaminated well may be monitored for pollutants once a year, whereas a well that had been found to be polluted has to be monitored more often.

A monitoring network book has to be prepared. In it the information for every well should include the rock cross section and other information collected during drilling, the results of the one-time detailed examination, the nature of the groundwater system, the reasons why the well is included in the monitoring program, and the suggested monitoring schedule.

For each well in the network, red lines have to be defined, and clear instructions should be given as to what should be done whenever the red lines are violated.

The outcome of monitoring activities should be continually processed and evaluated, and the monitoring operation should be reevaluated and reorganized according to the management and research requirements.

18.9 Active Data Banks

Groundwater research and monitoring operations produce large amounts of data of many kinds. These data have to be gathered in one place. The following groups of data should be included in a groundwater data bank:

Stratigraphic and lithological cross sections of each well, based on drilling data and regional geology, and structural geology of every spring.

Physical data—temperature, depth and altitude of the water table, discharge rate, or pumping rate.
Chemical data, including dissolved common ions, trace elements, pH, and organic compounds.
Contaminants—following the list of drinking water standards.
Microorganisms and pathogenic findings.

Data banks should be centralized; that is, all the data pertaining to a certain region should be accessible within the frame of a single data bank. The sources of data are:

Monitoring networks of water authorities, health authorities, and other organizations.
Historical data from archives of various authorities, and consultants' reports.
Data from research institutes, papers published in professional journals, and theses.

Data banks are justified only if they are active. They must not just gather information, but their staff must put the information into use in the following ways:

Storing the data in a user-friendly accessible way so that researchers, planners, management personal, and other customers can readily retrieve the data they need.
The data have to be stored in a form that is easy to understand and use (e.g., in the form of tables, maps, transects, depth profiles, fingerprint diagrams. composition diagrams, etc).
The data bank personnel have to disseminate the information through periodic and special reports, maps, and other graphical representations.
Data banks can issue alarms in real time to management personnel and decision makers.

Data banks should not be data sinks, they should be active and turn the data into a productive resource.

19
HYDROCHEMIST'S REPORTS

19.1 Why Reports?

Scientific research has to be summed up in a written format with appropriate circulation. The final stage of any research is a report, needed to:

Expose the work to critical comments by other specialists.
Convey the findings to the authorities that financed the investigation.
Convey the results and recommendations to the engineers and other researchers who have to execute the recommendations.
Document the findings for future investigators.
Share your experience with the international scientific community.

In addition to the listed reasons for preparing written reports there is a benefit in report writing that cannot be overestimated: the preparation of the report clears the researcher's mind, clears up inconsistencies, and focuses thoughts on the essential issues. Thus, preparing reports is an integral part of the scientific thinking process.

19.2 Types of Reports

Written reports vary in length from a single page to hundreds of pages. The length of a report depends on the type of report and the specific nature of each study.

A research proposal is, in most cases, the first document to be written in any investigation. It should focus on the problem to be addressed, work done so far, strategy of the investigation, suggested schedule, organizations involved, and required budget. A research proposal should be brief and contain only the most important points.

Progress reports are required periodically by the financing agencies. Beyond that they are also of prime importance to the researcher. Progress reports force the researcher to realize that data have to be processed and discussed continuously. Conclusions reached in progress reports influence the continuation of the study as priorities are reassessed. Progress reports have to contain a substance—results and preliminary discussion of the data. The length of a progress report is basically defined by the amount of data and the maturity of their processing.

Final reports are just that—final. In other words, they should contain all relevant information and be comprehensive. Final reports are the longest of all report types and, if needed, may contain hundreds of pages. Final reports may have an outline similar to scientific papers, but they are supported by all the raw data, mostly organized in appendices. Because of their length, final reports should contain an abstract.

Scientific papers are often an outcome of a final report. A scientific paper should be concise, a task that can be achieved by referring to the final report for details. A scientific paper should deal with a local problem, but in a mode that is applicable elsewhere. Thus, scientific papers have a methodological aspect that is of much importance. In optimal cases the specific study serves as a case study to demonstrate the methodology of the scientific approach.

No recipes exist for writing reports—each case has to be solved for itself. The following sections deal with the various parts of a report in a generalized form.

19.3 Internal Structure of Reports

19.3.1 Name or Title

The name of a project is, in a way, a one sentence summary of the most essential parts of the whole study. Thought has to be given to proper formulation of the name of the project, the report, or the scientific paper. This is best done by first listing the key words that should be included:

The main issue—pollution, recharge or discharge estimates, mixing of water
    types, groundwater dating, pumping consequences, general survey.
Methods used—pumping tests; chemical analyses and isotopic analyses; dating
    by tritium, ^{14}C, ^{36}Cl, or ^{4}He; trace elements; specific polluting compounds
    analyzed; monitoring; water table data, temperature surveys.
Geographical location of the study.

Below are examples of report titles, based on selected key words:

Chemical and isotopic survey of springs and wells in the Paradise Valley.
Flow direction in the Apollo aquifer, Utopialand, traced by a landfill leachate.
Leakage of the Fantasia Dam, traced by environmental water isotopes.
A groundwater transect through the Episode Valley: hydrological, chemical, and isotopic parameters.
Identification of recharge areas of the Maximum Springs, applying stable isotopes, tritium, carbon-14, and helium.
Paper factory as a possible source of salinization of the Alpha aquifer, Betaland.

19.3.2 Abstract

The abstract provides a brief summary of the major outcome of the research. It does not deal with the history of the project, the type of measurements conducted, or hypothesis that could not be confirmed. The abstract should avoid noninformative sentences. For example, what is wrong with the following abstract:

> An intensive study has been conducted on a large number of water sources in the study area. The paper deals with the various aspects of the main problems, leading to a series of practical recommendations.

This example contains no single item on information, and the abstract gives the reader no idea of what the paper is about. Noninformative words are "a large number" (instead of "21 wells"); "water sources" (instead of "a lake and 12 springs"); "study area" (instead of "Wisdom Valley, Bestland"); "the paper deals with" (should be omitted); "various aspects" (instead of "consequences of clearing of the Green Forest"); "leading to a series of" (should be omitted); "practical recommendations" (instead of "limiting forest clearing to no more than 10% of the area is recommended"). The abstract should contain the most important data, for example,

> The Curiosity aquifer is saline (1500–2200 mgTDI/l), Na-HCO$_3$ dominated, with temperatures of 59 ± 2°C, devoid of tritium, ^{14}C concentrations of 6–8 pmc, dD of –120 ± 10‰ and d^{18}O of –16 ± 2‰. Recharge from the Top Mountains is concluded, depth of circulation being >850 m, effective age –22,000 ± 3000 years. So far no invasion of the Filthy Landfill leachate has been noticed, based on the chemical and biological data.

19.3.3 Targets of the Study

Hydrochemical studies may be motivated by pure scientific curiosity, urgent management problems, or routine surveys. Behind each of these motivations stand one or several targets and they should be spelled out. There is a saying that "a good question is half its answer." By the same token it may be said that well-defined targets promise well-conducted research and focusing on the major issues. Which of the following sentences are well-defined hydrochemical research targets?

Checking hydraulic interconnections between the Grand Dam and the Greytown municipal well fields I and IV.
Conducting general observations on the local hydrological system.
Testing the applicability of ^{14}C dating techniques to the noncarbonate aquifers of the Glorious groups in the Papaland.

The first and last examples are well-defined hydrochemical research targets because they are specific and focused.

19.3.4 Definition of the Study Area

Terms like "the northern Great Artesian Basin," "surroundings of London," or "arid zones of Israel" may be suitable for the name or title of a project, but in the report or the paper itself the study area should be precisely defined. The aerial definition may be by coordinates or by names of geographical boundary objects. The study area should always be defined on a clearly readable map, along with a map that shows the study area in a larger geographical frame, for example, a country.

19.3.5 Hydrological Inventory

The report should have at its beginning a section describing the hydrological inventory of the study area: seacoast, lakes, marshes, rivers, dry riverbeds, springs, abandoned wells, operating wells, observation bore holes. Sources of available information on these water bodies should be discussed, such as published papers, reports issued by various authorities, research thesis submitted at universities, and data stored in archives. The exact titles can be included in the reference list of the report. To each source of information a sentence or two should be added on the nature and extent of the included information, such as water level measurements, discharge values, periodic quality checks, borehole and casing data, and chemical and isotopic analyses.

19.3.6 Human Activities and Installations Relevant to the Study

A clear and concise description should be provided on relevant human activities in the study area. These should include dams, drainage channels, recharge installations, sewage ponds, industrial effluent ponds, fluid waste disposal installations, pumped well fields, and the nature and extent of agricultural activity, including irrigation schemes and the use of fertilizers (types, quantities) and pesticides (types, quantities). Part of the information may be abstracted from detailed maps, but most information has to be obtained directly from local authorities, farmers, and industry.

19.3.7 Strategy of the Investigation

A good investigation is not done by shooting in all directions. It is performed with an outlined strategy. A clear distinction has to be made between knowns and unknowns, and between established methods that have already been tested and calibrated, and developing methods that have to be checked and calibrated.

Formulation of a preliminary conceptual model based on existing data is most helpful in defining the strategy of an investigation. Such a preliminary conceptual model should include alternative recharge areas, modes of flow, flow paths, mixing combinations, sources of heat, dissolved solids, and dissolved gases, and alternative routes of potential contamination. In the second stage, criteria should be defined by which incorrect alternatives can be ruled out and correct alternatives can be proven. In the third stage the research strategy is formulated, focusing efforts on necessary observations and collection of required data. The following is an example of an investigation strategy, taken from the introduction of a paper on research of CO_2-rich springs:

> The strategy of data processing was (a) searching for water groups of common chemical composition, (b) searching for geological or geographic logic of the established chemical groups, turning them into geochemical water groups, (c) in each group the distribution pattern of ion concentration variations was studied. Mixing lines emerged, and the properties of the end members were qualitatively defined, (d) once it was found that HCO_3 is the dominant anion, accompanied by different cations, possible water-rock interactions were explored to explain the identified geochemical water groups, (e) the stable hydrogen and oxygen data were processed to look for the role of local recharge (lying on the meteoric line) versus waters with modified isotopic compositions (e.g., ^{18}O shifts). For water lying on the meteoric line the average recharge altitude was calculated from the isotopic composition, (f) the tritium data were interpreted in terms of pre- and post-nuclear bomb test water components, (g) the temperature data were interpreted in terms of depth of circulation of the groundwater end members, (h) $d^{13}C$ values, obtained for the CO_2 emanating in a few mofettes, and bubbling in a few springs, were applied to define constraints on the possible origin, i.e., decomposition of organic material, heat-induced decomposition of carbonatic

rocks, or mantle contribution. This choice of CO_2 origins was narrowed by noble gas data: nonatmospheric He is observed in the springs, explained as accompanying the ascending CO_2. $^3He/^4He$ ratios of 0.1 of the atmospheric ratio ruled out a mantle origin, leaving both the He and CO_2 as of crustal origin.

19.3.8 Results

Data are the major product of every investigation. Our interpretations, conceptual models, and recommendations may all be challenged by other investigators, or turn out to be wrong by our own subsequent work. Observations and measurement results, however, may always be incorporated in new models and coupled with additional observations. Hence, results are a most important part of a report or paper.

Results should be presented in a way that makes them usable by other investigators. Data are best presented in well-organized tables (section 6.1), along with relevant technical data, such as who measured each parameter, measuring error, limit of detection, resolution, results of duplicate and triplicate measurements, and measuring methods applied. A major part of data tables may be included in the appendices of a report.

19.3.9 Data Processing and Interpretation

Hydrochemical data can be processed in a variety of ways and, accordingly, can be presented in many different ways, including fingerprint diagrams (section 6.2), composition diagrams (section 6.3), histograms, maps, transects (section 7.9), and a host of other graphic modes.

Presentation of processed data is closely linked to interpretation since the obtained patterns reflect distinct boundary conditions, such as, the number of water types occurring, mixing of water types and mixing percentages (natural and polluted systems), hydraulic interconnection and hydraulic isolation, positive and negative correlations between parameters, and directional changes in properties (on transects and maps) indicating directions of flow and modes of intermixing.

Much thought should be devoted to determining the best way to present processed data in a report. The order should be defined by the study strategy and the logic of the case presentation. In a poorly organized report, use is often made of statements such as "this point is discussed later on" or "the relevant data are presented in a later section." In a well-organized report topics are discussed in a progressive order. The data processing and interpretation sections of a report lead each to a subconclusion or define a constraint, to be all taken into account in the formulation of the overall conceptual model.

19.3.10 Exact Terms, Logic, and Informative Statements

Conclusions, boundary conditions, and conceptual models have to be formulated with the aid of exact and relevant terms.

- What is wrong in the following statement?
  'The shallow wells contain fresh water and therefore contain post-bomb tritium.'

  The term "fresh water" relates to *low TDI concentrations*, whereas post-bomb tritium relates to a *recent age*. The presence of post-bomb tritium *indicated* the water is of a recent age, and not vice versa.
- What is wrong in the following sentence?
  'The low temperature of the springs indicates dominance of a recent water component.'

  Low temperature may indicate dominance of a cold water end member, but *temperature* does not relate to *age*.
- What is wrong in the following phrase?
  'The water is old, as indicated by its high salinity.'

  High salinity may be caused by many processes (evaporation, intermixing with saline water, dissolution from rocks) and is not an indication of a high groundwater age.

Logic has to be maintained in every sentence of a report. *Cause* and *outcome* should not be mixed, neither should *observation* and *hypothesis*.

- Correct the following statement:
  'The new deep wells tap mixed waters and therefore have discordant ages.'

  Observation (discordant ages) leads to the *conclusion* (the water in the wells is a mixture) and not vice versa.
- Correct the following statement:
  'Piston flow is assumed to dominate in the Good Hope wells that have seasonally varying temperatures.'

  The order should be reversed: *observation* (seasonally varying temperatures) leads to the *conclusion* (dominance of piston flow). The term "assumed" is out of place.

Statements in a report should always be informative as much as possible.

Examples
- How can the following sentence be made informative?
  'Group A waters are warm and group B waters are cold.'

  Warm and *cold* are relative terms, but providing the relevant temperatures will make the sentence informative: Group A waters are warm (52–64°C) and group B waters are cold (15–17°C).

- Is the following sentence good enough to be included in a hydrochemist's report?
  'Magnesium is correlated to bicarbonate in two out of the three water types prevailing in the area.'
  This sentence should read something like: Magnesium reveals a linear positive correlation in water groups I and III of the Grand Scheme project area.

19.3.11 Conceptual Model

Conceptual models (section 1.5) should be presented in a clear, logical, and concise way. Each part of the model should be based on a specific sub-conclusion or constraint, reached in previous parts of the report. A conceptual model should discuss alternatives, and explain criteria used to select the preferred possibility.

A conceptual model should, as much as possible, be formulated in terms that will lead to conclusions and recommendations. Therefore, discussion of a conceptual model should also include definition of weak points and open possibilities. The conceptual model is, in a way, the heart of every investigation, and much thought should be devoted to its construction and proper presentation.

19.3.12 Quantitative (or Mathematical) Model

Parts of a conceptual model, or all of it, may in many cases be quantified, using mass or energy balances, chemical equilibrium and degree of saturation calculations, flow velocities calculated by hydraulic parameters, or velocities calculated via isotopic age indicators. In presenting the mathematical calculations, *knowns* and *unknowns* have to be clearly defined, and basic assumptions have to be discussed. Each calculation should be accompanied by an evaluation of its degree of confidence.

19.3.13 Scientific Conclusions

Models may be followed by a discussion of the scientific conclusions reached in the study; for example, which elements and compounds tend to behave as conservative parameters; which ions participate in secondary water-rock interactions; or reliability of isotopically based calculations, such as recharge altitude (section 9.8) or $\delta^{13}C$ values as a tool to correct observed ^{14}C values for water-rock interaction.

Scientific conclusions are of a local nature but of a general value, because they can be applied to studies in other regions as well.

19.3.14 Operational Recommendations

Operational recommendations can be of a practical aspect, for example, rate of suggested pumping, or the spacing of new wells and their required depths. Operational recommendations can also relate to further research needed, for example, new parameters to be measured, special pumping tests, or new wells. Recommendations have to be clear, and their reasoning should be explained. Does the following recommendation fit the described specifications?

"A few deeper wells should be drilled to get a better understanding of the system."

This recommendation is totally worthless: "A few"—how many?, what are the technical specifications?, where?; "to get better understanding"—how? by closing large gaps between existing wells? by better installation of the wells? by proper pumping tests?

Recommendations for further research should be given only if they are fully justified. Vague words such as "deeper," "more," or "better" should be replaced by explicit definitions. Recommendations should also be weighed in light of the costs involved.

20
ANSWERS

Answer 1.1: It seems easy to grasp that along the seacoast all rivers flow into the sea. The observation that all rivers run into the sea describes runoff. Water flows down along the topographic relief, and the sea is the terminal base of drainage.

Answer 1.2: The law of mass conservation. When we pour water into a vessel, the level of the water rises—not so in the case of the sea. So, there must be an outlet of water from the ocean.

Answer 1.3: Two segments of the water cycle were concealed: the evaporation of water from the ocean, forming clouds that transport water into the continents, and the flow of water underground.

Answer 1.4: Once the water (runoff or underground drainage) is returned to the sea. There, all water is well mixed and the accumulated information is wiped out.

Answer 1.5: The temperature that prevailed at the base of the aerated zone at the time of recharge (via the concentration of the atmospheric noble gases); the depth of circulation or storage (via the water temperature); the types of rocks passed (via the chemical composition); the time the water stayed underground, i.e., its age (via the decay of tritium and ^{14}C and the accumulation of radiogenic ^{4}He and $^{40}Argon$).

Answer 2.1: 1—C; 2—F; 3—B; 4—D; 5—E; 6—A.

Answer 2.2: Every well has to be dug out or drilled (rock and soil material has to be taken out). Water exists underground only as a fluid stored in rock voids (pores, fissures, and small cavities). The well space acts as a collecting tube for the water in the rocks—the rocks are yielding water to the well.

Answer 2.3: In an extra-rainy year infiltrating rain is pushed beneath the depth of evapotranspiration activity, and a significantly large portion flows into the saturated zone and is turned into recharge. In contrast, in an extra-dry year water infiltrates to a shallow depth (about the root zone) and all of it is returned to the atmosphere by evapotranspiration.

Answer 2.4: To assess recharge we also need the range of annual precipitation intensities or the variance. There are reports of recharge in arid areas that is sustained by once-in-a-decade heavy rain events. This is an example of the problems that are connected with the selection of single representative parametric values—they cannot adequately describe an entire system.

Answer 2.5: The first water occurrence indicates the existence of a phreatic aquifer, and the second occurrence of water may indicate the existence of an underlying confined aquifer. However, the drillers could have difficulties in telling water occurrences apart if the drilling itself is conducted with water. Additional data are always needed—water table, water temperature, and water chemistry. Differences between the properties of the deeper and shallower waters will establish that two separated aquifers exist, and that the deeper one is confined.

Answer 2.6: The spring has a temperature of 15°C, whereas the local average annual temperature is 19°C. Thus, the spring water "remembers" the low snowmelt temperature. Due to the rapid flow in karstic conduits, temperature equilibration is incomplete. In addition, the discharge varies significantly during the year, in high correlation to the snowmelt season.

Answer 2.7: The time that elapsed since the water was recharged into the saturated zone. Water that is recharged into an unconfined system at its inland reaches attains a relatively high age by the time it arrives at the base of drainage at the sea, but along its flow path it gets mixed with local recharge that is younger. Hence, in a phreatic aquifer, water age does not have a singular value (the *average* age will be on the order of a few decades). In trapped aquifers, the age will indicate the time that elapsed since the system was isolated from active recharge. This will be a singular age, and of much interest concerning the paleohydrology of the study area.

Answer 3.1: High-quality drinking water is low in dissolved solids, less than 800 mg/l.

Answer 3.2: 2—A, D, G; 3—A, D, G; 4—A, D, F; 5—C, E, H; 6—occasionally A, occasionally C; occasionally D, occasionally E, G; 7—C, H; 8—B, D, F; 9—B, D, F.

Answer 3.3: Lithology, fractures and open fissures, dissolution cavities, rock cementation, layering of rocks, occurrence of intrusive bodies, and the nature of faults and folded structures.

Answer 3.4: Similar chemical and isotopic compositions of the waters of wells I and II support the hypothesis that these wells are hydraulically interconnected. Differences in the water properties of wells III and IV support the hypothesis that these wells are hydraulically separated.

Answer 3.5: Atmospheric (section 3.7.1).

Answer 4.1: The answer is given in Fig. 4.1.

Answer 4.2: Masl stands for meters above sea level (section 4.1). The altitude of the water table was 553 masl (see also Fig. 4.3).

Answer 4.3: The arrows in Fig. 4.4 show the direction of flow, along the steepest gradient. However, some flow goes also from the area of the well with a 153 masl water table to the area of the well with a 137 masl water level and so on. These side flows are slower because they occur along smaller gradients (compare to the structure of the equation of Darcy's law, section 2.10).

Answer 4.4: The fault in Fig. 4.6 was plotted on the basis of abrupt changes in recorded lithologies between wells 6 and 7 (Table, Fig. 4.6a). An alternative interpretation: The calcareous sandstone was deposited on a limestone relief with a steep paleo-erosional face between the locations of wells 6 and 7.

Answer 4.5: The water table curve in Fig. 4.8 is not correlated to the precipitation record (no match of peaks). The reason: the well is recharged by various sources not included in the precipitation record—recharge is by the time of snowmelt and not by the time of snow fall. In addition, water demands are not reflected in the precipitation record. In other cases the water table–precipitation record correlation may be good, indicating a different hydrological history.

Answer 4.6: The river water flows to the wells mainly via conduits or zones of preferred water transmission.

Answer 4.7: Because the hydraulic gradient is zero (the two wells plot on the same water table contour).

Answer 4.8: Locally, water from aquifer I could leak into aquifer II, provided the pressure in aquifer II was significantly lower than the pressure in aquifer I.

Answer 4.9: Provided two aquifers (e.g., one below the other) have the same temperature, breaching during the pumping test would not be noticed on the temperature-discharge curve. So, what may be recommended?

Answer 4.10: (1) In October 1960 the temperature of the river was 64°F and that of the closest wells was 50°F (the wells were colder); (2) in January 1961 the river was 32°F and the closest wells were 60°F (the wells were warmer); (3) in March 1961 the river was 41°F and the closest wells were 45°F (the wells were slightly warmer); (4) in September 1961 the river was 77°F and the closest wells were 50°F (the wells were colder). Thus, the temperature of the wells followed the temperature of the river, but with a time lag indicating the river recharged the wells.

Answer 4.11: The contours in Fig. 4.22 depict the annual change of groundwater as measured in wells. Generally, the values drop as the distance from the river increases (from >30°F near the river to <10°F a few hundred meters away). Two explanations come to mind: (1) recharge to the wells decreases with distance and, hence, the temperature is closer to the local average annual ambient temperature, or (2) the distant wells are discharged as well but the water travels a longer time, allowing for temperature equilibration with the aquifer rocks. The authors prefer the first explanation. This shows that additional data is needed, can you give an example?

Answer 4.12: The changes over the measured period were small (only 1.5°C, from 120 to 167 l/min, and from 523 to 578 µs/cm. The question whether these temperature, discharge, and conductivity variations are significant is answered by comparison of the changes to the accuracy of the measurements. In the discussed case measurements were done to ± 0.2°C, ± 5 l/min, and ± 15 µs/cm. Thus, the observed changes were of significance. This shows that the accuracy of the data has to be indicated in each case (sections 5.7–5.10).

Answer 5.1: 24.3; 2+; (24.3:2)/1000 = 12.1 mg/l.

Answer 5.2: 1—C; 2—F; 3—B; 4—H; 5—A; 6—J; 7—D; 8—E; 9—I; 10—G.

Answer 5.3: Milliequivalent per liter. It is used to describe the concentration of ions dissolved in water. An *equivalent* is the number of grams that equal the respective ionic weight, divided by the valency. 35.5 mg/l.

Answer 5.4: The sum of cations equals the sum of anions in this (synthetic) case. Hence, the reaction error is 0 meq/l, or 0%.

Answer 5.5: The reproducibility is a measure of how well a measurement can be repeated. The mean value of the given measurements is 12.0°C, the mean deviation is 0.1, hence, the reproducibility is ± 0.1°C, or ± 0.8%.

Answer 5.6: (a) A quick glance reveals that the value of 202.02 mg/l is outstanding, and most probably results from a typing error and should be 102.02 mg/l (it was checked and admitted by the laboratory); (b) if the resolution is 2 mg/l, there is no reason for the laboratory to report the data with two decimal places. The data should be reported as 101, 99, 97, 103, and 102 mg/l; (c) the average value is 100 ± 2 mg/l (an analytical error of 2%), which in most cases is acceptable.

Answer 6.1: This limited set of data provided the researchers with the preliminary conclusion that the wells abstracted water from at least two aquifers of different chlorine concentration, and that the produced mixing ratio varied according to pumping intensity and/or alternations of rainy and dry seasons. Thus, all available water sources in the region were included in the study in order to detect the intermixing groundwaters. Later research revealed that the wells tapped up to three aquifers: the Kalahari sand (hosting fresh recently recharged water), underlain by basalt (hosting saline water), and beneath the Cave sandstone aquifer (hosting fossil slightly saline groundwater) (Mazor et al., 1977).

Answer 6.2: Lines extrapolating to the zero points are observed for Li, Na, and K, plotted as a function of Cl. The temperature is roughly positively correlated to the concentration of these ions. Thus, if only these parameters were measured, the conclusion could be reached that one deals with a warm (and deep) saline water end member that is diluted by very fresh and shallow (cold) local recharge water. Yet, the correlation lines of SO_4, Ca, and Mg extrapolate to the vertical axis, indicating that both water end members are mineralized to different degrees. The correlation lines of HCO_3, Ca, and Sr are horizontal, revealing that the intermixing end members have the same concentrations of these ions. Thus, the full picture leads to the conclusion that both end members come from aquifers that contain gypsum (contributing the bulk of SO_4 and Ca and smaller amounts of Mg that is present in most gypsum rocks), but the two aquifers differ in that the deeper one contains some halite and the shallower one has no halite.

Answer 6.3: Such a fingerprint is given in Fig. 20.1. The lines of the four springs are nearly parallel, suggesting dilution of a slightly saline end member with various amounts of fresh water. Both end members have similar HCO_3 concentrations. More of this case study is given in section 11.10.

Answer 6.4: The data span from 0.001 meq/l for the lowest value in the table up to 5814 meq/l for the highest value. To accommodate all these data, seven logarithmic cycles are needed. Is your interpretation of the obtained patterns in agreement with the discussion of these data given in section 6.6?

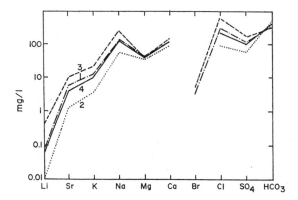

Fig. 20.1

Answer 6.5: The Dead Sea values stand out as being much higher than the values of the springs. Dividing these values by 10 made better use of the space in the fingerprint diagram, and the reduction in the number of needed logarithmic cycles resulted in better resolution of the other lines in the diagram. The diagram can be further improved by multiplying the lithium and potassium data by 10, this will cut another logarithmic cycle.

Answer 6.6: The researcher concluded that pumping of the aquifer caused vertical migration of salty groundwater from a deeper semiartesian aquifer. Plugging the lower parts of the wells in this field and/or reduced pumping rates may be recommended for proper management of the freshwater aquifer.

Answer 6.7: The sum of cations nearly equals the sum of the anions, hence the diagram is in meq/l. The data, except SO_4, plot in clusters, revealing the spring had a constant composition (with random analytical errors). Thus, the spring is fed by a single homogeneous aquifer. The SO_4 data show significant variations, in contrast to the other ions. Therefore the quality of the SO_4 data should be questioned and a new analysis of SO_4 will be needed before any conclusions can be drawn based on this ion.

Answer 6.8: The wells tap two distinct waters: (1) wells W-A, W-L, and W-K tap relatively fresh water with a composition of Na = Mg > Ca >> K >> Li and HCO_3 > Cl > SO_4 > Br; (2) wells W-B and W-H tap significantly more saline water with a composition of Na > Ca > Mg > K >> Li and Cl > SO_4 > HCO_3 >> Br. The wells tap two different aquifers and no intermixing seems to occur. Hence, the casings of the wells seem to be well sealing in their nonslotted sections.

Answer 6.9: No, since the increase in cations is balanced by an increase in HCO_3, meaning the underlying clay is most probably not being reached and the water quality is still good.

Answer 6.10: Evaporites, as well as clay and shale, since these rocks commonly contain some halite and would significantly raise the chlorine concentration; limestone and dolomite are not plausible since the sodium concentration is rather high and not balanced by chlorine. The aquifer rocks are most probably igneous rocks or their debris.

Answer 6.11: Extreme degree of evaporation of surface water and soil water in the dry season, resulting in the concentration of rainborne salts in playas and salinas. During extra-heavy rain events these salts are redissolved and transported by the recharge water into the saturated zone. This process is common in certain arid zones.

Answer 6.12: The well clearly struck gypsum, as a dominant rock or hosted in an evaporitic complex. Valuable information was missed by not having collected water samples from shallower depths during drilling. Careful study of the local rock section, and water quality examination in neighboring wells, may result in the recommendation to plug the lower part of the well and try to reach water from shallower, nonevaporitic rocks.

Answer 7.1: The aim was to understand the sources of recharge to the local through-flow system. Hence, an undisturbed well was needed, and the park area provided it. At first glance the study may seem to have needed only water level measurements, but data concerning all possible recharge and consumption sources were needed as well—summed up as precipitation in Fig. 4.8, and much more information of that year is provided in the upper part of the figure.

Answer 7.2: One has to understand the problems that bother the spa owner in order to plan the study, but in any case the question arises whether the collected samples properly represent the setup? For example, the water has to be collected at the well head or spring, before the temperature changes and the gases escape; water from two wells may be mixed before entering the specific bathing installation; etc. A visit to the location of study and careful sample collection are essential.

Answer 7.3: By preference of wells that meet criteria of the following kind: (1) available information on the well depth, construction, lithology and stratigraphy, and initial temperature, water level, and chemistry (avoid "pirate" wells); (2) records of measurements conducted in the past; (3) spatial representation of maximum potential aquifers; (4) high yields, if possible; (5) coverage of a maximal range of salinity and temperatures. Based on the results of this first stage of planning, decide whether you will present the data as contour maps or as transects, and select the final wells accordingly.

Answer 7.4: Date of drilling, water horizons encountered as drilling progressed, their temperature and composition, intervals of loss of circulation fluids, depth of the well, the lithological and stratigraphic profile, perforation intervals, initial hydraulic head and temperature of the pumped water, rise of the water table after the water was first encountered by the drillers (indicating confinement), and more. This is indispensable information that justifies every effort needed to get the drillers' reports.

Answers 7.5: Experience reveals that the properties of water often change from the first minute of the pump operation until a steady state is reached sometime later. The explanation is that the pumping operation draws water from the entire water column—from the upper part where recharge and contaminants arrive, and possibly from several aquifers passed during drilling. The ideal study of a well starts with its monitoring before pumping starts and continuous monitoring thereafter until the properties stabilize. If the properties do not change, this indicates one deals with a single and uniform body of groundwater, whereas if the water properties change it indicates different water types are pumped and their properties can be deciphered. If such a detailed study has already been performed once, the following work can be conducted on continuously pumped wells, providing well-average parametric values.

Answer 7.6: The fewer the available monitoring points, the more detailed and profound has to be their study. For example, the study should cover a very wide list of parameters measured in the laboratory, a well should be profiled after a long enough time of no pumping, each well has to be monitored without pumping and during pumping, and a frequent time series should be obtained.

Answer 9.1: The $\delta^{18}O$ values are $-6‰$ for freshwater end member and $+4.5‰$ for the Dead Sea, hence the difference, $10.5‰$, equals 100%. The $\delta^{18}O$ value of Hamei Yesha is $0‰$, and the contribution of Dead Sea brine in it is

$$\frac{(6-0) \times 100}{10.5} = 57\%$$

The corresponding δD values are $-30‰$ for the fresh water, $0‰$ for the Dead Sea, $-13‰$ for Hamei Yesha, and the calculated contribution of Dead Sea brine to the mineral spring is 55%. The values of 57% and 55% are very close, con-

Answers

sidering the analytical errors and the inaccuracy of reading the data from a graph.

Answer 9.2: (a) 1264 masl; (b) 1404 masl.

Answer 9.3: δD measurements should serve as a good check.

Answer 9.4: Colder climate and/or heavier rains (causing an amount effect).

Answer 9.5: Calibration of the local isotopic altitude effect. Springs or shallow wells located (a) at different altitudes, and (b) at the base of small ridges, to be sure they are recharged by local precipitation. One should exclude springs or wells located at faults, deep wells, or thermal waters—all of these may have remote recharge, and therefore will not represent an identifiable recharge altitude.

Answer 9.6: They are recharged by partially evaporated water. One can exclude salinization due to complete drying, as in playas or salinas, because in these locations rainwater evaporates completely, the salts accumulate, and extra-strong rain events dissolve the salts and transport them into the saturated groundwater zone. In the last case, salinity will be elevated, but the δD and $\delta^{18}O$ values will be relatively light.

Answer 9.7: The HCO_3 concentration decreases as a function of salinity increase. The deeper waters reflect, by their elevated chlorine concentration, that the wells tapped seawater, and this is supported by the lower HCO_3 concentrations.

Answer 9.8: By plotting the data on linear axes rather than on a semilogarithmic plot.

Answer 10.1: The pre-1952 tritium concentration in precipitation was about 5 TU (decayed to less than 2.5 TU in 1965). The available data (Fig. 10.3) indicate values above 100 TU since 1961. Hence, the water was recharged after 1952 and prior to 1961. This is an *effective age*. Hydrologically it means rapid recharge.

Answer 10.2: Pre-1952.

Answer 10.3: The tritium can be used in this case both as an age indicator and as a tracer (0.3 TU reveals a pre-1952 effective age, and therefore no post-1952 water reached it from the sewage pond (which equilibrated to the prevailing atmospheric tritium value). The tritium concentration in the pond could be measured, or assumed from the local tritium curves. Suppose that the pond contained 300 TU, then its contribution to the well could be at most,

$$\frac{0.3 \times 100}{300} = 0.1\%.$$

Answer 10.4: The conventional method is based on the comparison of the measured tritium concentration in the water sample with the recorded tritium concentrations in precipitation, taking the latter as the possible initial concentration in the water. The tritium-^{3}He method is based on the ratio of two parameters directly measured in the water sample: the tritium and ^{3}He concentrations.

Answer 11.1: Post-bomb (post-1952).

Answer 11.2: Pre-bomb, but probably quite recent, since 70 pmc is in fact a common initial ^{14}C concentration. An age of a few decades to a few hundred years (but not thousands).

Answer 11.3: More than 50,000 years ago (Fig. 11.1).

Answer 11.4: This is a "forbidden" combination of high tritium and low ^{14}C, leading to discordant ages: 6 TU indicates a post-1952 age and 6 pmc indicates an age of several tens of thousands years. This is a mixture of post-1952 water with very old water (0 TU, ~0 pmc).

Answer 11.5: The Parly-Chennons well is tapping a phreatic aquifer with post-1952 water.

Answer 11.6: The Migennes well has already been discussed in the text. It is reported to be a confined well, but the tritium-^{14}C combination indicates two water types are pumped in this well as a mixture. This nature is distinct from the comparison of the repeated measurements (10/67 and 7/69). In the last sample the percentage of recent water was greater than in the first sample. Corrosion and rupturing of the well casing may have deteriorated the well, so it pumps an increasing amount of shallow recent water. However, such a well may cause a short circuit between the two aquifers: due to different hydraulic pressures, water from one aquifer may flow uncontrolled into the other aquifer. Repair of the well may be necessary.

Answer 11.7: The Montier-en Der well is another example of a well reported to be confined, but the June 1969 sample revealed that recent (shallow) water is pumped as well.

Answer 11.8: The Issy well pumps very old water. The δ^{13}C value, δ^{13}C = $-15‰$, reveals no "dilution" of carbon isotopes occurred in the aquifer beyond the initial water-rock interactions. Hence, an initial groundwater ^{14}C of 60 pmc (at least) may be assumed, and the present value of 3.7 pmc indicates four half-lives have passed (reducing the ^{14}C from 60 pmc to 30, 15, 7, and 3.5 pmc respectively). Thus, the water has an age of $5.730 \times 4 = 23,000$ years.

Answer 11.9: Both figures depict tritium in groundwater as a function of ^{14}C, revealing the fact that groundwater that contains bomb tritium also has high ^{14}C concentrations, indicating that bomb radiocarbon is present as well.

Answer 12.1: About 700,000 years, as read from the radioactive decay curve given in Fig. 12.1.

Answer 12.2: The ^{36}Cl concentration, read from Fig. 12.10, is 120×10^7 atoms/l. The initial ^{36}Cl concentration, read from Fig. 12.10 by the technique explained in Fig. 12.4, was 220×10^7 atoms/l. The respective P_{36} is $120 \times 100/220 = 55\%$. Thus, the age, read from Fig. 12.1, is about 2.9×10^5 years.

Answer 12.3: Two half-lives. Hence, the age is 600,000 years.

Answer 12.4: The water samples underwent different degrees of evapotranspiration prior to their recharge into the groundwater system. This, in turn, indicates that recharge was at different surface cells, typified by different degrees of evapotranspiration.

Answer 12.5: Bromine—to check for possible dissolution of halite (why?); tritium and other anthropogenic contaminants—to exclude samples that contain bomb ^{36}Cl (for what reason?); ^{14}C—to select a number of samples that are very young in terms of the ^{36}Cl half-life (what for?); Cl—to establish the initial ^{36}Cl concentration in studied water samples (how?).

Answer 12.6: The most important and most informative wells have to be selected for full analyses. The criteria will include the following: (a) at least 5 samples to establish the local intake line (how will they be selected?); (b) the rest of the samples will be selected so that they include representatives of all water groups present (i.e., waters from different stratigraphic units, and hence possibly of different aquifers, and waters from the widest range of salinity); (c) wells with large yield are most important, as they represent larger portions of the exploited resource; (d) wells with much historical data (time series) are preferable, since the chance to understand their systems is greater if more auxiliary information is available.

The well selection may be based on available data, and complimentary preliminary measurements, most important being chlorine, and in unconfined groundwaters, tritium.

Answer 13.1: Using Fig. 13.1, the following values were abstracted for the solubility of the noble gases in air-saturated water, at sea level, expressed in cc STP/cc water:

Temperature	He	Ne	Ar	Kr	Xe
15°C	4.6×10^{-8}	2.0×10^{-7}	3.5×10^{-4}	8.0×10^{-8}	11.0×10^{-9}
22°C	4.5×10^{-8}	1.9×10^{-7}	3.1×10^{-4}	7.0×10^{-8}	9.0×10^{-9}
51°C	4.7×10^{-8}	1.7×10^{-7}	2.1×10^{-4}	4.1×10^{-8}	5.0×10^{-9}

Answer 13.2:
(a) The altitude correction factor at 1500 masl is 1.20 (Fig. 13.1).

(b)	He	Ne	Ar	Kr	Xe
solubility (5°C, 0 m)	4.9×10^{-8}	2.2×10^{-7}	4.4×10^{-4}	11×10^{-8}	16.5×10^{-9}

(c)	He	Ne	Ar	Kr	Xe
solubility (5°C, 1500 m)	4.1×10^{-8}	1.8×10^{-7}	3.7×10^{-4}	9.2×10^{-8}	13.8×10^{-9}

All values are in cc STP/cc water.

Answer 13.3: (a) The average recharge altitude is $(250 + 650)/2 = 450$ masl; the respective altitude correction factor is 1.060; the Ar concentration normalized to sea level is $(3.2 \times 10^{-4}) \times 1.060 = 3.4 \times 10^{-4}$ cc STP/cc water; the corresponding intake temperature is 15°C. (b) 4°C, (c) 10°C.

Answer 13.4: As it is a coastal spring, and nothing is known about the recharge area, no altitude correction is to be applied. The intake temperature may be directly read from the argon concentration from Fig. 13.1. The outcome: 12°C. Hence, the water was heated 32°C - 12°C = 20°C. As the local heat gradient is not known, we shall use the common value of 3°C/100 m. Accordingly, the minimum depth of circulation is $20 \times 100/3 = 660$ m.

Answer 13.5: The neon curve in Fig. 13.1 is seen to vary little with temperature, and therefore it is of little use in intake temperature calculation. The argon curve is, in contrast, sensitive to temperature changes. An intake (base of aerated zone) temperature of 8°C is obtained. Hence, 24°C - 8°C = 16°C is the temperature gain obtained by circulation, reflecting a distant and high intake—thus indicating recharge from the mountain. If so, a respective altitude correction should be applied to the raw data, resulting in an intake temperature lower than 8°C, further supporting mountain (and snowmelt) recharge.

Answer 14.1: Placing the data into the equation, we get

$t = 26 \times 10^{-8}/10^{-7} \times 0.8 \ (1.21 \times 2 + 0.287 \times 2) \times 10^{-6} \times 2.5 \times 10$

$= 43{,}000$ years.

This is in the same order of magnitude of the reported ^{14}C ages of the waters of group C.

Answer 14.2: Tritium, ^{14}C, and ^{36}Cl provide the base of three decaying dating methods, whereas ^{4}He is the base of a cumulative dating method (and so is the yet poorly developed ^{40}Ar method). In the first group the age-indicating isotope concentrations are easy to measure at the young end of the age range, and in the second group the concentrations are easier to measure the older the samples are.

Answer 14.3: The maximum helium concentration expectable in newly recharged water is read from the curve in Fig. 13.1 to be 5×10^{-8} cc STP/cc water. Excesses over this concentration are of radiogenic helium. In groundwaters that have an age of over 10,000 years, this excess is significant.

Answer 14.4: There are two basic arguments: (1) in most rocks the helium age is found to be significantly lower that the corresponding age defined by all other isotopic methods, indicating much of the radiogenic helium left the rock, and (2) as the partial pressure of helium in the rock crystals increases with age, the amount expelled increases as well, until a steady state is reached at which the newly produced helium is fully balanced by the helium expelled into the associated groundwater.

Answer 14.5: (a) Gas separation induced by pressure decrease during ascent in a well—corrected for via the atmospheric Ne, Ar, Kr, and Xe that are lost by the same degree as the He, but their initial concentrations can be well estimated from assumed recharge altitude and ambient temperature; (b) mixing with recent (commonly shallow) groundwater, either in a spring or in a poorly cased well—detectable by disagreeing age indications, for example, detectable tritium or ^{14}C together with appreciable concentrations of radiogenic He. Mixing corrections are discussed in sections 6.6 and 6.7.

Answer 14.6: The five diagrams report case studies in which the radiogenic helium concentration increases with depth (increase in temperature reflects increase of depth). In general, deeper groundwaters reveal higher helium concentrations, interpreted as higher ages. The deeper groundwater systems in each study location are confined, and in general the older ones were deeper buried during basin subsidence. Deviations from the general positive correlation may reflect varing concentrations of uranium and thorium and different rock:water ratios.

Answer 14.7: The following patterns are seen in Fig. 14.2: (a) the line segments of Ne to Xe have a similar general pattern for all water samples, resembling the shape of the line of water saturated with air at 15°C (ASW 15°C); thus, the respetive Ne, Ar, Kr, and Xe concentrations originate from initially dissolved air. (b) The concentrations of the dissolved atmospheric Ne, Ar, Kr, and Xe vary from one sample to the other over one and one-half orders of magnitude, with different degrees of gas losses or gains. (c) In contrast, the He data are all significantly higher than in the air-saturated water, indicating the presence of radiogenic He in all the studied samples.

Answer 14.8: The data were normalized to the concentrations in the approximated recharge water, presented by the ASW line (see the location of this line in the diagram). This mode of noble gas data representation reveals that a few samples almost retained all their initial atmospheric noble gases, others retained as little as 20%, and two samples reveal enrichments, possibly from air added during sample collection (for such reasons duplicate and even triplicate samples should be collected from each well and measured).

Answer 15.1: The Curiosity 1 well taps stagnant groundwater, because at a depth of 105m - 15m = 95 m below sea level, groundwater is in the zone of zero hydraulic potential, and the water is expected to be stagnant. The age of around 17,000 years indicates recharge during the last glacial maximum. At that time the sea level was substantially lower (by how much can in most places be learned from other studies). The ancient water table was accordingly lower. The following climatic warming was accompanied by ice melting and a rise in the sea to its present level. Thus, the presently active through-flow system, tapped by the shallow well, lies above the ancient groundwater.

Answer 15.2: Probably not. As the $\delta^{13}C$ value was $-8.1‰$, it was concluded that the water reacted mainly with the local marine limestone, and an initial ^{14}C concentration of 65 pmc was applied for the age calculation, with no $\delta^{13}C$ correction (section 11.7). The similarity of the $\delta^{13}C$ values in the ancient and recent groundwaters revealed that, most probably, no change in the plant community occurred.

Answer 15.3: Yes, the water sampled from the deep well could be a mixture of water from different depths if the casing was corroded (section 11.10). In such a case the sampled water would be a mixture of different ages, and the calculated age would be meaningless. To check this possibility data on the concentration of tritium and other anthropogenic pollutants is needed. If the latter are not present, the water is most probably not a mixture.

Answer 15.4: Yes, the chemical and isotopic composition should be established in the deep well and in the adjacent shallow well. This should be done in repeatedly collected samples under different pumping rates (with the aid of a portable pump). The following possibilities are envisaged: (a) The chemical composition of the water in the two wells is significantly different, in which case mixing can be checked. (b) The dissolved ion concentrations in the deep well will be similar to those in the shallow well, in which case the chemistry will not serve as a mixing indicator. But, (c) the isotopic composition is expected to differ—a lighter composition is expected in the older water as the age coincides with the last ice age. In the last case the stable isotopes may serve as mixing indicators: if the chemical composition does not change, and the lighter isotopic composition is repeated in the samples collected in the deep well under different pumping rates, the deep water is conclusively not a mixture.

Answer 15.5: Yes, it seems ideal, as it provides an unmixed sample of an established age, coinciding with the last glacial maximum (i.e., a phase of an extreme climate). A higher concentration of the heavier atmospheric noble gases is expected, and if found, it will support the colder climate reconstruction.

Answer 15.6: Since the present climate is temperate, and the location is in the coastal plain, it seems safe to conclude that even during the glacial maximum no snow line came near the paleorecharge location. Hence, no snowmelt was involved, and the deduced paleotemperature is the average ambient one. To be safe in this point, it is worth studying the setup of the shallow well. If the present recharge is through a porous aerated zone, and not karstic, then the conclusion that one deals with the average ambient paleotemperature is supported.

Answer 16.1: The observed nitrate contamination was attributed to extensive use of nitrate fertilizers. This hypothesis has to be checked using the following steps:

Conduct a survey of adjacent agricultural activities—where are fertilizers used? What types of fertilizers are used? How long have they been in use?

From the above, make a list of compounds that are introduced into the ground along with each type of applied fertilizer (e.g., K, SO_4). These compounds should be analyzed, along with nitrate, in accessible wells in depth profiles of the type shown in Fig. 16.1.

Answer 16.2: No, the injected brine might have contained petroleum compounds that could be adsorbed in the aquifer rocks and be slowly released into the groundwater. Hence, petroleum contaminants, and not just chlorine, should be analyzed in the damaged groundwater.

Answer 16.3: The monitoring operation cannot be stopped, and the following steps have to be undertaken: (a) The water in the ash pit should be measured periodically for possible changes. (b) The water in the ash pit should be analyzed for a large number of possible toxic compounds, especially trace metals. (c) The detailed analyses of the pit water should be used to search for more sensitive tracers. (d) The conclusion, based on wells A and C, that the pit water was well confined, was based on the assumption that water moves in a granular medium with no conduit-controlled flow. This might either have been wrong from the beginning, or conduit flow may have evolved due to water leaking from the pit. Hence, monitoring is always needed. The composition of the water in the ash pit looked harmless from the data given in Table 16.3. It is the additional analyses, recommended for this water, that will determine whether it is really harmless, or whether some toxic compounds exist, necessitating more careful monitoring.

Answer 16.4: The leachate monitored by the piezometers located in the pit revealed that a threat to local groundwater exists (What are the concentrations of the various ions?). A plume of leachate entering the local groundwater is observed in the direction of the water level gradient. Contamination is noticeable at a distance of how far (use the scale of the maps)? The mentioned observation that metals found in the leachate do not show up in the monitoring wells is explainable by adsorption on clay material in the soil. However, this means that a hallow of metals is formed in the soil and this hallow may reach local groundwater in the future. Therefore, metals observed in the leachate should be measured periodically in the monitoring wells.

REFERENCES

Textbooks for further reading are marked by an asterisk (*), and special texts published by the International Atomic Energy Agency are given at the end of this list.

Adar, E. (1984) Quantification of aquifer recharge distribution using environmental isotopes and regional hydrochemistry (Ph.D. diss.) Dept. of Hydrology and Water Resources, University of Arizona, Tucson

Allison, G.B. and Hughes, M.W. (1975) The use of environmental tritium to estimate recharge to a South-Australian aquifer. *J. of Hydrology* **26**, 245-254.

Andersen, L.J., Kelstrup, V., and Kristiansen, H. (1980) Chemical profiles in the Karup water-table aquifer, Denmark. In: *Nuclear Techniques in Groundwater Pollution Research*, IAEA, Vienna, 47-60.

Andrews, C.B. and Anderson, M.P. (1978) Impact of a power plant on the groundwater system of a wetland. *Groundwater* **16**, 105-111.

Andrews, J.N., Balderer, W., Bath, A.H., Clausen, H.B., Evans, G.V., Florkowski, T., Goldbrunner, J.E., Ivanovich, M., and Loosli, H. (1984) Environmental isotope study in two aquifer systems: In: *Isotope Hydrology 1983*, IAEA, Vienna 535-576.

Andrews, J.N., Goldbrunner, J.E., Darling, W.G., Hooker, P.J., Wilson, G.B., Youngman, M.J., Eichinger, L.,. Rauert, W., and Stichler, W. (1985) A radiochemical, hydrochemical and dissolved gas study of groundwaters in the Molasse basin of upper Austria. *Ear and Planetary Sci. Lett.* **73**, 317-332.

Andrews, J.N. and Lee, D.J. (1979) Inert gases in groundwater from the Bunter sandstone of England as indictors of age and paleoclimatic trends. *J. of Hydrology* **41**, 233-252.

Andrews, J.N., Youngman, M.J., Goldbrunner, J.E., and Darling, W.G. (1987) The

Arad, A., Kafri, U., Halicz, L., and Brenner, I. (1984) Chemical composition of some trace and minor elements in natural groundwaters in Israel. *Geol. Survey Israel*, Rep. 29/84.

Back, W. and Hanshaw, B.B. (1970) Comparison of chemical hydrogeology of the carbonate peninsulas of Florida and Yucatan. *J. of Hydrology* **10**, 330-368.

Bath, A.H., Edmunds, W.M., and Andrews, J.N. (1979) Paleoclimatic trends deduced from the hydrochemistry of a Triassic sandstone aquifer, United Kingdom. In: *Isotope Hydrology 1978*, IAEA, Vienna, 545-568.

Bebout, D.G. and Gutierrez, D.R. (1981) Geopressured geothermal resources in Texas and Louisiana—geological constraints. *Proc. 5th Conf. Geopressured-Geothermal Energy*, Baton Rouge, Louisiana, 13-24.

Bentley, H.W. and Davis, S.N. (1980) Isotope geochemistry as a tool for determining regional groundwater flow. *Proc. 1980 National Terminal Storage Program Information Meeting*, Columbus, Ohio, 35-41.

Bentley, H.W, Davis, S.N., Elmore, D., and Swanick, G.B. (1986b): Chlorine-36 dating of very old groundwater 2. Milk River Aquifer, Canada. *Water Resourc. Res.* **22**, 2003-2016.

Bentley, H.W., Phillips, F.M., and Stanley, N.D. (1986c) Chlorine-36 in the terrestrial environment. In: Handbook of Environmental Isotope Geochemistry, vol. 1B, 442-475.

Bentley, H.W., Phillips, F.M., Stanley, N.D., Habermehl, M.A., Airey, P.L., Calf, G.E., Elmore, D., Gove, H.E. and Torgersen, T. (1986a) Chlorine-36 dating of very old groundwater 1. The Great Artesian Basin, Australia. *Water Resourc. Res.* **22**, 1991-2001.

Bird, J.R., Calf, G.E., Davie, F.F., Fifield, L.K., Ophel, T.R., Evans, W.R., Kellett, J.R., and Habermehl, M.A. (1989) The role of ^{36}Cl and ^{14}C measurements in Australian groundwater studies. *Radiocarbon* **31**(3).

Borman, F.H. and Likens, G.E. (1970) The nutrient cycles of an ecosystem. *Sci. A.* **223**(4), 92-101.

Borole, D.V., Gupta, S.K., Krishnaswami, S., Datta, P.S., and Desai, B.I. (1979) Uranium isotopic investigations and radiocarbon measurements of river-groundwater systems, Sabaramati Basin, Gujarat, India. In: *Isotope Hydrology 1978*, IAEA, Vienna, **1**, 181-201.

Bortolami, G.C., Ricci, B., Suzella, G.F., and Zuppi, G.M. (1978) Isotope hydrology of the Val Coraoglia, Maritime Alps, Piedmont, Italy. In *Isotope Hydrology*, IAEA, Vienna, 327-350.

Bosch, A. (1990) Atmospheric and non-atmospheric noble gases in the geohydroderm (Ph.D. diss.), Weizmann Institute of Science, Rehovot, Israel.

Bradley, J.S. and Powly, D.E. (1994) Pressure compartments in sedimentary basins: a review. In: Basin Compartments and Seals, Ortoleva, P.J. (ed.); American Association of Petroleum Geologists, Memoir 61, Tulsa, Oklahoma, 3-26.

*Brassington, R. (1988) *Field Hydrology*. Open University Press, Milton Keynes; Halsted Press John Wiley, New York.

Bredenkamp, D.B., Schutte, J.M., and Du Toit, G.J. (1974) Recharge of a dolomitic aquifer as determined from tritium profiles. In: *Isotope Techniques in Groundwater Hydrology 1974*. IAEA, Vienna, **1**, 73-96.

References

Broecker, W.S. (1986) Oxygen isotope constraints on surface ocean temperatures. *Quaternary Res.* **26**, 121-134.

Calf, G.E., Bird, J.R., and Evans, W.R. (1988) Origins of chloride variation in the Murray Basin using environmental Cl-36 (abstract). Conference on Studies in Hydrology, Murray Basin, Canberra, 23-26 May 1988.

*Carter, M.W., Moghissi, A.A., and Kahn, B. (eds.) (1979) *Management of Low-Level Radioactive Wastes*. Pergamon Press, Oxford.

Castany, G. (1960) Quelques aspects nouveaux de l'hydrogéologie du basin parisien. *C.R. ann. Comité Fr. Geol. Geoph.*

Cherkauer, D.S. (1980) The effect of flyash disposal on a shallow groundwater system. *Ground Water* **18**, 544-550.

Conrad, G., Jouzel, J., Merlivat, L., and Puyoo, S. (1979) La nappe de la craie en Haute-Normandie (France) et ses relations avec les eaux superficielles. In: *Isotope Hydrology 1978*, IAEA, Vienna, **1**, 265-287.

Cotecchia, V., Tuzioli, G.S., and Magri, G. (1974) Isotopic measurements in research on seawater ingression in the carbonate aquifer of the Salentine Peninsula, Southern Italy. In: *Isotope Techniques in Groundwater Hydrology 1974*, IAEA, Vienna, **1**, 445-463.

Craig, H. (1961a) Isotopic variations in meteoric waters. *Science* **133**, 1702-1703.

Craig, H. (1961b) Standard for reporting concentrations of deuterium and oxygen-18 in natural waters. *Science* **133**, 1833-1834.

Dansgaard, W. (1964) Stable isotopes in precipitation. *Tellus* **16**, 436-469.

Davie R.F., Fifeld, L.K., Bird, J.R., Ophel, T.R., and Calf, G.E. (1988) The application of ^{36}Cl measurements to groundwater modeling (abstract 47-50). Conference on studies in Hydrology, Murray Basin, Canberra, 23-26 May 1988.

Davie R.F., Kellett, J.R., Fifield, L.K. Evans, W.R., Calf, G.E., Bird, J.R., Topham, S., and Ophel, T.R. (1989) Chlorine-36 measurements in the Murray Basin: preliminary results from the Victorian and South Australian Mallee region. *J. Australian Geophys.* **11**, 261-272.

*Davis, S. and DeWiest, R.J.M. (1966) *Hydrogeology*. John Wiley, New York.

Davis, S.N. and Bentley, H.W. (1982): Dating groundwater, a short review. In: *Nuclear and Chemical Dating Techniques: Interpreting the Environmental Record*, Currie L.A. (ed), *American Chemical Society Symp. Series* **176**, 187-222.

Deák, J. (1979) Enviromental isotopes and water chemical studies for groundwater research in Hungary. In: *Isotope Hydrology 1978*, IAEA, Vienna, **1**, 221-249.

De Marsily, G. (1986) *Quantitative Hydrology*. Academic Press, New York.

Deak, J., Stute, M., Rudolph, J., and Sonntag, C. (1987) Determination of the flow regime of Quaternary and Pliocene layers in the Great Hungarian Plain (Hungary) by D, ^{18}O, ^{14}C and noble gas measurements. In: *Isotope Techniques in Water Resources Development*, IAEA, Vienna, 335-350.

Dennis, H.W. (1973) Salt pollution of a shallow aquifer, Indianapolis, Indiana. *Ground Water* **11**, 18-22

Dowing, R.A., Smith, D.B., Pearson, F.J., Monksouse, R.A., and Otlet, R.L. (1977) The age of groundwater and its relevance to the flow mechanism. *J. of Hydrology* **33**, 201-216.

*Drever, J.I. (1982) *The Geochemistry of Natural Waters*. Prentice-Hall, Englewood Cliffs, N.J.

*Eriksson, E. (1985) *Principles and Application of Hydrochemistry*. Chapman and Hall, London.

Evin, J. and Vuillaume, Y. (1970) Etude par le radiocarbone de la Nappecaptive de L'Albien du Bassin de Paris. In: *Isotope Hydrology 1970*, IAEA, Vienna, 315-332.

Fontes, J.Ch. (1983) *Dating of Groundwater. Guide on Nuclear Techniques in Hydrology*. Technical Reports Series no. 91, IAEA, Vienna, 285.

Fontes, J. Ch., Bortolami, G.C., and Zuppi, G.M. (1979) Isotope hydrology of the Mont Blanc Massif. In: *Isotope Hydrology 1978*, IAEA. Vienna, 411-436.

Foster, K.E. and Fogel, M.M. (1973) Mathematical modelling of soil temperature. *Progressive Agriculture in Arizona* **25**, 10-12.

*Freeze, R.D. and Cherry, A. (1979) *Groundwater*. Prentice-Hall, Englewood Cliffs, N.J.

*Fritz P. and Fontes, J. Ch. (eds) (1980 and 1986) *Handbook of Environmental Isotope Geochemistry*. Vols 1 and 2.

Fridman, V., Mazor, E., Becker, A., Avraham, D., and Adar, E. (1995) Stagnant aquifer concept: III. Stagnant mini-aquifers in the stage of formation, Makhtesh Ramon, Israel. *J. of Hydrology* **173**, 263-282.

Fritz, P., Drimmie, R.J., and Render, F.W. (1974) Stable isotope contents of a major prairie aquifer in central Manitoba, Canada. In: *Isotope Techniques in Grounwdater Hydrology 1974*, IAEA, Vienna, **1**, 379-398.

Fritz, P., Hennings, C.S., Suzolo, O., and Salati, E. (1979) Isotope hydrology in northern Chile. In *Isotope Hydrology 1978*, IAEA, Vienna, **2**, 525-544.

Fritz, P., Matthess, G., and Brown, R.M. (1976) Deuterium and oxygen-18 as indicators of leachwater movement from a sanitary landfill. In: *Interpretation of Environmental Isotope and Hydrochemical Data in Groundwater Hydrology*, IAEA, Vienna, 131-142.

Fryberger, J.F. (1975) Investigation and rehabilitation of a brine contaminated aquifer. *Ground Water* **13**, 155-160.

Gat, J.R. (1971) Comments on the stable isotope method in regional groundwater investigations. *Water Resourc. Res.* **7**, 980-993.

Gat, J.R. and Dansgaard, W. (1972) Stable isotope survey of the fresh water occurences in Israel and the Northern Jordan Rift Valley. *J. of Hydrology* **16**, 177-212.

Gat, J.R., Mazor, E., and Tzur, Y. (1969) The stable isotope composition of mineral waters in the Jordan Rift Valley, Israel. *J. of Hydrology* **7**, 334-352.

Geyh, M.A. (1972) Basic studies in hydrology and ^{14}C and ^{3}H measurements. *24th Int. Geol. Cong.*, Montreal, **2**, 227-234.

Geyh, M.A. (1980) Interpretation of environmental isotopic groundwater data, arid and semi-arid zones. In: *Aridzone Hydrology: Investigations with Isotope Techniques*, IAEA, Vienna, 31-46.

Geyh, M.A. and Wirth, K. (1980) ^{14}C ages of confined groundwater from the Gwandu aquifer, Sokoto basin, Northern Nigeria. *J. of Hydrology* **48**, 281-288.

Goldenberg, L.C., Margaritz, M., and Mandel, S. (1983) Experimental investigation of irrevesible changes of hydraulic conductivity on the seawater-freshwater interface in coastal aquifers. *Water Resourc. Res.* **19**, 77-85.

Gonfiantini, R., Dincer, T., and Derekoy, A.M. (1974) Environmental isotope hydrology in the Honda region, Algeria. In *Isotope Techniques in Groundwater Hydrology 1974*, IAEA, Vienna, **1**, 293-316.

Goolsby, D.A. (1971) Hydrochemical effects of injecting wastes into a limestone aquifer near Pensacola, Florida. *Ground Water* **9**, 13-19.

Habermehl, M.A. (1980) The Great Artesian Basin, Australia. *BMR J. Australian Geol. Geophys.* **5**, 9-38.

Heaton, T.H.E. (1984) Rates and sources of ^{4}He accumulation in groundwater. *Hydrology Sci. J.* **29**, 29-47.

Herczeg, A.L., Simpson, H.J., and Mazor, E. (1993): Transport of soluble salts in a large semiarid basin: River Murray, Australia. *J. of Hydrology* **144**, 59-84.

Heaton, T.H.E. and Vogel, J.C. (1981) 'Excess air' in groundwater. *J. of Hydrology* **50**, 201-216.

*Hem, J.D. (1985) *Study and Interpretation of the Chemical Characteristics of Natural Water*. U.S. Geological Survey, Water Supply Paper **254**.

Herzberg, O. and Mazor, E. (1979) Hydrological applications of noble gases and temperature measurements in underground water systems, examples from Israel. *J. of Hydrology* **41**, 217-231.

Hubbert, M.K. (1940) The theory of ground-water motion. *J. of Geology* **48**(8), 785-944.

Love, A.J., Hreczeg, A.L., Armstrong, D., Stadter, F., and Mazor, E. (1993) Groundwater flow regime within the Gambier Embayment of the Otway Basin, Australia: evidence from hydraulics and hydrochemistry. *J. of Hydrology* **143**, 297-338.

Hufen, T.H., Lau, L.S., and Buddemeier, R.W. (1974) Radiocarbon, ^{13}C and tritium in water samples from basaltic aquifers and carbonate aquifers on the island of Oahu, Hawaii. In *Isotope Techniques in Groundwater Hydrology 1974*, IAEA, Vienna, **2**, 111-127.

*Hutton, L.G. (1983) *Field Testing of Water in Developing Countries*. Water Research Centre, Medmeham, England.

Jorgensen, D.G. (1968) An aquifer test used to investigate a quality of water anomaly. *Ground Water* **6**, 18-20.

Kroitoru, L. (1987) The characterization of flow sysetms in carbonatic rocks defined by the groundwater parameters: Central Israel (Ph.D. diss.), Feinberg Graduate School of the Weizmann Institute of Science, Rehovot, Israel.

Kroitoru, L., Carmi, I., and Mazor, E. (1987) Groundwater ^{14}C activity as affected by initial water-rock interactions in a carbonatic terrain with deep water tables: Judean Mountains, Israel. *Int. Symp. on the Use of Isotope Techniques in Water Resources Development*, IAEA, Vienna, extended abstract, 134-136.

Leontiadis, I.L., Payne, B.R., Letsios, A., Papagianni, N., Kakarelis, D., and Chadjiagorakis, D. (1983) Isotope hydrology study of Kato Nevroko of Dramas. In: *Isotope Hydrology 1983*, IAEA, Vienna, 193-206.

Levy, Y. (1987) The Dead Sea, hydrographic, geochemical and sedimentological changes during the last 25 years (1959-1984). *Geological Survey of Israel* (in Hebrew).

Lewallen, M.J. (1971) Pesticide contamination of a shallow bored well in southeastern Coastal Plains. *Ground Water* **9**, 45-49.

Loosli, H.H. and Oeschger, H. (1978) ^{39}Ar, ^{14}C and ^{85}Kr measurements in groundwater samples. In: *Isotope Hydrology 1978*, IAEA, Vienna, **2**, 931-997.

*Matthes, G. (1982) *The Properties of Groundwater*. John Wiley, New York.

Marine, I.W. (1979) The use of naturally occurring helium to estimate groundwater velocities for studies of geologic storage of radioactive waste. *Waste Resourc. Res.* **15**, 1130-1136.

Martinsen, R.S., (1994) Summary of published literature on anomalous pressure: implications for the study of pressure compartments. In: *Basin Compartments and Seals*, Ortoleva, P.J. (ed.) AAPG Memoir 61, American Association of Petroleum Geologists, Tulsa, Oklahoma, 477.

Marty, B., Criaud, A., and Fouillac, C. (1988) Low enthalpy geothermal fluids from the Paris sedimentary basin—1. Characteristics and origin of gases. *Geothermics* **17**, 619-633.

Mazor, E. (1972) Paleotemperatures and other hydrological parameters deduced from noble gases dissolved in groundwaters: Jordan Rift Valley, Israel. *Geochim. et Cosmochim. Acta* **36**, 1321-1336.

Mazor, E. (1975) Atmospheric and radiogenic noble gases in thermal waters: their potential application to prospecting and steam production studies. *Proc. 2nd UN Symp. on Development and Use of Geothermal Resources*, San Francisco, **1**, 793-801.

Mazor, E. (1976) Multitracing and multisampling in hydrological studies. In: *Interpretation of Environmental Isotope and Hydrochemical Data in Groundwater Hydrology*, IAEA, Vienna, 7-36.

Mazor, E. (1978) Mineral waters of the Kinneret basin and possible origin. In: *A Monography on Lake Kinneret*, Serruya, C. (ed.), W. Junk N.V. Publishers, The Hague, 103-120.

Mazor, E. (1979a) Dilute water-rock reactions in shallow aquifers of the Kalahari flatland. *Proc. 3rd Water-Rock Interaction Symp.*, Edmonton, Canada, 14-15.

Mazor, E. (1979b) Noble gases in a section across the vapour dominated geothermal field in Larderello, Italy. *Pure and Applied Geophys.* **117**, 262-275.

Mazor, E. (1982) Rain recharge in the Kalahari—a note on some approaches to the problem. *J. of Hydrology* **55**, 137-144.

Mazor, E. (1985) Mixing in natural and modified groundwater systems: detection and implications on quality and management. In: *Scientific Basis for Water Resources Management*. IAHS Publ. no. 153, 241-251.

Mazor, E. (1986) Chemical, isotopic and physical data used to check basic (commonly assumed) hydrological interrelations. *Proc. Int. Conf. on Groundwater Systems Under Stress*, Brisbane, Australia, May 1986, 1-14.

Mazor, E. (1992a) He as a semi-quantitative tool for groundwater dating in the range of 10^4 to 10^8 years. In: *Isotopes of Noble Gases as Tracers in Environmental Studies*, IAEA, Vienna, 163-178.

Mazor, E. (1992b) Reinterpretation of Cl-36 data: physical processes, hydraulic interconnections, and age estimates in groundwater systems. *Applied Geochem.* **7**, 351-360.

Mazor, E. (1993a) Chlorine-36 data and basic concepts of hydrology—comments on F.M. Phillips' comment, with special reference to the Great Artesian Basin. *Applied Geochem.* **8**, 649-651.

Mazor, E. (1993b) Interrelations between groundwater dating, paleoclimate and paleohydrology. In: *Applications of Isotope Techniques in the Study of Past and Current Environmental Changes in the Hydrosphere and the Atmosphere*, IAEA, Vienna, 249-257.

Mazor, E. (1993c) The need for auxiliary data to address the long list of processes recruited to interpret ^{36}Cl data in groundwater systems—a reply to the comment by J.C. Fontes and J.N. Andrews. *Applied Geochem.* **8**, 667-669.

Mazor, E. (1993d) Some basic principles of ^{36}Cl hydrology—a reply to the discussion by Kellett, Evans, Allan, and Fifield. *Applied Geochem.* **8**, 659-662.

Mazor, E. (1995) Stagnant aquifer concept I. Large scale artesian systems—Great Artesian Basin, Australia. *J. of Hydrology* **174**, 219-240.

Mazor, E. and Bosch, A. (1987) Noble gases in formation fluids from deep sedimentary basins: a review. *Applied Geochem.* **2**, 621-627.

Mazor, E. and Bosch, A. (1990) Dynamics of groundwater in deep basins: He-4 dating, hydraulic discontinuities and rates of drainage. International Conference on Groundwater in Large Sedimentary Basins, Perth, Western Australia, 9-13 July 1990. Department of Primary Industries and Energy Proc., 380-389.

Mazor, E. and Bosch, A. (1992a): He as a semi-quantitative tool for groundwate dating in the range of 10^4 to 10^8 years. In: *Isotopes of Noble Gases as Tracers in Environmental Studies*, IAEA, Vienna, 163-178.

Mazor, E. and Bosch, A. (1992b): Physical processes in geothermal systems derived by noble gases. In: *Isotopes of Noble Gases as Tracers in Environmental Studies*, IAEA, Vienna, 203-218.

Mazor, E., Drever, J.I., Finely, J., Huntoon, P.W., and Lundy, D.A. (1992) Hydrochemical implications of groundwater mixing: an example from the southern Laramie Basin, Wyoming. *Water Resourc. Res.* **29**, 193-205.

Mazor, E., Dubois, J.D., soom-Flück, J., and Wexsteen, P., (1990) Time-data series as a major diagnostic tool in groundwater hydrology: case studies from Switzerland II. International Conference on Water Resources in Mountainous Regions, IAH and IAHS, Lausanne, Switzerland, Memoires of the 22nd Congress of IAH, vol. XXII, 281-288.

Mazor, E. and George, R. (1992) Marine airborne salts applied to trace evapotranspiration, local recharge and lateral groundwater flow in Western Australia. *J. of Hydrology* **139**, 63-77.

Mazor, E., Gilad, D., and Fridman, V. (1995) Stagnant aquifer concept: II. Small-scale artesian systems—Hazeva, Dead Sea Rift Valley, Israel. *J. of Hydrology* **173**, 241-261.

Mazor, E., Jaffe, F.C., Flück, J., and Dubois, J.D. (1986) Tritium corrected C-14 and atmospheric noble gas corrected He-4, applied to deduce ages of mixed groundwaters: examples from the Baden region, Switzerland. *Geochim. et Cosmochim. Acta* **50**, 1611-1618.

Mazor, E., Kaufman, A., and Carmi, I. (1973) Hammet Gader (Israel): geochemistry of a mixed thermal spring complex. *J. of Hydrology* **18**, 289-303.

Mazor, E. and Kroitoru, L. (1987) Phreatic-confined discontinuities and restricted flow in confined groundwater systems. *Int Symp. on the Use of Isotope Techniques in Water Resources Development*, IAEA, Vienna, extended abstract, 130-131.

Mazor E. and Kroitoru, L. (1992) Beer Sheva-Yarkon—Taninim: one groundwater group or several separate ones? *Israel Bull. Earth Sci.* **39**, 67-70.

Mazor, E. and Mero, F. (1969a) Geochemical tracing of mineral and fresh water sources in the Lake Tiberias basin, Israel. *J. of Hydrology* **7**, 276-317.

Mazor, E. and Mero, F. (1969b) The origin of the Tiberias-Noit mineral water association in the Tiberias-Dead Sea Rift Valley, Israel. *J. of Hydrology* **7**, 318-333.

Mazor E., Nadler, A., and Harpaz, Y. (1973) Notes on the geochemical tracing of the Kaneh-Samar spring complex, Dead Sea basin. *Israel J. of Earth Sci.* **22**, 255-262.

Mazor, E. and Nativ, R. (1992) Hydraulic calculation of groundwater flow velocity and age: examination of the basic premises. *J. of Hydrology* **138**, 211-222.

Mazor, E. and Nativ R., (1994): Stagnant groundwaters stored in isolated aquifers: implications related to hydraulic calculations and isotopic dating—a reply to a comment by T. Torgersen. *J. of Hydrology*, **154**, 409-418.

Mazor, E., Rosenthal, E., and Eckstein, J. (1969) Geochemical tracing of mineral water sources in the South-Western Dead Sea Basin, Israel. *J. of Hydrology* **7**, 246-275.

Mazor, E. and Truesdell, A.H. (1984) Dynamics of a geothermal field traced by noble gases: Cerro Prieto, Mexico. *Geothermics*, **13**, 91-102.

Mazor, E., Vautax, F.D., and Jaffé, F.C. (1985) Tracing groundwater components by chemical, isotopic and physical parameters, example: Schinziach, Switzerland. *J. of Hydrology* **76**, 233-246.

Mazor, E., Vautaz, F.D., and Jaffe, F.C., (1990) Time-data series as a major diagnostic tool in groundwater hydrology: case studies from Switzerland I. International Conference on Water Resources in Mountainous Regions, IAH and IAHS, Lausanne, Switzerland, Memoirs of the 22nd Congress of AIH, vol. XXII, 271-280.

Mazor, E. and Verhagen, B. Th. (1983) Dissolved ions, stable isotopes and radioactive isotopes and noble gases in thermal waters of South Africa. *J. of Hydrology* **63**, 315-329.

Mazor, E., Verhagen, B. Th., and Negrenenu, E. (1974) Hot springs of the igneous terrain of Swaiziland. In *Isotope Techniques in Groundwater Hydrology 1974*, IAEA, Vienna, **2**, 29-47.

Mazor, E., Verhagen, B. Th., Sellschop, J.P.F., Jones, M.T., Robins, N.E., Hutton, L., and Jennings, C.M.H. (1977) Northern Kalahari groundwaters: hydrologic, isotopic and chemical studies at Orapa, Botswana. *J. of Hydrology* **34**, 203-234.

Mazor, E., Verhagen, B. Th., Sellschop, J.P.F., Jones, M.T., and Hutton, L.C. (1981) Sodium exchange in a NaCl waste disposal case (Lobatse, Botswana): implications to mineral water studies. *Environ. Geol.* **3**. 195-199.

Mazor, E. and Wasserburg, G.J. (1965) Helium, neon, argon, krypton and xenon in gas emanation in Yellowstone and Lassen Volcanic National Parks. *Geochim. et Cosmochim. Acta* **29**, 443-454.

*Ozima, M. and Podosek, F.A. (1983) *Noble Gas Geochemistry*. Cambridge University Press, Cambridge.

Payne, B.R., Quijano, L., and Latorred, C.D. (1980) Study of the leakage between two aquifers in Hermosillo, Mexico, using environmental isotopes. In: *Arid Zone Hydrology: Investigations with Isotope Techniques*, IAEA, Vienna, 113–130.

Payne, B.R. and Yurtsever, Y. (1974) Environmental isotopes as a hydrogeological tool in Nicaragua. In *Isotope Techniques in Groundwater Hydrology 1974*, IAEA, Vienna, **1**, 193–202.

Pearson, Jr., F.J. and Hanshaw, B.B. (1970) Sources of dissolved carbonate species in groundwater and their effects on carbon-14 dating. In: *Isotope Hydrology 1970*, IAEA, Vienna, 271–286.

Pearson, F.J. and Swarzenski, W.V. (1974) ^{14}C evidence for the origin of arid region groundwater, northeastern province, Kenya. In: *Isotope Techniques in Groundwater Hydrology 1974*, IAEA, Vienna, **2**, 95–109.

Pearson, G.N. (1974) Tritium data from groundwater in the Kristianstad plain, southern Sweden. In: *Isotope Techniques in Groundwater Hydrology 1974*, IAEA, Vienna, **1**, 45–56.

Phillips, F.M., Benrley, H., Davis, S.N., Elmoer, D., and Swanick, G.B. (1986) Chlorine-36 dating of very old groundwater 2. Milk River Aquifer, Alberta, Canada. *Water Resourc. Res.* **22**, 2003–2016.

Pickens, J.F., Cherry J.A., Grisak, G.E., Merritt, W.F., and Risto, B.A. (1978) A multilevel device for ground-water sampling and piezometric monitoring. *Ground Water* **16**, 322–327.

Rafter, T.A. (1974) The dating of fossil man in Australia. *Proc. Symp. Hydrogeochemistry and Biogeochemistry*, Tokyo, Japan.

Rudolph, J., Rath, H.K., and Sonntag, C. (1983) Noble gases and stable isotopes in ^{14}C-dated paleowaters from central Europe and the Sahara. In: *Isotope Hydrology 1983*, IAEA, Vienna, 467–477.

Saffinga, P.G. and Keeney, D.R. (1977) Nitrate and chloride in ground water irrigated agriculture in Central Wisconsin. *Ground Water* **15**, 170–177.

Salati, E., Matsui, E., Leal, J.M., and Fritz, P. (1980) Utilization of natural isotopes in the study of salinization of the waters in the Pajeu River Valley, northeast Brazil. In: *Arid-Zone Hydrology: Investigations with Isotope Techniques*, IAEA, Vienna, 133–151.

Schlosser, P., Stute, M., Dörr, H., Sonntag, C., and Münich, K.O. (1988) Tritium/^{3}He dating of shallow groundwater. *Earth and Planetary Sci. Lett.* **89**, 353–362.

Schoch-Fischer, H., Rozanski, K., Jacob, H.J., Sonntag, C., Jouzel, I., Ostlund, G., and Geyh, M.A. (1983) Hydrometeorological factors controlling the time variation of D, ^{18}O and ^3H in atmospheric water vapour and precipitation in the northern westwind belt. In: *Isotope Hydrology 1983*, IAEA, Vienna, 3–30.

Schoeller, H. (1954) *Arid Zone Hydrology—Recent Developments*. UNESCO, Paris.

Shampine, W.J., Dincer, T., and Noory, M. (1979) An evaluation of isotope concentrations in the groundwater of Saudi Arabia. In: *Isotope Hydrology 1978*, IAEA, Vienna, **2**, 443–463.

Shuster, E.T. and White, W.B. (1971) Seasonal fluctuations in the chemistry of limestone springs: a possible means for characterizing carbonate aquifers. *J. of Hydrology* **14**, 93–128.

Siegenthaler, U. and Oeschger, H. (1980) Correlation of ^{18}O in precipitation with temperature and altitude. *Nature* **285**, 314–317.

Sonntag, C., Klitzsch, E., Lohnert, E.P., Ee-Shazly, E.M., Munnich, K.O., Junghans, Ch., Thorweihe, U., Weistroffer, K., and Swailem, F.M. (1979) Paleoclimatic information from deuterium and oxygen-18 in carbon-14 dated north Saharian groundwaters. In: *Isotope Hydrology 1978*, IAEA, Vienna, **2**, 569–581.

*Stahl, W., Aust, H., and Dounas, A. (1974) Origin of artesian and thermal waters determined by oxygen, hydrogen and carbon isotope analyses of water samples from the Sperkhios Valley, Greece. In: *Isotope Techniques in Groundwater Hydrology 1974*, IAEA, Vienna, **1**, 317–339.

Stute, M., Clark, J.F., Schlosser, P., Broecker, W.S., and Bonani, G. (1995b) A 10,000-yr continental paleotemperature record derived from noble gases dissolved in groundewater from the San Juan basin, New Mexico. *Quarternary Res.* **43**, 209–220.

Stute, M., Forster, H., Frischkorn, A., Serejo, Clark, J.F., Schlosser, P., Broecker, W.S., and Bonani, G. (1995a) Cooling of tropical Brazil (5°C) during the last glacial maximum. *Science* **269**, 379–383.

Tamers, M.A. and Scharpenseel, H.W. (1970) Sequential sampling of radiocarbon in groundwater. In: *Isotope Hydrology 1970*, IAEA, Vienna, 241–257.

*Todd, D.K. (1980) *Groundwater Hydrology*, 2nd ed. John Wiley, New York.

Torgersen, T. and Clarke, W.B. (1985) Helium accumulation in groundwater, 1: An evaluation of sources and the continental flux of crustal ^{4}He in the Great Artesian Basin, Australia. *Geochim. et Cosmochim. Acta* **49**, 1211–1218.

Toth, J. (1963) A theoretical analysis of groundwater flow in small drainage basins. *J. Geophys. Res.* **68**, 4795–4812.

Tremblay, J.J., D'Cruz, J., and Anger, H. (1973) Salt water intrusion in the Summerside area, P.E.I. *Ground Water* **11**, 21–27.

Verhagen, B.Th., Mazor, E., and Sellschop, J.P.F. (1974) Radiocarbon and tritium evidence for direct recharge to groundwaters in the northern Kalahari. *Nature* **249**, 643–644.

Verhagen, B.Th., Smith, P.E., McGeorge, I., and Dzimebowski, Z. (1979) Tritium profiles in Kalahari sands as a measure of rain-water recharge. In *Isotope Hydrology 1978*, IAEA, Vienna, **2**, 733–751.

Vogel, J.C. (1970) Carbon-14 dating of groundwater. In: *Isotope Hydrology 1970*, IAEA, Vienna, 225–239.

Vogel, J.C. and Van Urk, H. (1975) Isotopic composition of groundwater in semi-arid regions of southern Africa. *J. of Hydrology*, **25**, 23–36.

Vogel, J.C., Thilo, L., and Van Dijken, M. (1974) Determination of groundwater recharge with tritium. *J. of Hydrology* **23**, 131–140.

Vuataz, F.D. (1982) Hydrologie, géochimie et géothermie des eaux thermales de Suisse et des regions Alpines limitrophes. *Matériaux pour la géologie de la Suisse - hydrologie* No. 29. Kummerly & Frey Geographischer Verlag, Berne.

*Walton, W.C. (1988) *Groundwater Pumping Tests Design and Analysis*. H.K. Lewis, London.

Wilmoth, B.M. (1972) Salty groundwater and meteoric flushing of contaminated aquifers in West Virginia. *Ground Water* **10**, 99–105.

Wilson, L.G. and DeCook, K.J. (1968) Field observations on changes in the subsurface water regime during influent seepage in the Santa Cruz River. *Water Resourc. Res.* **4**, 1219–1233.

Winslow, J.D., Stewart, Jr., H.G., Johnston, R.H., and Crain, L.J. (1965) Groundwater resources of eastern Schenectudy County, New York, with emphasis on infiltration from the Mohawk River. State of New York Conservation Department Water Resources Commission Bull. **57**, 148.

Yurtsever, Y. and Payne, B.R. (1979) Application of environmenetal isotopes to groundwater investigation in Qatar. In: *Isotope Hydrology 1978*, IAEA, Vienna, **2**, 465–490.

Zaikowsky A., Kosanke, B.J., and Hubbard, N. (1987) Noble gas composition of deep brines from the Palo Duro Basin, Texas. *Geochim. et Cosmochim. Acta* **51**, 73–84.

Zartman, R.E., Wasserburg, G.J., and Reynolds, J.H. (1961) Helium, argon and carbon in some natural gases. *J. Geophys. Res.* **66**, 277–306.

Special Publications of the International Atomic Energy Agency P.O. Box 100, Vienna, Austria

Application of Isotope Techniques in Hydrology, 31 pp. 1962.
Arid-Zone Hydrology: Investigations with Isotope Techniques, 265 pp., 1980.
Concentration in Precipitation (1969–1983), vols. 1–7.
Environmental Isotope Data No. 1: World Survey of Isotope Concentration in Precipitation (1969–1983), vols. 1–7.
Interpretation of Environmental Isotope and Hydrochemical Data in Groundwater Hydrology, 228 pp., 1976.
Isotopes in Hydrology, 740 pp., 1967.
Isotope Hydrology 1978, vol. 1: 440 pp., vol. II: 540 pp., 1979.
Isotope Hydrology 1983, 874 pp., 1984.
Isotope Techniques in Hydrology, (Bibliography) vol. I: 1857–1965, 228 pp., 1968; vol. II: 1968–1971, 233 pp., 1973.
Isotope Techniques in the Hydrological Assessment of Potential Sites for the Disposal of High-Level Radioactive Wastes, 164 pp., 1983.
Isotope Techniques in the Study of the Hydrology of Fractured and Fissured Rocks, 306 pp., 1986.
Nuclear Techniques in Groundwater Pollution Research, 286 pp., 1980.
Paleoclimates and Paleowaters: A Collection of Environmental Isotope Studies, 216 pp., 1983.
Stable Isotope Hydrology: Deuterium and Oxygen-18 in the Water Cycle, 339 pp., 1981.
Statistical Treatment of Environmental Isotope Data in Precipitation, 276 pp., 1981.
Tritium and Other Environmental Isotopes in the Hydrological Cycle, 83 pp., 1967.

INDEX

Accuracy, 96
Aerated zone, 12
Age indicators, 26, 59, 260, 298
Algeria, 175, 187, 188
Alkalinity, 163
Altitude effect, 180
Amount effect, 177
Aquiclude, 16
Aquifer, 15
  confined, 23, 54, 236
  phreatic, 22, 236
  stagnant, 25, 34, 38, 61
  unconfined, 22, 31
Artesian, 23, 228
Atomic weight, 91, 92
Australia, 44, 137, 226, 257, 260, 262, 264, 299, 305
Austria, 247, 299, 308

Base of drainage, 30, 33, 36
  terminal, 28, 33, 38
Basin, subsidence, 29
Botswana, 171, 351
Brazil, 173, 179, 187

Canada, 174, 267

Carbon
  carbon-13, 224
  carbon-14, 165, 217
Chili, 175, 225, 226
Chlorine, 36, 57, 138, 140, 255
Clay, 17, 49
CO_2, 126, 140
Coal ash contamination, 340
Compartments, 39
Composition diagrams, 113, 120
Concentration units, 94
Conglomerate, 15
Critical angle of flow, 34

Darcy's Law, 19, 27
Data bank, 372
Data base, 9
Denmark, 325
Depth of circulation, 6, 76
Depth profiles, 149
Detection limit, 96
Discharge, 18, 84
Dolomite, 16
Drilling, 8, 147
Dyke, 56

Efflorescence, 166

411

Electrical conductivity, 160
Elements, 90
England, 240, 246, 288, 299, 303
Equivalent, 94
Errors, 97, 103
Evaporation, 18
Evapotranspiration, 18, 32, 139, 257

Filtering, 164
Fingerprints, 108
Flow
  basin wide, 44
  direction, 71
  discontinuity, 236
  in conduits, 207
  in fissures, 50
  in pores, 50, 282
  lateral, 32
  modes of, 29, 37, 45
  piston, 13, 203, 250
  structural controls, 52, 54, 55
  through-, 22, 31, 33, 37
  velocity, 26, 70 73, 81
  vertical, 32, 34, 36
France, 179, 210, 229, 235

Germany, 219, 225, 343
Gradient, 71, 117
Greece, 191
Groundwater
  composition, 1, 47, 60, 126, 128
  concealed, 1
  cycle, 3
  depth of circulation, 76
  hot, 83
  information, 3
  trapping, 36
Gypsum, 48

HCO_3, 60, 126
Hawaii, 221
Helium-dating, 293
Historical data, 150
Hungary, 252
Hydraulic
  barriers, 37

[Hydraulic]
  connectivity, 6, 35, 50, 57, 59, 116, 118
  potential, 32, 54
Hydrogen isotopes, 91, 168
Hydrograph, 66, 69

Igneous rocks, 16, 17
Impermeability, 19, 20
India, 234
Infiltration, 14, 50
Ions, 92, 94
  conservative, 57
  exchange study, 348
Isotopes, 90
  carbon, 91, 126, 217
  chlorine, 255
  effects, 176
  fractionation, 169
  hydrogen, 91, 168
  oxygen, 92, 168
Israel, 95, 123, 141, 148, 171, 194, 237, 249, 274, 281
Italy, 184, 185, 189

Kalahari desert, 134, 203, 207, 220
Karstic, 25, 250, 284

Landfill contamination, 340, 368
Limestone, 16
L-shape flow model, 31

Mediterranean Sea, drying up, 314
Meteoric isotope line, 172
Mixing, 6, 59, 78, 121, 126, 134, 209, 248, 302, 312
Models
  mathematical, 7
  phenomenological, 7
Monitoring network, 371
Multisampling, 190

Nicaragua, 182, 183
Nigeria, 246
Nitrate contamination, 323, 324, 326
Noble gases, 93, 270

Oil brine contamination, 335
Overflow, 30
Oxygen
  dissolved, 162
  isotopes, 92, 168

Paleoclimate, 60, 317
Paleohydrology, 313
Paleotemperature, 287, 318
Permeability, 19, 21
Pesticide contamination, 327
Petroleum, 39
pH, 161
Piezometric head, 24
Pollution, 15, 320
Porosity, 22
Pressurized, 39
Pumping test, 73, 149

Quality of water, 5, 47, 49

Reaction error, 97, 103
Recharge, 14, 17, 33, 49, 66, 68, 79, 202, 205, 313, 317
Reproducibility, 96, 99, 103
Rift Valley, 56
Runoff, 32

Salt contamination, 346
Salt rock, 48, 138
Sample collection, 145
Sandstone, 15
Saturated zone, 14
Saturation, 136
Sea derived salts, 136

Sea level changes, 314
Seawater intrusion, 131, 186, 360
Shale, 17, 49
Sill, 57
Smell, 159
Soil, 12
  profiles, 203, 205, 206
South Africa, 185, 204, 206, 228
Spa, 363
Stagnant systems, 28, 30, 33, 37, 54, 61, 312, 364
Street runoff, 367
Sustainable development, 356
Swaziland, 283
Sweden, 202
Switzerland, 87, 132, 135, 178, 181, 210

Table organization, 106
Taste, 159
Temperature, 74, 75, 78, 160, 280, 286
  effect, 176
Time series, 9, 81, 85, 129, 151
Transpiration, 19
Tritium, 196, 214

Unconfined, 37
Undersaturation, 136
United States, 68, 69, 78, 79, 82, 130, 208, 246, 329, 330, 332, 336, 339, 342, 350
U-shape flow model, 40, 42

Valences, 92

Water table, 1, 14, 62, 64, 117